The Origin of Higher Clades

Osteology, Myology, Phylogeny and Evolution of Bony Fishes and the Rise of Tetrapods

The Origin of Higher Clades

Osteology, Myology, Phylogeny and Evolution of Bony Fishes and the Rise of Tetrapods

Rui Diogo
Department of Anthropology
The George Washington University
Washington, DC
USA

CRC Press
Taylor & Francis Group
Boca Raton London New York

CRC Press is an imprint of the
Taylor & Francis Group, an **informa** business
A SCIENCE PUBLISHERS BOOK

First issued in hardback 2018

SCIENCE PUBLISHERS
An imprint of Edenbridge Ltd., British Isles.
Post Office Box 699
Enfield, New Hampshire 03748
United States of America

Website: *http://www.scipub.net*

sales@scipub.net (marketing department)
editor@scipub.net (editorial department)
info@scipub.net (for all other enquiries)

Library of Congress Cataloging-in-Publication Data

Diogo, Rui
The origin of higher clades: osteology, myology, phylogeny, and evolution of bony fishes and the rise of tetrapods/Rui Diogo.
p. cm.
Includes bibliographical references and index.
ISBN 978-1-57808-530-9 (Paperback)
1. Osteichthyes--Evolution. I. Title.

QL618.2D56 2007
597.13'8--dc22

2007028539

ISBN 13: 978-1-138-45355-5 (hbk)
ISBN 13: 978-1-57808-530-9 (pbk)

Paperback edition 2008

Published by Science Publishers, Enfield, NH, USA
An imprint of Edenbridge Ltd.

Preface

The Osteichthyes, including bony fishes and tetrapods, is a highly speciose group of animals, comprising more than 42000 living species. The extraordinary taxonomic diversity of osteichthyans is associated with a remarkable variety of morphological features and adaptations to very different habitats, from the deep-sea to high mountains. Osteichthyans therefore provide a very interesting case study to analyze the origin and morphological macroevolution of higher-clades. In this book, I provide a new insight on the osteology, myology, phylogeny and evolution of this fascinating group, which is based on my own research and on a survey of the literature. Chapters 1 and 2 provide a short introduction to the main aims of the book and to the methodology and methods used. Chapter 3 deals with an extensive cladistic analysis of osteichthyan higher-level interrelationships based on a phylogenetic comparison of 356 characters in 80 extant and fossil terminal taxa representing all major groups of Osteichthyes. This cladistic analysis includes various terminal taxa and osteological characters, and namely a large number of myological characters, not included in previous analyses. Chapter 4 provides a general discussion on issues such as the comparative anatomy, homologies and evolution of osteichthyan cranial and pectoral muscles, the development of zebrafish cephalic muscles and the implications for evolutionary developmental studies, the origin, homologies and evolution of one of the most peculiar and enigmatic structural complexes of osteichthyans, the Weberian apparatus, and the use of myological versus osteological characters in phylogenetic reconstructions. I hope that this work may stimulate, and pave the way for, future studies on the comparative anatomy, functional morphology, phylogeny and evolution of osteichthyans and of vertebrates in general, which, as stressed throughout the book, should ideally take into account the precious information obtained from the study of muscular features.

Dedicated to **MICHEL CHARDON, to his outstanding knowledge, to his friendship, and to his humbleness**

Acknowledgements

First of all, I want to thank P. Vandewalle and M. Chardon for accepting me in the Laboratory of Functional and Evolutionary Morphology in 1998 and thus for giving me the opportunity to begin my research on the anatomy, functional morphology, phylogeny and evolution of vertebrates and of bony fishes in particular. I also want to thank E. Parmentier. His persistence, the remarkable ability that he has to solve all types of challenges, and the courage he has to get deeply involved in different scientific areas were really inspiring for me.

I am also thankful to R. Vari, as well as his colleagues S. Weitzman, J. Williams and S. Jewett from the National Museum of Natural History, for accepting me in that amazing museum during two academic years and for providing numerous specimens analyzed in this work. I also want to thank I.. Doadrio, who received me in the Museo Nacional de Ciencias Naturales de Madrid, and has made available many specimens of the vast fish collection of this museum, which is mainly the fruit of his hard work. Another bright scientist who received me in his lab for several months was S. Hughes, whom I thank very, very much. In his lab, at the prestigious MRC Centre for Developmental Neurobiology of the King's College of London, I took my first steps in Evolutionary Developmental Biology ("Evo-Devo"). I enjoyed much his availability, his interest, and his continuous questioning and curiosity. I also want to take this occasion to thank B. Wood for inviting me to continue my research at the Anthropology Department of the George Washington University, where I shall have the opportunity to expand my work to other osteichthyan groups, and particularly to primates.

A special thanks to the late G. Teugels, as well to J. Snoeks and E. Vreven (Musée Royal de l'Afrique Centrale), P. Laleyé (Université Nationale du Bénin), Z. Peng and S. He (Academy of Sciences of China at Wuhan), T. Grande (Field Museum of Natural History), D. Catania (California Academy of Sciences), M. Stiassny (American Museum of Natural History), M. Sabaj and J. Lundberg (Academy of Natural Sciences of Philadelphia), W. Fink, D. Nelson and H. Ng (Museum of Zoology, University of Michigan), R. Bills and P. Skelton (South African Institute for Aquatic Biodiversity), L. Page and M. Retzer (Illinois Natural History Survey), P. Pruvost and G. Duhamel (Museum National d'Histoire Naturelle) and R. Walsh and F. Slaby (George Washington University) for kindly providing a large part of the specimens analyzed. I also want to acknowledge T. Abreu, A. Zanata, F. Meunier, D. Adriaens, F. Wagemans, M. de Pinna, P. Skelton, F. Poyato-Ariza, T. Grande, H. Gebhardt, M. Ebach, A. Wyss, J. Waters, G. Cuny, L. Cavin, F. Santini, J. Briggs, L. Gahagan, M. Gayet, J. Alves-Gomes, G. Lecointre, L. Soares-Porto, P. Bockmann, B. Hall, F. Galis, T. Roberts, G. Arratia, L. Taverne, E. Trajano, C. Ferraris, M. Brito, R. Reis, R. Winterbottom, C. Borden, B. Richmond and many other colleagues for their helpful advice and assistance and for their discussions on osteichthyan anatomy, functional morphology, phylogeny and/or evolution in the last years. A special thanks to V. Abdala, with whom I have discussed many of the parts of this work, and with whom I hope to undertake the numerous projects we have in mind concerning vertebrate musculature, as well as to J. Joss (Macquarie University) and A. Gosztonyi (Centro Nacional Patagónico) for providing me the large dipnoan specimens analyzed, and to J. Fernández and other people from the Museo Nacional de Ciencias Naturales de Madrid for providing me the nice salamander and lizard specimens examined.

My special thanks to all my friends, particularly to Pedro Brito, Claudia Oliveira, Henry Evrard and Diego Alarcon Reina. Thank you very much, Alejandrita Pelito Lindo, and thanks to our amazing and adorable Tots Pelluda. Very special thanks to my parents, Valter and Fatima, to my brothers, Hugo and Luis, and to my late grandfathers Raul and Amélia.

Thank you very much for the confidence in my work and for the close cooperation in the several projects we have together. Finally, thanks to all those who have been involved in administering the various grants and other awards that I received during the last years, without which this work would really not have been possible.

List of Abbreviations*

II, III, IV, V, VII, IIX, IX, X	foramens/nerves of Miles's 1977 original drawing
A0, A1, A1-OST	**adductor mandibulae A0, A1 and A1-OST**
A1-OST-L, A1-OST-M	**lateral and mesial sections of adductor mandibulae A1-OST**
A2	**adductor mandibulae A2**
A2-D, A2-PVM, A2-V	**dorsal, posteroventromesial and ventral sections of adductor mandibulae**
A3', A3''	**adductor mandibulae A3' and A3''**
AB-PRO	**abductor profundus**
AB-SUP	**abductor superficialis**
abs	anterior bulla of swimbladder
AC	**anconaeus coracoideus**
AD-AP	**adductor arcus palatini**
AD-HYO	**adductor hyomandibulae**
AD-OP	**adductor operculi**
AD-PRO	**adductor profundus**
AD-SUP	**adductor superficialis**
ADM	**adductor mandibulae**
AED1	**abductor et extensor digiti I**
AHL, AHM	**anconaeus humeralis lateralis and medialis**
am	ampulla
am-m	macula of ampulla
AME	**adductor mandibulae externus**
ana	anterior neural arch
ang	angular

*Myological structures are shown in bold

angart	anguloarticular
angrart	anguloretroarticular
anocl	anocleithrum
aorb	groove and foramen for orbital artery of Miles, 1977
apal	autopalatine
ar-chp	articulatory area for posterior ceratohyal
ar-hm	articulatory area for hyomandibula
ar-mnd	articulatory area for mandible
ar-neu	articulatory area for neurocranium
ar-op	articulatory area for opercular bone
ar-pq	articulatory area for palatoquadrate
ar-q	articulatory area for quadrate
ar-sym	articulatory area for symplectic
ARR-3	**arrector 3**
ARR-D	**arrector dorsalis**
ARR-D-1,2	**sections of arrector dorsalis**
ARR-V	**arrector ventralis**
ARR-V-1,2	**sections of arrector ventralis**
art	articular
artrart	articuloretroarticular
asi	atria sinus imparis
ASM	**anconaeus scapularis medialis**
atpm	anterior transversal peritoneal membrane
AW, Aω	**adductor mandibulae Aw**
AW-D, AW-V	**bundles of adductor mandibulae Aw**
b	cranial bone B of Miles, 1977
bb	basibranchial
BC	**basicranial muscle**
BH	**branchiohyoideus**
BM	**branchiomandibularis**
boc	basioccipital
boc-phapr	pharyngeal process of basioccipital
BRM	**branchial muscles**
bsph	basisphenoid
C	**cucullaris**
c-apal-eth	cartilage between autopalatine and ethmoid region

c-eth	ethmoid cartilage
c-ia	interatrial cartilage
c-mapa	cartilage between maxilla and autopalatine and/or dermopalatine
c-Meck	Meckel's cartilage
c-peth	pre-ethmoid cartilage
cam	camera aerea Weberiana
can	anterior semicircular canal
cart	cartilage
cb1	ceratobranchial 1
CBL	**coracobrachialis longus**
cc	complex centrum
CCL	**contrahentium caput longum**
CCO	**constrictor colli**
cctr	canalis communicans transversus
CD	**contrahentes digitorum**
CEH	**ceratohyoideus**
CERV	**cervicomandibularis**
ch, ch-a, ch-p	ceratohyal, anterior ceratohyal and posterior ceratohyal
cho	horizontal semicircular canal
cl	cleithrum
cl-hp	humeral process of cleithrum
cla	claustrum
clav	clavicle
CM	**coracomandibularis**
co	concha of scaphium
com	coronomeckelian bone
COP	**constrictor operculi**
cor	coracoid
cor-vmp	ventromesial process of coracoid
CORAD	**coracoradialis**
coro	coronoid bone
crb	cranial rib
CRB-PECG	**muscle between cranial rib and pectoral girdle**
cus	utriculo-saccular canal
dI, dII, dIII, dIV, dV	digits I, II, III, IV and V

den	dentary bone
den-alp	anterolateral process of dentary bone
df	deep fossa
DIL-OP	**dilatator operculi**
DM	**depressor mandibulae**
DM-A, DM-P	**anterior and posterior parts of depressor mandibulae**
dmtte	dorsomesial area of thin tunica externa ("median slit")
dpal	dermopalatine
DS	**dorsalis scapulae**
dsph	dermosphenotic
EACR	**extensor antebrachii et carpi radialis**
EACU	**extensor antebrachii et carpi ulnaris**
ECR	**extensor carpi radialis**
ect	ectopterygoid
ECU	**extensor carpi radialis**
EDB	**extensores digitorum breves**
EDC	**extensor digitorum communis**
EDL	**extensor digitorum longus**
ehy	epihyal
ELD4	**extensor lateralis digiti IV**
ent	entopterygoid
EP	**epaxialis**
EPIST	**episternocleidomastoideus**
EPITR	**epitrochleoanconeus**
epoc	epioccipital
et	epipterygoid
exoc	exoccipital
exs	extrascapular
extracl	extracleithrum
f	cranial bone F of Miles, 1977
FACR	**flexor antebrachii et carpi radialis**
FACU	**flexor antebrachii et carpi ulnaris**
FAL	**flexor accessorius lateralis**
FAM	**flexor accessorius medialis**
FBP	**flexores breves profundi**
FBS	**flexores breves superficiales**

FCR	**flexor carpi radialis**
FDC	**flexor digitorum communis**
FDL	**flexor digitorum longus**
FLEP	**flexor plate**
fr	frontal
fte	ichthyocoll fibers of tunica externa inserting on transformator tripodis
GG	**genioglossus**
GG-L, GG-M	**genioglossus lateralis and medialis**
GH	**geniohyoideus**
gplate	gular plate
GT	**geniothoracicus**
HAB	**humeroantebrachialis**
hc	hyoid cornu
HG	**hyoglossus**
HH	**hyohyoideus**
HH-AB	**hyohyoideus abductor**
HH-AD	**hyohyoidei adductores**
HH-INF	**hyohyoideus inferior**
HH-SUP	**hyohyoideus inferior**
hm	hyomandibula
hum	humerus
hyh, hyh-d, hyh-v	hypohyal, dorsal hypohyal and ventral hypohyal
HYP	**hypaxialis**
i	cranial bone I of Miles, 1977
iclav	interclavicle
ih	interhyal
IMC	**intermetacarpales**
inc	intercalarium
inc-ap, inc-asc	articular and ascendens processes of intercalarium
int	intercalar
INTE	**interhyoideus**
INTE-L, INTE-M	**lateral and mesial divisions of interhyoideus**
INTM	**intermandibularis**
INTM-A, INTM-P	**anterior and posterior bundles of intermandibularis**

iop	interopercle
k-m	cranial bone K-M of Miles, 1977
keth	kinethmoid
j	jugal
l-A, B, C, D, E, F, G	ligaments A, B, C, D, E, F, G
l-ans	anterior ligament of os suspensorium
l-Bau	Baudelot's ligament
l-ch-mnd	ligament between ceratohyal and mandible
l-chp-mnd	ligament between posterior ceratohyal and mandible
l-cl-pecra1	ligament between cleithrum and pectoral ray 1
l-crb-scl	ligament between cranial rib and supracleithrum
l-ent-leth	ligament between entopterygoid and lateral ethmoid
l-hmsusp	hyosuspensory ligament of Miles, 1977
l-in	intercostal (intervertebral) ligament
l-io	interossicular ligament
l-iop-mnd	ligament between interopercle and mandible
l-meth-apal	ligament between mesethmoid and autopalatine
l-meth-prmx	ligament between mesethmoid and premaxilla
l-mx-mx	ligament between the two maxillae
l-pop-mnd	ligament between preopercle and mandible
l-post-epoc	ligament between posttemporal and epioccipital
l-post-epoc-1, 2	ligaments 1 and 2 between posttemporal and epioccipital
l-post-neupos	ligament between posttemporal and posterior margin of neurocranium
l-pri	primordial ligament
l-prmx-apal	ligament between premaxilla and autopalatine
l-rbr-mnd	ligament between branchiostegal rays and mandible
l-s	suspensor ligament
l-susp-neur	ligament between suspensorium and neurocranium
LA	**labial muscle**
lab	labyrinth

lac	lacrimal
lag	lagena
lagcap	lagenar capsule
lca	lateral cutaneous area
LD	**latissimus dorsi**
leth	lateral ethmoid
LEV-5	**levator arcus branchialis V**
LEV-AO	**levator anguli oris**
LEV-AP	**levator arcus palatini**
LEV-AP-1, LEV-AP-2	**sections of levator arcus palatini**
LEV-H	**levator hyoideus**
LEV-HYO	**levator hyomandibulae**
LEV-OP	**levator operculi**
LMS3, LMS4	**levator maxillae superioris 3 and 4**
mcor-ar	mesocoracoid arch
ment	mentomeckelian bone
mesopte	mesopterygium
metapte	metapterygium
meth	mesethmoid
MH	**mandibulohyoideus**
mnd	mandible
mp	metapterygoid
mx	maxilla
mx-b	maxillary barbel
n	nasal
na	neural arch
na1, 2, 3, 4, 5	neural arches 1, 2, 3, 4, 5
na3-adp	anterodorsal process of neural arch 3
naoc	occipital neural arch
neu	neurocranium
nsp	neural spine
OH	**omohyoideus**
OM	**ocular muscles**
op	opercular bone
opcart	opercular cartilage
OPE	**opercularis**
opmem	opercular membrane
osph	orbitosphenoid

osus	os suspensorium
ot-oc	otic-occipital
P	**pectoralis**
pa	parietal
pa-exs	parieto-extrascapular
PAC	**pronator accessorius**
pal	palatine
palm-ses	palmar sesamoid
paq	palatoquadrate
para	parasphenoid
part	prearticular
PCH	**procoracohumeralis**
pcl	postcleithrum
pe	perilymphatic space
pec-fin	pectoral fin
pec-ra	pectoral rays
pec-ra-1, 2	pectoral rays 1, 2
pec-splint	pectoral splint
pif	pineal foramen
PM-MA, PM-MI	**palatomandibularis major and minor**
po	postorbital
po-ch	posterior ("hydrostatic") chamber of the swimbladder
pof	postfrontal
pop	preopercle
post	posttemporal
pp	parapophysis
pp1, 2, 3, 4, 5	parapophyses of vertebrae 1, 2, 3, 4, 5
PPR	**pronator profundus**
PR-H	**protractor hyoidei**
PR-H-D, PR-H-V	**ventral and dorsal sections of protractor hyoidei**
PR-MUP	**protractor of "Müllerian" process**
PR-PEC	**protractor pectoralis**
PR-PM	**protractor hyomandibulae**
pra	proximal radial
prf	prefrontal
prmx	premaxilla

propte	propterygium
prot	prootic
ps	perilymphatic space
PSE-SUP	**pseudotemporalis superficialis**
psp	postsplenial
psph	pterosphenoid
pt	pterotic
pte	pterygoid
PTM	**pterygomandibularis**
PTR	**pronator teres**
pvm	prevomer
pvm-tlp	prevomeral tooth plate
q	quadrate
qju	quadratojugal
r-br	branchiostegal rays
r-br-I, II, IV	branchiostegal rays I, II, IV
rad	radius
rart	retroarticular
RC	**rectus cervicis**
RE-AO	**retractor anguli oris**
RE-HM	**retractor hyomandibulae**
rib3, 4, 5	rib of vertebrae 3, 4, 5
rm-mb	mesial branch of ramus mandibularis
rsph	rhinosphenoid
S1, S2, S3, S4, S5	**"supinator muscles" of Millot and Anthony, 1958**
sa	saccule
SAR1	**subarcualis rectus 1**
sale	supple area of tunica externa
sb	swimbladder
sc	scaphium
sc-ap, sc-asc	articular and ascendens processes of scaphium
sca	scapula
sca-cor	scapulo-coracoid
scl	supracleithrum
SCO	**supracoracoideus**
sdo	supradorsal

se	sinus endolymphaticus
sem	septomaxilla
sepl, sept	longitudinal and transversal septa of the swimbladder
SH	**sternohyoideus**
SH-PRO, SH-SUP	**sternohyoideus profundus and superficialis**
smx	supramaxilla
sne	supraneural
sne1, 2, 3	supraneurals 1, 2, 3
soc	supraoccipital
sop	subopercular bone
sopcart	subopercular cartilage
sp	splenial bone
spe	sphenethmoid
sph	sphenotic
sppsp	splenialpostsplenial
spv	saccus paraventralis
sq	squamosal
st	supratemporal
stp	stapes
sucom	supratemporal commissure
sura	surangular
sym	symplectic
T-A1,T-A2	**tendons of adductor mandibulae A1 and A2**
T-AW-V	**tendon of adductor mandibulae Aw-V**
T-FDL	**tendons of flexor digitorum longus**
T-SH	**tendons of sternohyoideus**
tab	tabular
te	tunica externa of swimbladder
tf	transformator tripodis
tri	tripus
tri-ap	articular process of tripus
u	utricle
uh	urohyal
ul	ulna
v1, 2, 3, 4, 5	vertebrae 1, 2, 3, 4, 5
vm	vomer
x	cranial bone X of Miles, 1977
y1+y2	cranial bone Y1+Y2 of Miles, 1977

Contents

Chapter 1

Introduction and Aims

The Osteichthyes, including bony fishes and tetrapods, is a highly speciose group of animals comprising more than 42,000 living species. Two main osteichthyan groups are usually recognized: the Sarcopterygii (lobefins and tetrapods), with an estimate of more than 24,000 living species (e.g., Stiassny et al., 2004), and the Actinopterygii (rayfins), including more than 28,000 extant species (e.g., Nelson, 2006). The extraordinary taxonomic diversity of osteichthyans is associated with a remarkable variety of morphological features and adaptations to very different habitats, from the deep sea to high mountains. In this brief Introduction, I will not provide a detailed historical account of all the numerous works dealing with osteichthyan phylogeny. Such information can be found in overviews such as Arratia (2000: relationships among major teleostean groups), Clack (2002: relationships among major groups of early tetrapods), Stiassny et al. (2004: relationships among major groups of gnathostome fishes), Cloutier and Arratia (2004: relationships among early actinopterygians), and Nelson (2006: relationships among numerous fish groups). I prefer simply to give the reader a general idea of the phylogenetic scenario that is nowadays most commonly accepted in textbooks concerning the relationships between the major osteichthyan groups, which is shown in Fig. 1. Further details about this subject will be given in Chapter 3, in which I will discuss each of these groups separately and compare the phylogenetic results obtained in this work with those of previous studies.

The extant vertebrates that are usually considered to be the closest relatives of osteichthyans are the chondrichthyans (Fig. 1). However, it should be stressed that according to most authors there is a group of fossil fishes that is even more closely related to osteichthyans: the †Acanthodii, which, together with the Osteichthyes, form a group usually named Teleostomi (e.g., Kardong, 2002). In addition, it should be noted that apart from the Teleostomi and Chondrichthyes, there is another group that is

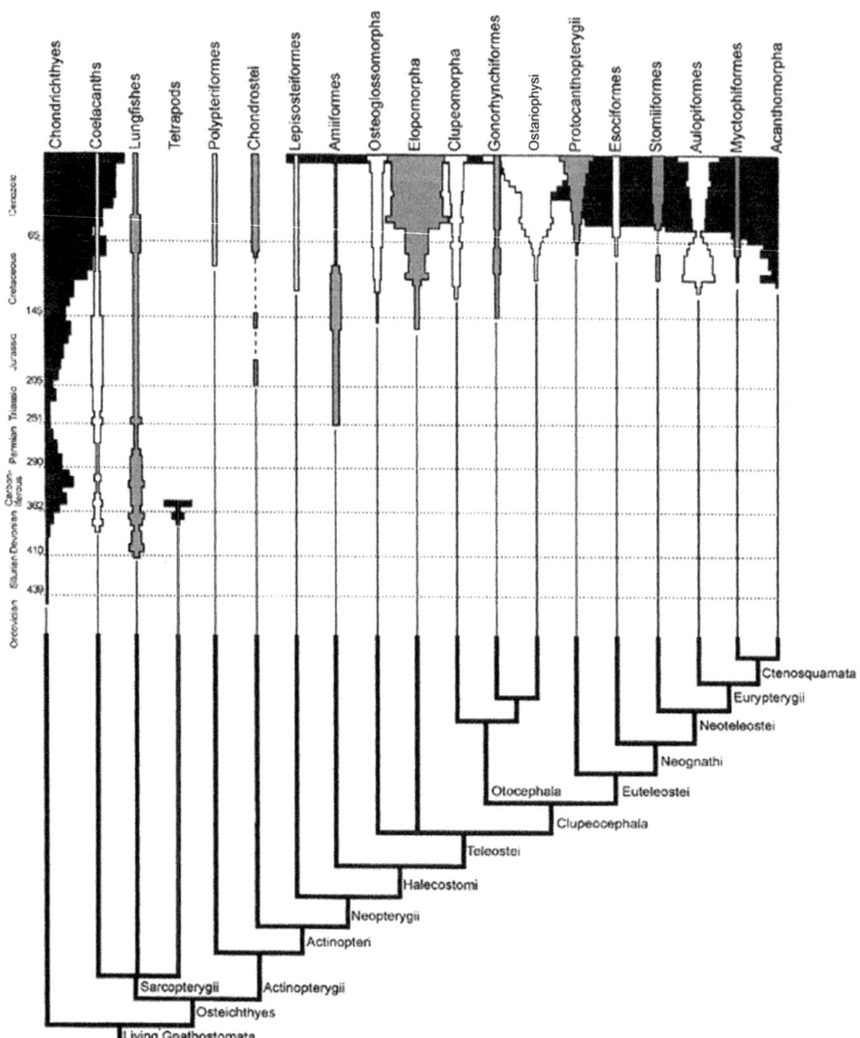

Figure 1. Relationships between the major extant gnathostome groups, modified from Stiassny et al. (2004); past and present counts of nominal families by column width through time (numbers are millions of years; tetrapod diversity truncated, chondrichthyan diversity truncated to the left and acanthomorph diversity truncated to the right); familial diversity is charted and this does not necessarily reflect known species diversity (for more details, see text).

usually included in the gnathostomes and that is usually considered the sister-group of teleostomes + chondrichthyans: the †Placodermi (e.g., Kardong, 2002). There are only two groups of extant sarcopterygian fishes, the coelacanths (Actinistia) and lungfishes (Dipnoi) (Fig. 1). The Polypteridae (included in the Cladistia) are commonly considered the most basal extant actinopterygian taxon (Fig. 1). The Acipenseridae and Polyodontidae (included in the Chondrostei) are usually considered the sister-group of a clade including the Lepisosteidae (included in the Ginglymodi) and the Amiidae (included in Halecomorphi) plus the Teleostei (Fig. 1). Regarding the Teleostei, four main living clades are usually recognized in recent works: the Elopomorpha, Osteoglossomorpha, Otocephala (Clupeomorpha + Ostariophysi) and Euteleostei (Fig. 1).

In order to provide more detail on the various subgroups of these four teleostean clades, I will refer to a cladogram provided by Springer and Johnson (2004), which, in my opinion, adequately summarizes the scenario that is probably accepted by most researchers nowadays. A simplified version of this cladogram is shown in Fig. 2. As can be seen, among these four teleostean groups the Osteoglossomorpha appears as the most basal one, the Elopomorpha appearing as the sister-group of Otocephala + Euteleostei (Fig. 2). Within the Euteleostei, the Esociformes are placed closely related to the Neoteleostei, although many authors consider the esociforms as part of the "Protacanthopterygii" (see below). Other fishes usually included in the "Protacanthopterygii" are the Alepocephaloidea, the Argentinoidea, the Salmoniformes, the Osmeroidea, and the Galaxioidea.

As stressed by Springer and Johnson (2004) and Stiassny et al. (2004), on whose works Figs. 1 and 2 are based, although the scenarios shown in those figures are widely accepted nowadays, they are far from being agreed upon by all specialists. For instance, Filleul (2000) and Filleul and Lavoué (2001) argued that the Elopomorpha is in fact not a monophyletic unit. Ishiguro et al. (2003) and other authors maintained that the Otocephala, as currently recognized (Ostariophysi + Clupeomorpha), is also not monophyletic, since certain otocephalans are more closely related to the "protacanthopterygian" alepocephaloids than to other otocephalans. Also, contrary to what is accepted by most authors (Fig. 2), Ishiguro et al. (2003) suggested that the non-alepocephaloid "protacanthopterygians" (sensu these authors, that is, the Esociformes, Salmoniformes, Osmeriformes and Argentinoidea) form a monophyletic, valid "Protacanthopterygii" clade. To give another example, Arratia (1997, 1999) argued that the most basal extant teleostean group is the Elopomorpha, and not the Osteoglossomorpha, as shown in Fig. 2. In fact, as can be seen in Fig. 1, in contrast with Springer and Johnson (2004), Stiassny et al. (2004) opted to place the Osteoglossomorpha, Elopomorpha and remaining teleosts in an unresolved trichotomy. And, as can also be seen in that figure, this is not the only trichotomy appearing in Stiassny et al.'s (2004) cladogram.

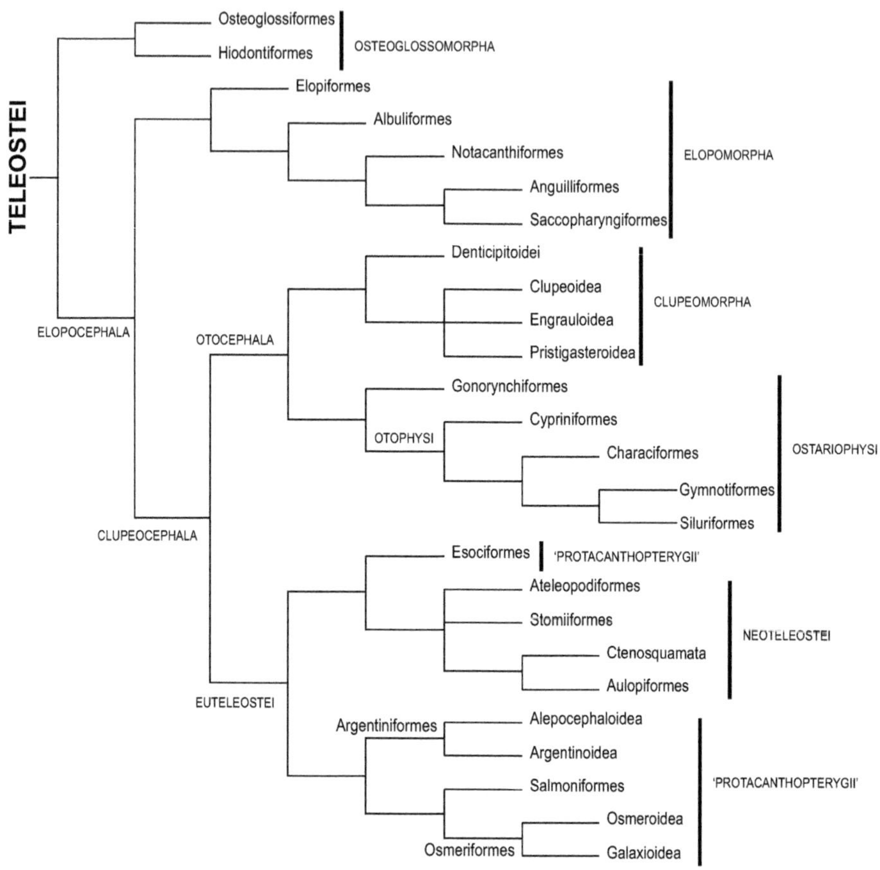

Figure 2 Relationships between the major extant teleostean groups, modified from Springer and Johnson (2004); the "protacanthopterygian" groups shown in the tree correspond to those of Ishiguro et al. (2003) (for more details, see text).

The other trichotomy appearing in that cladogram concerns one of the most discussed topics in osteichthyan phylogeny: that concerning the identity of the closest living relatives of the Tetrapoda (Fig. 1). This topic has been, and continues to be, the subject of much controversy. In general textbooks such as those by Lecointre and Le Guyader (2001), Kardong (2002) and Dawkins (2004), the tetrapods often appear more closely related to lungfishes than to the coelacanths. This view has been defended, at least partly, in many morphological and molecular works, such as those by Rosen et al. (1981), Patterson (1981), Forey (1980, 1991), Cloutier and Ahlberg (1996), Zardoya et al. (1998), Meyer and Zardoya (2003), and Brinkmann et al. (2004). However, researchers such as Zhu and Schultze (1997, 2001), on the basis of anatomical studies, defended a closer relationship between tetrapods and coelacanths than between tetrapods and lungfishes. And Zardoya and Meyer (1996) and Zardoya et al. (1998), on the basis of molecular analyses, have inclusively suggested that coelacanths and lungfishes may be sister-groups. A completely different scenario has been defended in a series of molecular works by Arnason and colleagues (Rasmussen and Arnason, 1999a,b; Arnason et al., 2001, 2004): that tetrapods are the sister-group of a clade including taxa such as lungfishes, cladistians, coelacanths, sharks and teleosts. Nevertheless, it should be said that the methodology and the results of these latter works have been severely criticized and questioned by numerous researchers (see below).

In light of all this controversy, Stiassny et al. (2004) opted to place coelacanths, lungfishes and tetrapods in an unresolved trichotomy inside the sarcopterygian clade (Fig. 1). In other words, they considered that the data currently available does not allow us to suitably answer two fundamental questions concerning two of the most highly diverse osteichthyan groups, the Tetrapoda and Teleostei: Which is the closest living relative of tetrapods? And which is the most basal extant teleostean clade? And, as stressed by Stiassny et al. (2004), "although most workers have followed Patterson (1973) in the recognition of *Amia* as the closest living relative of the Teleostei, there remains some controversy" about the phylogenetic position of this taxon. Actually, according to Grande (2005), an extensive study in progress done by Grande and Bemis supports the hypothesis that, contrary to what is often accepted nowadays (see Fig. 1), the Halecomorphi (including the Amiidae) may be more closely related to the Ginglymodi (including the Lepisosteidae) than to the Teleostei. Such a view has also been supported by molecular studies by Inoue et al. (2003) and Kikugawa et al. (2004). Thus, another fundamental question concerning one of these two highly diverse osteichthyan groups remains disputed: which are the closest living relatives of teleosts, the amiids of the genus *Amia*, or both these fishes and the lepisosteids of the genera *Lepisosteus* and *Atractosteus*, that is, the members of the three extant genera of an eventual clade Halecomorphi + Ginglymodi? Together with what was mentioned above about the controversies regarding the monophyly/non-monophyly of the Elopomorpha, the Otocephala, and the "Protacanthopterygii", these are just a few examples to illustrate that, in fact, despite the progress achieved in osteichthyan phylogeny in the last decades, some crucial questions concerning this subject do remain unresolved and highly debated.

Le et al. (1993) and Meyer and Zardoya (2003) have stated that in order to help resolve such crucial questions it is extremely important to promote "new morphological character analyses". In fact, apart from the obvious need for new studies using, and combining, different types of molecular data and, very important, including more terminal taxa than the few species usually included in most molecular studies, I also consider that there is an imperative need for new, fresh morphological cladistic analyses to help clarify osteichthyan higher-level phylogeny. When one reads certain recent molecular works, it may appear that morphologists have already played all their cards regarding the resolution of major issues on the phylogeny of groups such as osteichthyans. In my opinion, this is clearly not the case.

First of all, many studies have focused on the anatomy of representatives of the major osteichthyan groups, but much of the vast amount of anatomical data available has unfortunately not been used to promote explicit cladistic analyses (Diogo, 2004a). To put it more simply, a certain researcher may describe in detail a region of the body of a certain taxon A, and may even compare it with the same region of the body of a certain taxon B. In many cases, however, this anatomical data is ultimately not used to promote a cladistic study in which are included not only taxa A and B but also other taxa, and in which all the data available is presented in the form of phylogenetic characters that then allow the building of a phylogenetic matrix that is, in turn, analyzed under an explicit cladistic procedure.

Another important point is that among the unfortunately few explicit morphological cladistic analyses published so far on osteichthyans, the great majority are focused on a single family, a single subfamily, or even a single genus (Diogo, 2004a). Such studies are of course needed and much welcomed. But the fact is that explicit cladistic studies using a high number of characters and a high number of representatives of various osteichthyan orders are very rare. And including terminal taxa from osteichthyan groups as varied as, for example, the Teleostei, the Halecomorphi, the Ginglymodi, the Chondrostei, the Cladistia, the Actinistia, the Dipnoi, and the tetrapods, as in the present work, is even more rare. As stressed by Le et al. (1993) and Ishiguro et al. (2003), the consequence of this is that certain major osteichthyan clades that are generally accepted by morphologists, and by other researchers as well, have in reality never been supported by explicit morphological cladistic analyses. I will provide some examples of this in Chapter 3.

Another aspect pointed out by Diogo (2004a,b) is that most morphological cladistic analyses that have been published so far on osteichthyans concern mainly osteological and/or external characters. Very few of them include a significant number of myological characters, not even those dealing exclusively with extant terminal taxa. At least to my knowledge, there is not a single morphological cladistic analysis published on osteichthyan higher-level phylogeny that has included a great number of both osteological and muscular characters.

In summary, in my opinion morphologists have clearly not played all their cards for the resolution of major issues regarding osteichthyan phylogeny: (1) there is a vast amount of anatomical data already available in the literature that could be used in explicit cladistic analyses; (2) an effort could also be made to promote cladistic analyses including representatives of various major osteichthyan groups, in order to test whether the higher clades often accepted in the literature are, or are not, supported by these analyses; and (3) an effort could be made to include other types of anatomical characters, for example, myological ones, in such cladistic analyses, since much useful anatomical information is being lost in using mainly osteological characters.

One of the main aims of the present work is precisely to provide a cladistic analysis that includes terminal taxa from osteichthyan groups as varied as the Teleostei, Halecomorphi, Ginglymodi, Chondrostei, Cladistia, Actinistia and Tetrapoda, and that includes a large number of both osteological and myological characters. This may make it possible, for instance, to check whether the major osteichthyan clades shown in Figs. 1 and 2 are in fact supported by such a cladistic analysis. The inclusion of a great number of both osteological and myological characters may also allow us to compare how these different types of characters behave in the phylogenetic study, for example, to see which characters provide better support for the major clades obtained in the study or which characters appear to be more homoplasic within these clades. Such issues have unfortunately not been much discussed in the literature (Diogo, 2004a,b). According to Diogo (2004a,b), myological characters may play an important role in phylogenetic reconstructions, particularly in those concerning the relationships of higher clades. One of our aims here is thus to discuss whether or not the results of the cladistic analysis of the present work, which is precisely focused on the higher-level phylogeny of a major group such as the osteichthyans, support such a view. The inclusion of muscular characters in a study such as the present one may also pave the way for a more detailed, integrative reflection on the functional morphology and evolution of, for example, the pectoral girdle or the head (Winterbottom; 1993; Galis, 1996; Borden, 1998, 1999; Diogo, 2004a). It is important to emphasize that structural complexes are constituted by a set of integrated bones, but also muscles, cartilage and ligaments. For this reason, as well as for the other reasons that will be given in these first two chapters, the characters examined in the present work concern not only the configuration of bones, but also that of numerous muscles, cartilage and ligaments.

It should be stressed that the cladistic analysis provided in this work refers essentially to the higher-level relationships among the major basal groups of the sarcopterygian and actinopterygian lineages, and not to the interrelationships between the various taxa of highly diverse, derived extant groups such as the actinopterygian neoteleosts and the sarcopterygian tetrapods. As explained above, this cladistic analysis does include some representatives of these two groups, but with the main aim of searching for the higher-level relationships between these groups and other major osteichthyan clades. A detailed morphological cladistic analysis of the relationships among the numerous teleostean subgroups and among the numerous tetrapod taxa would require at least two books such as this one. But I do hope that the present work will pave the way for such analyses. For instance, the results obtained here may help to clarify which of the taxa examined may be more closely related to neoteleosts and to tetrapods and may thus help in choosing appropriate outgroups and polarizing characters in future works on the interrelationships among the various subunits of these two groups.

These results can also help to clarify the origin and homologies of certain structures found in the members of these groups, for example, the peculiar tongue muscles found in most tetrapods. One of the main goals of the book is to provide a new insight into the homologies and evolution of certain key osteological and myological structures examined, taking into account the phylogenetic results obtained. I consider that it is particularly important to give a fresh look to the homologies and evolution of osteichthyan muscular structures, which have been much less studied and discussed than the osteological ones. The most extensive, detailed comparative works that have been done on osteichthyan muscles and that were actually based on a direct observation of these muscles in osteichthyan taxa as varied as teleosts, lepisosteids, amiids, acipenseriforms, polypterids, dipnoans, and tetrapods, and not mainly on a recompilation from the literature, were done several decades ago (e.g., Luther, 1913-1914; Edgeworth, 1935; Kesteven, 1942-1945). Among these works, I have a special admiration for Edgeworth's (1935) excellent book, which continues to be, in my opinion, one of the most fundamental studies on vertebrate muscles. However, one should keep in mind that this book was written more than 70 years ago. Therefore, Edgeworth did not have access to information now available, which may, or may not, contradict some of the hypotheses proposed in his book. For example, he had no access to the data now available on the muscles of the coelacanth *Latimeria chalumnae*, since this fish had not been discovered. He had no access to the data provided by evolutionary developmental biologists about muscle development. Also, some of his hypotheses were based on phylogenetic scenarios that have been contradicted by numerous works. For instance, according to Edgeworth

(1935), the chondrichthyans, actinopterygians and tetrapods in the sense of the present work were derived from an "early dipnoan stock"; birds are the sister-group of a taxon including all his "reptiles"; his clade Reptilia + Aves and his clade Mammalia originated independently from amphibians; and the Teleostei is the sister-group of an assemblage including all non-teleostean actinopterygians (his "Ganoidei"). It is also worth noting that Edgeworth (1935) analyzed and interpreted his work in the light of some evolutionary notions that, in view of the scientific data available at present, may sound somewhat odd. For instance, in one of his last chapters he stated: "animals are psychological units characterized by memory and purpose, by striving after ends in view; the variations in the development of homologous structures and in the structures by which the same function is carried out show that this psychological factor is of great importance from the very first; this immaterial, non-spatial, teleological factor, the mind, can initiate and inhibit physico-chemical processes; life development of individual organisms and evolution are primarily due to this power" (Edgeworth, 1935: 227). I thus consider that it is opportune to revise the work of Luther (1913-1914), Edgeworth (1935), Kesteven (1942-1945), and other authors in order to check to what extent the hypotheses advanced in those works are, or are not, supported by the new data available in the literature and by the observations and phylogenetic results of the present work.

Chapter 2

Methodology and Material

The phylogenetic methodology used in the present work basically follows that used in Diogo (2004a), in which more details about this methodology can thus be found. The procedure employed for proposing hypotheses of relationships is cladistics; parsimony was employed to find the hypothesis best supported by the analyzed data, using both the Hennig86 (Farris, 1988) and Nona&Winclada (Nixon, 2002) computer programs. Tree manipulations and diagnostics were done with the help of Nona&Winclada. Autapomorphies for the different taxa examined were actively searched for and included in the analysis. All multistate characters were ordered. However, unlike in the earlier work (Diogo, 2004a), in the present analysis I did not proceed to a "normalization" of these multistate characters. That is, the weight of the characters with three states (there are no characters with more than three states in the present analysis) was not multiplied by a factor of 0.5, as was done earlier (Diogo, 2004a). It should be noted that I did check whether the phylogenetic results obtained in the present work were different from those that would have been obtained using the "normalization" procedure of Diogo (2004a), and no significant differences were found.

Special attention was given to the process of determining the polarity of the different character states used on the cladistic analysis. Thus, an extensive, detailed survey of the literature was done in order to obtain as much information as possible on taxa other than those included in the analysis, both living and fossil, with a particular focus on the †Acanthodii and on the Chondrichthyes, as well as on other vertebrate groups such as the †Placodermi. Regarding the 80 terminal taxa included in the cladistic analysis, I will briefly summarize, below, some of the main reasons for choosing these particular taxa.

First of all, it is important to note that although the majority (73) of these 80 terminal taxa concern extant groups, some of them (7) concern fossils;

roughly, it can be said that for every 10 extant terminal taxa included in the analysis, one fossil terminal taxon was included. The main reason for using more extant taxa is that, as mentioned above, a great number of the phylogenetic characters used in the analysis refer to the configuration of muscles, cartilage and ligaments. Thus, the seven fossil taxa that were used, †*Chanoides macropoma*, †*Clupavus maroccanus*, †*Santanichthys diasii*, †*Lusitanichthys characiformis*, †*Sorbininardus apuliensis*, †*Tiktaalik roseae* and †*Acanthostega gunnari*, were chosen for a precise reason. Regarding the five former fossil taxa, which are often included in the teleostean clade Ostariophysi, Gayet (1981, 1985, 1986a), Taverne (1977a, 1995, 1999), Filleul and Maisey (2004), and others have argued that these are particularly "problematic" fossil taxa that, if included in an explicit cladistic analysis together with other ostariophysans and non-ostariophysans, could well show that the four extant otophysan orders, and eventually the ostariophysans, as currently recognized, do not form monophyletic groups. Since the ostariophysan fishes play a central role for a proper understanding of the higher-level phylogeny and evolution of one of the most diverse osteichthyan groups, the Teleostei, and since the testing of the monophyly of the otophysan, ostariophysan and otocephalan fishes is one of the aims of the present work, these five fossil taxa were included in the analysis. The inclusion of these fossils in an explicit cladistic analysis is also crucial in clarifying a major issue in the evolution of teleosts: whether the characteristic Weberian apparatus of extant otophysans was, or was not, acquired just once within the evolutionary history of these fishes (see, e.g., Gayet, 1981, 1985; Fink and Fink, 1981, 1996; Taverne, 1995; Filleul and Maisey, 2004). The reason for including these five fossil taxa and not, for instance, other "problematic" ostariophysan fossils sensu, for example, Gayet (1981, 1985, 1986a) such as †*Salminops ibericus*, is that these five are particularly well conserved, which is not the case with the latter. Regarding †*Acanthostega gunnari*, this taxon was selected to represent the so-called "early tetrapods" because among those fossils that are usually placed in the very base of the Tetrapoda, this is one of the best preserved and best described (e.g., Clack, 2002, 2006). The recently discovered, and relatively well-described, †*Tiktaalik roseae* was chosen because, according to Daeschler et al. (2006) and Shubin et al. (2006), it may well be one of the bony fishes more closely related to tetrapods. Thus, to put it in a rough and simplistic way, †*Tiktaalik roseae* was selected for being a relatively well-described bony fish seemingly "lying near the tetrapods", and †*Acanthostega gunnari* was chosen for being a well-described tetrapod seemingly "lying near the bony fishes". It is thus interesting to see in which position these fossils will appear on the trees obtained in the cladistic analysis of the present work. If the phylogenetic hypothesis of Daeschler et al. (2006), Shubin et al. (2006), and other authors is supported, these two fossils may help to partly cover the gap between extant tetrapod and non-tetrapod osteichthyans.

Concerning the 73 extant taxa included in the cladistic analysis, *Latimeria* and *Amia* were selected because they are the single extant representatives of the Actinistia and of the Halecomorphi, respectively. The Dipnoi comprises only three extant genera, which are usually placed in two different groups: the Neoceratodontidae, including *Neoceratodus*, and the Lepidosirenidae, including *Lepidosiren* and *Protopterus* (see, e.g., Schultze, 2004). *Neoceratodus* and *Lepidosiren* were thus chosen to represent each of these two groups. *Polypterus*, one of the two living cladistian genera (both from the family Polypteridae), was selected to represent the Cladistia. The Chondrostei includes two extant families, the Polyodontidae, with two living genera, and the Acipenseridae, with four living genera. These two families are represented in the present work by *Psephurus* and *Acipenser*, respectively. *Lepisosteus*, one of the two living genera of Ginglymodi (both from the family Lepisosteidae), was selected to represent this clade. As explained above, the main aim of the present work is not to examine the interrelationships between the numerous taxa of derived and highly diverse osteichthyan groups such as the Tetrapoda. The three tetrapod taxa incorporated in the analysis are thus selected for two main reasons: to help clarify the relationships between lower tetrapods and the other osteichthyan groups examined, and to help clarify the homologies and general evolution of the structures discussed in the book, from basal bony fishes to derived osteichthyan groups such as amphibians and amniotes. Since many of these structures are muscles, cartilage and ligaments, it was decided to include in the analysis at least some extant tetrapod taxa, for comparison with the other extant osteichthyan taxa examined. Thus, apart from †*Acanthostega gunnari*, a well-described tetrapod that is seemingly phylogenetically more plesiomorphic than the last common ancestor of amphibians and amniotes (e.g., Clack, 2002), it was determined to include two extant taxa representing these latter two main tetrapod clades: the salamander *Ambystoma*, representing extant amphibians, and the lizard *Timon*, representing extant lower amniotes. In fact, as explained by Kardong (2002) and others, although salamanders are not, in phylogenetic terms, the most basal extant amphibians, the configuration of their osteological and myological structures does seem to approximate more what may have been found in the first amphibians than for instance the configuration of living anurans and caecilians. This is precisely the reason salamanders are often chosen to represent extant amphibians in studies on the comparative anatomy of the major vertebrate groups (see, e.g., Tables 10.2 and 10.3 of Kardong, 2002). For a similar reason, lizards are often selected in such comparative studies to represent living lepidosaurs and eventually lower amniotes. For instance, Kardong (2002: 256) stated that "the first amniotes were small and would probably remind us of lizards in general appearance". For these reasons, and having in mind the main purposes of the book (see above), it was decided to include, in the present work, the salamander *Ambystoma* and the lizard *Timon* as key representatives of extant amphibians and of extant lower amniotes, respectively.

Regarding the living teleosts included in the cladistic analysis, representatives of each of the four osteoglossiform extant families sensu Hilton (2003) were included in the cladistic analysis: *Hiodon* (Hiodontidae), *Pantodon* (Osteoglossidae), *Xenomystus* (Notopteridae) and *Mormyrus* (Mormyridae). Representatives of all five extant elopomorph orders (see Fig. 1) were also included: *Elops* and *Megalops* (Elopiformes), *Albula* (Albuliformes), *Notacanthus* (Notacanthiformes), *Anguilla* and *Conger* (Anguilliformes) and *Eurypharynx* (Saccopharyngiformes). The five extant ostariophysan orders are also covered in the analysis, including all extant gonorynchiform genera, since both the interrelationships and phylogenetic position of gonorynchiforms have been particularly controversial: *Chanos, Gonorynchus, Phractolaemus, Kneria, Parakneria, Cromeria* and *Grasseichthys* (Gonorynchiformes), *Opsariichthys, Barbus, Danio, Cobitis* and *Catostomus* (Cypriniformes), *Xenocharax, Distichodus, Citharinus* and *Brycon* (Characiformes), *Sternopygus, Gymnotus* and *Brachyhypopomus* (Gymnotiformes) and *Diplomystes, Nematogenys, Trichomycterus, Callichthys, Cetopsis, Silurus, Pimelodus, Bagrus* and *Chrysichthys* (Siluriformes). The four major extant clupeomorph groups (see Fig. 1) are also represented: *Denticeps* (Denticipitoidei), *Ilisha* (Pristigasteroidea), *Ethmalosa* (Clupeoidea) and *Thryssa* and *Engraulis* (Engrauloidea). All the major extant groups of "Protacanthopterygii" sensu Ishiguro et al. (2003) (see Fig. 2) are represented: *Coregonus, Thymallus* and *Salmo* (Salmoniformes), *Stokellia, Retropinna* and *Galaxias* (Galaxioidea), *Osmerus* and *Plecoglossus* (Osmeroidea), *Searsia, Xenodermichthys* and *Alepocephalus* (Alepocephaloidea), *Argentina* and *Bathylagus* (Argentinoidea) and *Umbra* and *Esox* (Esociformes). Lastly, in order to test the monophyly/non-monophyly of the "Protacanthopterygii" (see above), four representatives of two of the most basal neoteleostean orders, that is, *Stomias* and *Astronesthes* (Stomiiformes) and *Aulopus* and *Chlorophthalmus* (Aulopiformes) (see Fig. 1), were also included in the cladistic analysis. More details concerning the choice of certain terminal taxa mentioned above will be given in Chapter 3.

The seven fossil taxa included in the cladistic analysis and *Latimeria* were not directly examined by the author and are thus coded following exclusively their descriptions in the literature (e.g., Millot and Anthony, 1958; Andrews, 1977; Taverne, 1977a, 1995, 1999; Gayet, 1981, 1985; Patterson, 1984; Fritzsch, 1987, 2003; Bemis and Northcutt, 1991; Schultze and Cloutier, 1991; Adamicka and Ahnelt, 1992; Northcutt and Bemis, 1993; Coates, 1996; Clack, 1998, 2002; Forey, 1998; Ahlberg and Clack, 1998; Bernstein, 2003; Filleul and Maisey, 2004; Daeschler et al., 2006; Shubin et al., 2006). However, I have personally observed the features referred to in the list of characters given in Chapter 3 for all the other 72 terminal taxa used in the analysis, excepting some very rare cases in which this was really not possible. For instance, the coding of *Neoceratodus* as CS-1 in the character "presence of a levator hyoideus in at least some developmental stages" is based on the literature, since I have dissected juvenile/adult members of this genus in which this muscle is missing, and not younger specimens in which, according to the literature, this muscle is present. In fact, it is important to note that, unless explicitly stated otherwise, the morphological features mentioned in that list of characters refer to the configuration found in wild-type adults. It should also be stressed that in the matrix shown in Table 1, inapplicable and missing character states for a certain taxon are indicated with '-' and '?', respectively. Unless otherwise stated, inapplicable character states correspond to cases in which, for example, a character refers to the shape of a bone that is missing in a certain taxon; missing character states correspond to those cases in which it was not possible to appropriately discern the respective state in a certain taxon (e.g., due to the poor preservation of the fossils described in the literature or of the extant specimens examined by the author).

There are documented cases, within actinopterygians, for example, of remarkable morphological variation within a single genus, a single species, and even within a single population of the same species (e.g., Hilton and Bemis, 1999). As will be mentioned throughout the list of 356 characters provided in Chapter 3, among these characters there are cases in which different wild-type, adult members (examined by the author and/or previously described in the literature) of a certain terminal taxon do seemingly exhibit different character states of a single character. Since in those cases it is not possible to assign these adult members to a single character state, the taxon is coded as '?'. It should, however, be noted that regarding the seven fossil taxa included in the cladistic analysis, I prefer using species, and not genera, as terminal taxa. The reason for this is that fossil species that are described in the literature under the same generic name are often anatomically very different, or at least the interpretations of the authors of these descriptions are rather different. That is, it may be that in those cases there is a high intrageneric variation, but it may well be that this is more related to the descriptions provided by the authors, or that the species should never have been included in the same genus. To give just one example, regarding certain morphological characters used in the present work, the description of †*Chanoides chardoni* given by Taverne (2005) is rather different from that of †*Chanoides macropoma* given by Patterson (1984). Since I did not have the opportunity to directly analyze any of these fossils, I prefer to prudently use †*Chanoides macropoma*, which was the first species of †*Chanoides* described in the literature, as a terminal taxon, and to thus basically follow the description of this species given by Patterson (1984). Apart from the seven fossil species, I also opted to use one extant species as a terminal taxon in this analysis, namely *Ambystoma ordinarium*. The reason for this is that the intrageneric variation found in the members of the genus *Ambystoma* is impressive, being in certain cases inclusively higher than the variation I have found between members of different osteichthyan families analyzed in the present work. For instance, some adult *Ambystoma* may be aquatic and non-metamorphosed, others may be metamorphosed and continue to inhabit mainly aquatic environments, and still others can be metamorphosed and terrestrial (e.g., Monath, 1965; Larsen and Guthrie, 1975; Lauder and Shaffer, 1985, 1988; Shaffer and Lauder, 1985a,b; Reilly and Lauder, 1989, 1991; Lauder and Reilly, 1990). I have personally dissected adult specimens of *Ambystoma* species other than *Ambystoma ordinarium*, such as *Ambystoma mexicanum* and *Ambystoma andersoni*, and there are remarkable differences between the specimens of these three species, such as the presence/absence of various muscles. Contrary to what I have done for the other 72 extant taxa included in the cladistic analysis, I have thus opted to use the species *Ambystoma ordinarium*, and not the genus *Ambystoma*, as a terminal taxon in that analysis.

The morphological features analyzed in the present work concern essentially the bones, cartilage, muscles and ligaments of the cephalic region (branchial apparatus excluded), anterior vertebrae and pectoral girdle. The ideal would be, obviously, to extend this analysis to all other regions of the body. However, a selection was necessarily made. This selection was the subject of a careful reflection. One of the reasons for choosing these three regions of the body was that many structures of the anterior part of the body of the taxa analyzed are, in reality, integrated in a major structural complex (see, e.g., Diogo, 2004a). The study of the muscles, bones, ligaments and cartilage of these regions thus allows us to make a broader, more integrative analysis on the whole anterior region of the body, its functional morphology, and its general evolution (see Introduction). It is important to note, nevertheless, that, in spite of this necessary selection, the cladistic analysis of the present work includes 356 phylogenetic characters, a number significantly higher than that used in most morphological cladistic studies done on osteichthyans.

The nomenclature of the anatomical structures examined and discussed in this work basically follows that of Diogo (2004a). A major exception concerns the names of certain pectoral girdle muscles, which differ from those employed by Diogo (2004a). As explained by Diogo (2004a), the names of the pectoral girdle muscles used in that work did not fully correspond to the nomenclature proposed by, for example, Winterbottom (1974). One of the main reasons for that was the presence, in catfishes, of a muscle that is named "arrector 3" in the present work, which was not described by Winterbottom (1974) and which was tentatively named arrector ventralis by Diogo et al. (2001a) and Diogo (2004a). In the present work, however, the nomenclature of the pectoral girdle muscles will follow, whenever it is possible, that of Winterbottom (1974). Thus, in order to facilitate comparisons with previous works such as Diogo et al. (2001a) and Diogo (2004a), it is worth noting that the "arrector ventralis", "arrector dorsalis", "abductor superficialis 1", "abductor superficialis 2", "adductor superficialis 1", "adductor superficialis 2" and "abductor profundus" of those previous works correspond respectively to the arrector 3, arrector ventralis, abductor superficialis, abductor profundus, adductor superficialis, adductor profundus and arrector dorsalis of the present work.

The specimens examined are from the Laboratory of Functional and Evolutionary Morphology of the University of Liège (LFEM), the Museo Nacional de Ciencias Naturales de Madrid (MNCN), the American Museum of Natural History (AMNH), the Academy of Natural Sciences of Philadelphia (ANSP), the Chinese Academy of Sciences at Wuhan (CASW), the California Academy of Sciences (CAS), the Field Museum of Natural History (FMNH), the Illinois Natural History Survey (INHS), the Museum National d'Histoire Naturelle de Paris (MNHN), the Centro Nacional Patagónico de Argentina (CONICET), the Macquarie University of Australia (MU), the Musée Royal de l'Afrique Centrale (MRAC), the Université Nationale du Bénin (UNB), the National Museum of Natural History (USNM), and the George Washington University (GWU). Anatomical observations were made after dissection of alcohol-fixed, formalin-embalmed, or trypsin-cleared and alizarine-stained specimens. Dissections and anatomical drawings were made using a Wild M5 dissecting microscope equipped with a camera lucida. The list of specimens examined is given below; the trypsine-cleared and alizarine-stained (c&s), formalin-embalmed (for) or alcohol-fixed (alc) condition of these specimens is given in parentheses following the number of specimens observed. It should be stressed that this list refers exclusively to the specimens examined for the purpose of the present work, and not to all osteichthyan taxa examined by the author (e.g., the author has analyzed numerous catfish taxa but only nine are included in the cladistic analysis of the present work; according to the previous research of the author these nine taxa adequately represent the siluriforms for the purpose of this work).

Non-teleostean actinopterygians: *Acipenser sturio*: MNCN 152172, 3 (alc). *Amia calva*: MNCN 35961, 1 (alc), 1 (c&s). *Lepisosteus osseus*: ANSP 107961, 2 (alc); ANSP 172630, 1 (alc); MNCN 246557, 1 (c&s). *Lepisosteus platyrhincus*: AMNH 74789, 2 (alc). *Polypterus bichir*: MNCN 1579, 7 (alc), 1 (c&s). *Psephurus gladius*: CASW, uncatalogued, 1 (alc).

Elopomorphs: *Albula vulpes*: MNCN 52124, 2 (alc). *Anguilla anguilla*: MNCN 41049, 3 (alc). *Elops lacerta*: LFEM, 2 (alc). *Elops saurus*: MNCN 48752, 2 (alc). *Conger conger*: MNCN 1530, 5 (alc). *Eurypharynx pelecanoides*: AMNH 44315, 1 (alc); AMNH 44344, 1 (alc). *Megalops cyprinoides*: MNCN 48858, 3 (alc). *Notacanthus bonaparte*: MNCN 107324, 3 (alc).

Osteoglossomorphs: *Hiodon tergisus*: MNCN 36019, 3 (alc). *Mormyrus niloticus*: LFEM, 1 (alc). *Mormyrus tapirus*: MNCN 80593, 3 (alc); MNCN 85283, 1 (alc). *Pantodon buchholzi*: MNCN 73493, 4 (alc). *Xenomystus nigri*: MNCN 227824, 25 (alc).

Clupeomorphs: *Denticeps clupeoides*: MRAC 76-032-P-1, 2 (alc). *Engraulis encrasicolus*: MNCN 68048, 2 (alc); MNCN 65097, 8 (alc); MNCN 1099, 3 (alc). *Engraulis* sp.: MNCN 48896, 3 (alc). *Ethmalosa fimbriata:* MNCN 48865, 3 (alc). *Ilisha fuerthii*: MNCN 49338, 8 (alc). *Thryssa setirostris*: MNCN 49294, 2 (alc).

Ostariophysans: *Bagrus bajad*: LFEM, 1 (alc), 1 (c&s). *Bagrus docmak*: MRAC 86-07-P-512, 1 (alc). *Barbus barbus*: LFEM, 1 (c&s). *Barbus guiraonis*: MNCN 245730, 3 (alc). *Brachyhypopomus brevirostris*: LFEM, 2 (alc). *Brachyhypopomus* sp.: INHS 89761, 2 (alc). *Brycon guatemalensis*: MNCN 180536, 3 (alc). *Brycon henni*: CAS 39499, 1 (alc). *Callichthys callichthys*: USNM 226210, 2 (alc). *Catostomus commersonii*: MNCN 36124, 10 (alc). *Citharinus* sp.: 86-016-P-72, 3 (alc). *Cetopsis coecutiens*: USNM 265628, 2 (alc). *Chanos chanos*: USNM 347536, 1 (alc), LFEM, 1 (alc). *Chrysichthys auratus*: UNB, 2 (alc). *Chrysichthys nigrodigitatus*: LFEM, 1 (c&s). *Cobitis paludica*: MNCN 248076, 7 (alc). *Cromeria nilotica*: MRAC P.141098, 2 (alc). *Danio rerio*: MNCN, 10 (alc). *Diplomystes chilensis*: LFEM, 3 (alc). *Distichodus notospilus*: MRAC A0-048-P-2630, 3 (alc). *Gonorynchus gonorynchus*: LFEM, 2 (alc). *Gonorynchus greyi:* FMNH 103977, 1 (alc). *Grasseichthys gabonensis*: MRAC 73-002-P-264, 3 (alc). *Gymnotus carapo*: INHS 35493, 2 (alc). MNCN 115675, 2 (alc). *Kneria wittei*: MRAC P-33512, 2 (alc). *Nematogenys inermis*: USNM

084346, 2 (alc). *Opsariichthys uncirostris*: MNCN 56668, 3 (alc). *Parakneria abbreviata*: MRAC 99-090-P-703, 3 (alc). *Phractolaemus ansorgii*: MRAC P.137982, 3 (alc). *Pimelodus blochii*: LFEM, 2 (alc), 1 (c&s). *Silurus aristotelis*: LFEM, 2 (alc). *Silurus glanis*: LFEM, 2 (alc). *Sternopygus macrurus*: CAS 48241, 1 (alc); INHS 62059, 2 (alc). *Trichomycterus areolatus*: LFEM, 2 (alc). *Xenocharax spilurus*: MRAC A0-048-P-2539, 3 (alc). [†*Chanoides macropoma*, †*Clupavus maroccanus*, †*Lusitanichthys characiformis*, †*Santanichthys diasii*, and †*Sorbininardus apuliensis*, not directly observed by the author, were also included in the cladistic analysis: see above.]

Euteleosts: *Alepocephalus rostratus*: MNCN 108199, 2 (alc). *Argentina brucei*: USNM 239005, 2 (alc). *Argentina sphyraena*: MNCN 001134, 12 (alc); MNCN 78530, 5 (alc). *Astronesthes niger*: MNCN 1102, 1 (alc). *Aulopus filamentosus:* MNCN 1170, 6 (alc). *Bathylagus euryops:* MNCN 124597, 1 (alc). *Bathylagus longirostris:* USNM 384823, 2 (alc). *Bathylagus tenuis:* MNHN 2005-1978, 2 (alc). *Chlorophthalmus agassizi:* MNCN 1193, 3 (alc); MNCN 1182, 5 (alc). *Coregonus lavaretus:* MNCN 75424, 1 (alc). *Coregonus tugun:* MNCN 75422, 2 (alc). *Esox lucius:* MNCN 197706, 5 (alc). *Galaxias maculatus:* USNM 344889, 2 (alc). *Osmerus eperlanus:* MNCN 193795, 11 (alc). *Osmerus mordax:* USNM 32565, 2 (alc). *Plecoglossus altivelis:* MNCN 192036, 1 (alc). *Retropinna retropinna:* AMNH 30890, 1 (alc). *Salmo trutta:* MNCN 136179, 2 (alc), 1 (c&s); MNCN 16373, 2 (alc); MNCN 40685, 2 (alc). *Salmo* sp.: MNCN 48863, 2 (alc). *Searsia koefoedi*: USNM 206896, 2 (alc). *Stokellia anisodon*: AMNH 31037, 1 (alc). *Stomias boa*: MNCN 74444, 8 (alc); MNCN 74456, 4 (alc). *Thymallus thymallus*: MNCN 115147, 1 (alc); MNCN 114992, 1 (alc). *Umbra limi*: MNCN 35672, 2 (alc); 36072, 2 (alc). *Umbra krameri*: MNCN 36659, 3 (alc). *Xenodermichthys copei*: MNCN 78950, 2 (alc); MNCN 1584, 2 (alc); USNM 215527, 2 (alc).

Non-tetrapod sarcopterygians: *Lepidosiren paradoxa*: CONICET, uncatalogued, 1 (alc). *Neoceratodus forsteri*: MU, uncatalogued, 2 (alc). [†*Tiktaalik roseae* and *Latimeria chalumnae*, not directly observed by the author, were also included in the cladistic analysis: see above.]

Tetrapods: *Ambystoma andersoni*: MNCN, uncatalogued, 2 (alc). *Ambystoma ordinarium*: MNCN, uncatalogued, 2 (alc). *Ambystoma mexicanum*: MNCN, uncatalogued, 2 (alc). *Homo sapiens*: GWU, uncatalogued, 6 (for). *Pan troglodytes*: GWU, uncatalogued, 2 (for). *Pongo pygmaeus*: GWU, uncatalogued, 1 (for). *Timon lepidus*: MNCN, 32544, 1 (alc), MNCN, uncatalogued, 1 (alc). [†*Acanthostega gunnari*, not directly observed by the author, was also included in the cladistic analysis: see above.]

Chapter 3

Phylogenetic Analysis

3.1 CLADISTIC ANALYSIS, DIAGNOSIS FOR CLADES OBTAINED, AND COMPARISON WITH PREVIOUS HYPOTHESES

A total of 356 phylogenetic characters were included in the cladistic analysis. These characters, listed in Section 3.2 below, were coded for each of the 80 terminal taxa included in the analysis, resulting in the data matrix shown in Table 1. The phylogenetic analysis of these characters resulted in 128 equally parsimonious trees with a length of 902 steps, a CI (Consistency Index) of 0.40, and an RI (Retention Index) of 0.77. These CI and RI values are considerably high, considering that the analysis includes 80 terminal taxa representing several major, different groups of a remarkably diverse and complex higher clade such as the Osteichthyes (see, e.g., Diogo, 2004a). Figure 3 shows the phylogenetic relationships between these 80 terminal taxa according to the "majority fools" tree (CI = 0.40, RI = 0.76) obtained by using the "majority fools" option of Nona&Winclada, which shows all clades that are supported by more than 50% of the equally parsimonious trees obtained and thus provides more information than that given by the use of the "strict consensus" option of this program. The numbers from 1 to 69 indicate the number of the clades, following the order given in the synapomorphy list provided below. The clades 14 and 27, marked with an asterisk, were supported by 75% of the equally parsimonious trees obtained; all remaining clades were supported by 100% of these trees. The relationships among the major osteichthyan groups included in the analysis, derived from the phylogenetic hypothesis presented in the tree of Fig. 3, are summarized in Fig. 4.

I have opted to follow, in this Section 3.1, a model somewhat similar to that of Lauder and Liem (1983). That is, apart from listing the synapomorphies supporting each clade obtained in the cladistic analysis, I

Figure 3 Phylogenetic relationships between the osteichthyan taxa included in the cladistic analysis of Chapter 3 according to the "majority fools" tree (CI = 0.39; RI = 0.76) obtained by applying the "majority fools" option of Nona&Winclada to the equally parsimonious trees (CI = 0.40; RI = 0.77) resulting from that analysis. The numbers from 1 to 69 indicate the number of the clades, following the order given in the synapomorphy list provided in the text. Clades 14 and 27, marked with an asterisk, were supported by 75% of the equally parsimonious trees obtained; all remaining clades were supported by 100% of these trees. The branch lengths illustrated are proportional to the number of unambiguous evolutionary transitions leading to the different nodes represented in the tree (for more details, see text).

Table 1 Data matrix of the 356 characters included in the cladistic analysis. Order of characters follows that of Section 3.2. "Inapplicable" and "missing" character states for a certain taxon are indicated with '-' and '?' respectively (for more details, see text). (Characters 001 to 120)

	000000000000000000000000000000000000000000000000000000000000000000000000000000000000000000000000000111111111111111111111
	000000000111111111122222222223333333333444444444455555555556666666666777777777788888888889999999999000000000011111111112
	123456789012345678901234567890123456789012345678901234567890123456789012345678901234567890123456789012345678901234567890
A. ORDINARIUM	000000--1110-00000000-0000--0---011111111111111111111111111?00010-0?00000000000001?010?00?00000?0010000000001100000--001
TIMON	000000--11?0-00000000-0000--0---010111111111111110011111111111100-0?00000000010011?000?00?00000?001010000000??00000--001
A. GUNNARI	?????????????????????????????????1?????????????????????????????00-0?00000000000001?0?0?00?00000?0010000000000100000???01
T. ROSEAE	?????????????????????????????????1?????????????????????????????00-0?000000000?00???0?0?00?00000?00100?00?00???0?0??????0
LEPIDOSIREN	100000--0000-00000010-0000--0---110000000000000000000000000000000-0?00000000000011?0?0?00?00000?00100000?00???0?0??--0?1
NEOCERATODUS	100000--0000-00000000-0000--0---110000000000000000000000000000000-0?00000000000011?0?0?00?00000?00100000?00???0?0??--0?1
LATIMERIA	000000--0000-00000000-0000--1---010000000000000000000000000000000?0?00000000000010?000?00?00000?00001100?00???0?0????0?0
POLYPTERUS	100000--0000-00001000-0000--?---000000000000000000000000000000000-0?00000010000010010000000000000000000000000100000--000
ACIPENSER	110000--0000-00001000-0000--?---000000000000000000000000000000000-0?0000010-0-0001?0?00000000000001000000?00???0?0??000?0
PSEPHURUS	100000--0000-00001000-0000--0---000000000000000000000000000000000-0?0000010-0-100100?00000000000001000000?00???0?0??000?0
LEPISOSTEUS	100000--0000-0001-000-0010001---000000000000000000000000000000000-0-000000000000010?00000000000000010000?00???0?0??--0?0
AMIA	000000--000??00001000?0010000---000000000000000000000000000000000-0-0000000000000000?000000000000010000000001000000??000
ELOPS	00101000000100001-011001101110000000000000000000000000000000000001000000000100000000000000000001000100000000000000000000
NOTACANTHUS	00201000000100001-011001101110000000000000000000000000000000000001010000000101001100000000000001010000000000000000?0001
ANGUILLA	102010000000-0001-01100110111000000000000000000000000000000000001-010000000100001000000000000000101000000000001000000000 1
CONGER	102010000000-0001-01100110111000000000000000000000000000000000001-0?0000000100001000001000000001010000000000010000000001
EURYPHARYNX	---1----0000---01-1--00??0??1???00000000000000000000000000000000?-0?000000010?00110000000000000100000000000001000000--001
MEGALOPS	00101000000100001-01100110111000000000000000000000000000000000000000000000001000000000000?000000100010000000000000000??000
ALBULA	100010--000100001-011001101110000000000000000000000000000000000001010000000100000000010000000001000100000000000000000000
HIODON	00101000000100001-0110011011100000000000000000000000000000000000000000000000?0100000000000?000001000100000000110000000000
PANTODON	002010000000-0001-01100111111000000000000000000000000000000000000000000000001000001 0??00000000001001100000000 0?0000000000
XENOMYSTUS	00???0??000??0001-001001111110010000000000000000000000000000000000100000000100000000000000000001000100000000110000000000
MORMYRUS	000010--000??1001-011001101110000000000000000000000000000000000001010000000101000100100000000000001100000000110000000000
AULOPUS	00201000000110001-011111102100000000000000000000000000000000000000100000000100000000000000000001010100000000010000 0--010
CHLOROPHTHALMUS	00201000000110001-01111110210000000000000000000000000000000000000010000000010000 1?0000000000000101010000000000000000--010
ASTRONESTHES	00201000000110001-01110110?110000000000000000000000000000000000000100000000100001000000000000 0?1010100000000010000 0??010
STOMIAS	00201000000110001-01110110?11000000000000000000000000000000000000010000000010000100000000000000?101010000000000000000000010
SALMO	00201000000110001-0111011011110000000000000000000000000000000000001000000001000000000000000000110101000000000000000000000
COREGONUS	00201000000100001-011101101111000000000000000000000000000000000000?00000000101000000000000000 0?1010100000000000000000000
THYMALLUS	00201000000110001-01110110111100000000000000000000000000000000000010000000010000100000000000000?101010000000000000000000 0
ESOX	00201000000110001-011101101111000000000000000000000000000000000000?00000000100001000000000000 0?10101000000000-0000000000
UMBRA	00201000000110001-0111011011110000000000000000000000000000000000001?0000000100001100000000000 01?01010000000001000000000 0
GALAXIAS	00201000000110001-01110110111?0000000000000000000000000000000000000000000000101001100000000000001010100000000010000000000
RETROPINNA	00201000000110001-01110110111100000000000000000000000000000000000-0-00000001000011000000000000010101000000001-0000000000
STOKELLIA	00201000000110001-00110110111100000000000000000000000000000000000-0-00000001000011000000000000010101000000001-0000000000
OSMERUS	002010000000-0001-011101101111000000000000000000000000000000000000000000000?00001100000000000011010100000000010000 0?0000
PLECOGLOSSUS	002010000000-0001-0111011011110000000000000000000000000000000000000000000001010011000000000000010101000000000100000000 00
ALEPOCEPHALUS	002010000000-0001-001101100010000000000000000000000000000000000000110000000101000001000000000011010100000000010000 0--000

Contd.

Table 1 (Contd.)

XENODERMICHTHYS	002010000000-0001-0011011000100000000000000000000000000000000000011000000001010000010000000000011010100000000100000--000
SEARSIA	002010000000-0001-0011011000100000000000000000000000000000000000011000000001000000010000000000011011100000000100000—0000
ARGENTINA	00201000000110001-0011011011100000000000000000000000000000000000001000000000100000001000000000000101010000000010000000000
BATHYLAGUS	00201000000110001-0111011011100000000000000000000000000000000000001000000000100000001000000000001101010000000010 0000--000
DENTICEPS	00201000000100001-01110110??1000000000000000000000000000000000000001000000000101000000000011001001?0101000000001000001?000
ENGRAULIS	00201000000100001-0111011011100000000000000000000000000000000000101000000001000000000000110010 01?0101000000001010001 1000
THRYSSA	?0201000000100001-0111011011100000000000000000000000000000000000101000000010?00000000000110010011010100000000 1010001?000
ILISHA	002010000001?0001-01110110111000000000000000000000000000000000000100000000010100000000001100100110101000000000 1010011000
ETHMALOSA	00201000000100001-011101101100000000000000000000000000000000000001100000000100000000000011001001101010000000010 00001?000
S. APULIENSIS	???0010?00000001?10001001000????0?01??10100?0?????0?00?????0
CHANOS	00201000000110001-01110110110000000000000000000000000000000000000000000001010011000000010001011010100000000?00100011000
GONORYNCHUS	00201001000?10001-01110110110000000000000000000000000000000000000000000001010011000?1001000001101010000000001010001-000
PHRACTOLAEMUS	00201100000110001-0011011011000000000000000000000000000000000000000011000011100110000000100000110101000000000001010001-000
GRASSEICHTHYS	00201000000110?01-01110110??000000000000000000000000000000000000000001000010100111000000100000 1?010100000000010?000??000
CROMERIA	00201000000110?01-01110110??0??000000000000000000000000000000000000000100001010011100000010000010010100000000010?000??000
PARAKNERIA	00201000000110101-0011011011000000000000000000000000000000000000000001000011010011000000010000010010100000000010000?-000
KNERIA	00201000000110101-0011011011000000000000000000000000000000000000000001000011010011000100010000010010100000000010000?-000
C. MAROCCANUS	???00?0000000001?10?01000000?????00?01?100?0??0???1100??00
S. DIASII	??00???00000?????0?0100?000?1?10000?01010010?10??11?00??00
C. MACROPOMA	??10100000000001?10001000000???00000?01110010?1001111011???0
L. CHARACIFORMIS	???0010?00000001?1010100?00???00?00?01010010?100011101???00
CATOSTOMUS	00201000000110001-0111011000100000000000000000000000000000000000100000000000101000100110001000000111010020120112120011100
COBITIS	00201000000110001-0111011000?000000000000000000000000000000000011-01000010010100010010000100000011010020120012120 01-100
OPSARIICHTHYS	00201000000110001-0111011000000000000000000000000000000000000000100000000000101000100100001000000111010020120012120011100
DANIO	00201000000110001-0111011000?00000000000000000000000000000000000100000000000101000100100001000000111010020120012121011100
BARBUS	00201000000110001-0111011000?00000000000000000000000000000000000100000000000101000100100001000000111010020120012121011100
XENOCHARAX	00201000000110001-00110110110000000000000000000000000000000000001100000001101000100000001110000101010021121112120011100
DISTICHODUS	10201000000110001-00110110110000000000000000000000000000000000001100000001101000100000001110000101010021121112120011100
CITHARINUS	00201000000110001-00110110110000000000000000000000000000000000001100000000101000100100001110000101010121121112120011100
BRYCON	00201000000100001-00110110110000000000000000000000000000000000001100000000101010100000001110000101010021121112120011100
BRACHYHYPOPOMUS	10201000000110001-00110110100000000000000000000000000000000000001110001100101000100000001000000000101002102011212 1011100
STERNOPYGUS	00201000000100011-01110110100000000000000000000000000000000000001110001000101000100000001000000000101002102011212 1011100
GYMNOTUS	00201000000100011-00110110100000000000000000000000000000000000001110001100101000100000001000000000101002102011212 1011100
DIPLOMYSTES	00201000000110001-001101100000000000000000000000000000000000000000111000000100000100110001000 01-0010100211201-2121011100
NEMATOGENYS	00201000000110001-011101100000000000000000000000000000000000000000111000000101000100110001000 01-0010100210201-212101-100
TRICHOMYCTERUS	00201000000110001-011101100000000000000000000000000000000000000000111000000101000100110001000 01-0010100010201-212101-100
CALLICHTHYS	00201000000110101-011101100000000000000000000000000000000000000000111000000101000100110001000 01-0010100010201-212101-100
CETOPSIS	00201010000110001-0111011000010000000000000000000000000000000000001110000001000001001101010000 1-0010100210201-212101-100
SILURUS	00201010000110001-0111011000011000000000000000000000000000000000001110010001000001001100010000 1-0010100211201-2121011100
CHRYSICHTHYS	00201010000110001-0011011000011000000000000000000000000000000000001110000001000001001100010000 1-0010100211201-2121011100
BAGRUS	00201010000110001-0011011000011000000000000000000000000000000000001110000001000001001100010000 1-0010100211201-2121011100
PIMELODUS	00201010000110001-0011011000011000000000000000000000000000000000001110000001010001001100010000 1-0010100211201-2121011100

Table 1 Data matrix of the 356 characters included in the cladistic analysis. Order of characters follows that of Section 3.2. "Inapplicable" and "missing" character states for a certain taxon are indicated with '-' and '?' respectively (for more details, see text). (Characters 121 to 240)

	111111111111111111111111111111111111111111111111111111111111111111111111111111112222222222222222222222222222222222222222
	222222222333333333344444444445555555555666666666677777777778888888888999999999900000000001111111111222222222233333333334
	123456789012345678901234567890123456789012345678901234567890123456789012345678901234567890123456789012345678901234567890
A. ORDINARIUM	1---0?1---0--1--??001----110111110000-000000111000-0----001---0-001-01-------00101------0-0--0000000000000000-000000000
TIMON	1---001---0--1--??001----110111000000-000001111010-0----001---00000-01-0-----11100??0---0-0--0000000000100-000-0-0000000
A. GUNNARI	1---000-?11--1--1?00????1?011100000???0000000??0000000??000000??
T. ROSEAE	00?0000??11?0000??00????1?011001000??000????0????0??00?000??
LEPIDOSIREN	1---110-000--00000001----100000010010-000001111000-0-----01---0-011-01-------00001------0-0--0---000--0-----------00--0
NEOCERATODUS	1---110-011--00000001----100000010000-000000101000-0-----01---0-011-01-------00001------0-0--0---000--0-----------00--0
LATIMERIA	1-00000??11?00001?001----101000010000-000000000000-0----00000-00000-00000-00000010000---1000?0---000--000-------00000000
POLYPTERUS	000000000000000100000?????00000010000-000000000000-0----0100000000000000-000000000000000-0--01000000000000000000000000
ACIPENSER	0100??01000-0000?0000??010000000000000-000000000000-0---000010-001-1-00110-0-0000000-0---0-0-000??000??0????????????00??0
PSEPHURUS	0000??01000-0000?0000??010000000000000-000000000000-0---000010-001-1-00110-0-0000000-0---0-0-000??000??0????????????00??0
LEPISOSTEUS	000001000100000000000-000000000??1000-000100000000-0----01000000001-00000-000000000000000-0--01000000000000000000000000
AMIA	000111000000000000000-000000000??1001100001000000-0----01001000000000000-0000000000000001000?0000001100000000010000000000
ELOPS	0001000010000000001001000000000011100-000000000000-0---1010010001-0000000-00000000000000100000000011100100001100000000
NOTACANTHUS	0---000-100-000000101----000000011101100100000000-0----01001100011-00000-0000000010000010000001000001?00100001-000000000
ANGUILLA	0---000-100-000000001----000000010100-000000000000-0----11001000001-00000-000000000-00001100000000101?0--0--011----00000
CONGER	0---000-100-000000001----000000010100-000000000000-0----11001000000000000-000000000-00001100000000101?0--0--011----00000
EURYPHARYNX	1---0?1---0-01--00001----?00000010100-000000000100-0---?001---001-1-00000-0-0000000-0---0-0--00010?01?0--0--???----00??0
MEGALOPS	00010000100000000100010000000000011100-000000000000-0---1010010000100000000-000000000000000100000000001110010000110000000000
ALBULA	00010000100000000000010010000000111010001000000000-0---1010010000100000000-000000000000000100000010001110010010 11?00000000
HIODON	000100001?001000010001000000000010100-000000000000-0---1010010001-0000000-0010000000100001000000000000100010000110000000000
PANTODON	00010000100100010110011110000000101010000?000000000-0---0010-10001-1-00000-000000000000000100000000000100100-00110-0000000
XENOMYSTUS	0001000010001000010001100000000010100-000000000000-0---0010-10001-0000000-000000000000000100000000000100010010110000000000
MORMYRUS	0001000010001001010001001?000000010101000100000000 1-0---0010010001-1-00000-00000000010000100000010000100110-001-0-0000000
AULOPUS	001100001000000000001----000000001010100100000000000-0---1010010001-0100000-00000000000000001000-101000111101001011100000000
CHLOROPHTHALMUS	001100001000000000001----000000001010100100000000000-0---1010010001-0100000-00100000001000010000-101000111101001011100000000
ASTRONESTHES	001?00001100000001000110000000000101010001000000000-0---1010010001-0000000-0010000000000001000?100000?11?0100001-100000000
STOMIAS	0011000011010000001001----000000001010100010000000000-0---1010010001-0000000-00100000000000100001000000?1110100101-100000000
SALMO	001100001000000000100011000000000010100-000000000000-0---1010010001-0000000-0000000000000001000000000011110100001110000000 00
COREGONUS	001100001000000000100011000000000010100-000000000000-0---1010010001-0000000-000000000000000100000010001 11?010000111000000 00
THYMALLUS	001100001000000000100011000000000010100-000000000000-0---1010010001-0000000-0000000000000001000000000011110100001110000000 00
ESOX	001100001000000000001----000000010100-000000000000-0----0100100001000000 0-00000000000000010000001000111001000011000000000
UMBRA	001100001000000000001----000000010100-000000000000-0----0100100001000000 0-001000000000001000000100001100100001000000000 0
GALAXIAS	00110000100100000?001----0000000010100-000000000000-0---?01001000??0000000-000000000011001000000100001100100101100000000 0
RETROPINNA	00110000000100000 1001----0000000010100-000000000000-0---1010010001-0000000-00000000000000010000000?00011001001011000000000
STOKELLIA	00110000000100000 1001----0000000010100-000000000000-0---1010010001-0000000-00000000000000010000001?00011001001011000000000
OSMERUS	001100001000000000100011000000000010100-000000000000-0---1010010001-0000000-0000000000000001000000000011110100101100000000 00

Contd.

Table 1 (Contd.)

PLECOGLOSSUS	0011000010000000010001100000000010100-000000000000-0---1010010001-1-00000-000000000000001000000100001100100?011000111?00
ALEPOCEPHALUS	00110000100000000110011000000000101010000000000000-0---0010010001-0000000-000000001000001000-10?00011100100001000000000
XENODERMICHTHYS	0011??0010000000001000-00000000010101000000000000-0---0010010001-0000000-000000000000001000-1000001110010001-000000000
SEARSIA	0011??001000000010000-0000000001010100000000000000-0---0010010001-0000000-000000000000001000-1000001110010001-001000000
ARGENTINA	0011000010000000000001100000000010100-000000000000-0---0110010001-0000000-001000001000001000010100001100100101010000010 0
BATHYLAGUS	001100001000000001101----0000000101010000000000000-0---0010010001-1-00000-001000001?00001000-1010000110010010 1-100000100
DENTICEPS	0000000010000000010001100000000010100-000000000000-0---1010010001-0000000-00100000000010100000?000010001000011000000?00
ENGRAULIS	00010000100000000100011000000000101010000000000000-0---0010010001-0000000-0000000010100010000000000111001001011000000000
THRYSSA	00000000100000000100011000000000101010000000000000-0---?010010001-0000000-000000001?000010000000000111001000011000000000
ILISHA	0001000010000000010001100000000010100-000000000000-0---1010010001-0000000-00000000110000100000000001100010010110000000000
ETHMALOSA	0001000010000000010001100000000010100-000000000000-0---1010010001-0000000-0000000010100010000001000111001001011000000100
S. APULIENSIS	000?00????0??00?0000?????0?000001??0??00???????01?0001??0100?01??000001??
CHANOS	000100?0??000000000001100000000010101010000000000000000010010001-0000000-0000000000000001000000100001100100101?000000100
GONORYNCHUS	001100?0??010000000001100000000010101110000000000000000110010001-1-00000-01100000000000010000001100011001001 01?010000100
PHRACTOLAEMUS	100000?0??00001001000110000000001010111000000000000011001001 0010001-1-00000-0000000000000010000001110010001 00?01?000000111
GRASSEICHTHYS	00?1001?--0?00000?00011000000000010101?100000000000000?0?1????01??01-1-00000-000000000-0000??000001?0001?00110?01?000000100
CROMERIA	00?100????0?00000?00011000000000101010100000000000000001001001 0101-1-00000-00000000010000100000001000001?00110101?000000100
PARAKNERIA	000100?0??0100000100011010000000101010100000000000000001001001 0001-1-00000-0000000001000010000000100001100110101000000000100
KNERIA	00?100?0??0100000100011010000000101010100000000000000001001001 0001-1-00000-000000000000100100000001000011001101010000000100
C. MAROCCANUS	000?00?0??00000001000110 00?000001??0???01?0011??0100?01??000001??
S. DIASII	000?00?0??0000000????????0?0000010?0???0100011??0100?01??000001??
C. MACROPOMA	000?00?0??00000001 0?0110?0?000001??0??0100011??0100?0???000001??
L. CHARACIFORMIS	000100?0??000000011001100000000001?10???0100011??0100?0???000001??
CATOSTOMUS	00000000000100000100011000000000101010100000000000000000010010001-1-00000-00100000001010010100001000011001000010000000100
COBITIS	00000000100100000000011000000000101010100000000000000000010010001-1-10001000000000001000010100001000011001000010000000100
OPSARIICHTHYS	00000000100100000000011000000000101010100000000000000000010010001-0000000-00100000100000102000010000110010000100000000100
DANIO	00000000100100000100011000000000101010100000000000000000010010001-0000000-0000000000000010200001000011001000001?000000100
BARBUS	00000000100100000100011000000000101010100000000000000000010010001-0000000-00000000000000102000010000110010000100000000100
XENOCHARAX	00010000000000000100011000000000101010100000000000000000-01001000010000000-00100000010100100000000000100010000110000000000
DISTICHODUS	00010000000000000100011000000000101010100000000000000000-01001100011-0000C-000000000101001000000100001000100001000000000000
CITHARINUS	0001000000000000010001100000000010100-100000000000000000-01001000010000001000000000101001000000100001000101001100000000000
BRYCON	0000000000000000010001100000000010100-100000000000000000-0100110001000000 0-00100000010100100000000000100010100110000000000
BRACHYHYPOPOMUS	00010000010100000000??000000000101011100000000000000000-01001001010000000-001000000100001000000110001000100001010000010 0
STERNOPYGUS	00010000010100000000000-000000000101011100000000000000000-01001001010000000-0010000001000010010001000010001000010100000000
GYMNOTUS	000100000000100000000??000000000101010100000000000000000-01001001010000000-0010000001000010010001100010001000010100000000
DIPLOMYSTES	1-000001010-00011001011000000000010100-10000000000000000-010-10000100000011?00000000-00001000000001000100010000100000000000
NEMATOGENYS	1-000001010-00011000011000000000010100-10000000000000000-110-1100001-000011?01000000-000011000001100010001000000000000000 0
TRICHOMYCTERUS	1-000001010-00011001011000000000010100-10000000000000000-110-1100001-000012?00000000-01011000000110001000100000000000000 00
CALLICHTHYS	1-000011--0-00011201011000000000010100-10000000000000000-110-1100001-000C12?00000000-00001000000110001000100000001000000100
CETOPSIS	1-000001010-00001001011000000000010100-10000000000000100-110-1000000000012?00000001-10001000000110001000100010 0?00000000
SILURUS	1-000001010-00011001011000000000010100-10000000000000000-010-1000000000012?00000000-0000100000011000100010001001 00000000
CHRYSICHTHYS	1-000001010-00011201011000000000010100-10000000000000000-010-1000000000012?00000000-000010000001100010001000000010000 0000
BAGRUS	1-000001010-00011211011000000000010100-10000000000000000-010-1000000000012?00000000-00001000000110001000100000001000 00000
PIMELODUS	1-000001010-00011201011000000000010100-10000000000000000-010-1000000000012?00000000-000010001001100010001000000010000 0000

Table 1 Data matrix of the 356 characters included in the cladistic analysis. Order of characters follows that of Section 3.2. "Inapplicable" and "missing" character states for a certain taxon are indicated with '-' and '?' respectively (for more details, see text). (Characters 241 to 356)

	22222222222222222222222222222222222222222222222222222222222222333333333333333333333333333333333333333333333333333333
	44444444455555555556666666666777777777788888888889999999999000000000011111111112222222222333333333344444444445555555
	12345678901234567890123456789012345678901234567890123456789012345678901234567890123456789012345678901234567890123456
A. ORDINARIUM	00101010000-0-000001001--00000101000?00-10-???1???-------1-011---0----0-0-1-1-01-100??01111-000-0-0000-0000000000000
TIMON	00111010000-0-000?000010000000001000?00-10-1--1----------00011---1----0---1-0-0?-1010101011-000-0-0000-0000000000000
A. GUNNARI	00110010000?0-0?0?00000000000?00000100010-1--0---00-----00011????----0-?-1-?-??-1010000011-000-0-00?0-0?0???0?????0
T. ROSEAE	00?0????000?0-????0000?0000000?00000?00010-1--0---00-----??0??????----?-?-1-?-??-001000?011-000-0-00?0-0?0???0?????0
LEPIDOSIREN	0111?01-000-0-0--?00000??01-00101000--0-0000001000-------00011---0----0---1-0-01-1001011101-000-0-0000-0000000000000
NEOCERATODUS	0111?01-000-0-0--?00000??00000101100000-000001000-------00011---0----0---1-0-01-1001001001-000-0-0000-0000000000000
LATIMERIA	0000000-000-000-00000000000000011100100000000000000-----000000000----?-0-1-1-01-00000000000000-0-0000-0000000000000
POLYPTERUS	0000000000000000000000000000000000100000000000000000-----001000000----0-0-1-1-00-0000000111-0000000000-0000000000000
ACIPENSER	0000?0???0000?0?000??0???000001000000001-0001--1----------000010000----0---010000-1?01???1---0000-00000-0000000000000
PSEPHURUS	0000?0???0000?0'000??0??'00000100000000-0001--1----------000010000----0---01000?-1?010001---0000-00000-0000000000000
LEPISOSTEUS	0000000000000000000000000000000010000000-000000000000-----01?010000-0000---00000001010001100000100000000000000000000
AMIA	000000000000000000000000000000111100000010000000000000000001001000000001---0000000001000110000000000001-0000000000000
ELOPS	00000001000000000000000000000011000100001000000?000000000100000000000100000100000000011-1101-0000000001000000000001000
NOTACANTHUS	000000??0?000?000100001000000011000000101000000100000000010000000000000000001000010011-1101-0000000001000000000001000
ANGUILLA	000000???000?00000??01??0000010100010010100000000000000001001000010000001001001-10011-1111-0000000001000000000001000
CONGER	000000???000?00000??01??0000010100010010100000000000000001001000000000001001001-10011-1111-0000000001000000000001000
EURYPHARYNX	000000???000?00000??0???0000101000000-11-1--1----------??0?1000?-????---1-0-01-10011-1111-000000000?0000000000001000
MEGALOPS	000000010000000000000000000000110000000010000000?0000000001000000000001000001000000011-11?1-0000000001000000000001000
ALBULA	0000000100000000010000010000001100000000100000000000000000100000001000000000100000?0011-1101-0000000001000000000001000
HIODON	00000110?00000000000000000000011000100001000000000000000001001000000000000001000010011-110??000000000100010000000000
PANTODON	00000?110?0000000000000000000001-0000001?-000000000????0100000000000000000001000010011-1110?000000000100010000000000
XENOMYSTUS	00001110?00000000000000000000011000000011000000000000000001001000010000000001000010011-1110?000000000100010000000000
MORMYRUS	0000101?-?000-0001001010000000001-00000001-00000000000000001001000000000110001000010011-1111-0000000001?0010000000000
AULOPUS	00000001000100000000000000000011000100101000000000000000001000000010000000001000010011-11101000000000100001000000000
CHLOROPHTHALMUS	000000010001000000000001000000011000000101000000000000000001000000010000000001000010011-11101000000000100001000000000
ASTRONESTHES	00000001?0010?000000?010000000011000000101000000000000000001000000010000000001010110011-111010000000000100001000000010
STOMIAS	0000000100000000000000010000000011000000101000000000000000001000100010000000001010110011-111010000000000100001000000010
SALMO	0000000100010000000000010000000011000000101000000000000000001000000010000000001000010011-111010000000000100000000000000
COREGONUS	0000100100000000000000010000000011000000101000000000000000001000000010000000001000001?011-111010000000000100000000000000
THYMALLUS	0000000100010000000000010000000011000000101000000000000000001000000010000000001000010011-111010000000000100000000000000
ESOX	000000010000000000100001001000011000?00101000000?0000000001000000010000000001000010011-111010000000000100000000000000
UMBRA	000000010000000000000010010000011000000101000000000000000001000000010000000001000010011-111010000000000100000000000000
GALAXIAS	000010010000000000001000--10000011000000101000000?0000000001000000010000000001000010011-111010000000000100000000000000
RETROPINNA	00001010?00000000000000000000011000000101000000000000000001000000010000000001000010011-111010000000000100000000000000

Contd.

Table 1 (Contd.)

STOKELLIA	000001010?0000000000000000000001100000010100000000000000001000000010000000001000010011-111010000000001000000000000000
OSMERUS	000000010000001000000000010000011000000101000000010000000010000000100000000010000100 11-11101000000?0100000000000000
PLECOGLOSSUS	00001001000000100000000001000001100000010100000001000000000100000001000000000100001101 1-1110101?00001010000000000000
ALEPOCEPHALUS	00000001000000100000000?00000000110000001010000000100000000010000000100000000C1000010011-1110100000000010000000100000000
XENODERMICHTHYS	0000100100000010000000100000000110000001010000000100000000101000000010000000001000010011-1110100000000010000000100000000
SEARSIA	0000100100000010000000000000000011000000101000000010000000001000000010000000001000010011-1110100000000010010001000000000
ARGENTINA	000000010000001001000010000000011000000101000000000000000010000010100000100010000110 11-1110100000000010100001000000000
BATHYLAGUS	000000010000001001000010000000?110000001010000000100000000010000010100000100010000100 11-1110100000000010000001000000000
DENTICEPS	000000010000000000000?000000001100010000100000000100000000100000001000000000100001?011-11101000000000100000000000000001
ENGRAULIS	000000010000000000000000000000110001000011000000000000000010000000000000000000100001001 1-11101000000000100000000000000001
THRYSSA	0000000100000000000000000000001100010000110000000000000000100000000000000000001000010011-111??00000000010000000000000001
ILISHA	0000000100000000000000000000001100010000100000200000000000100000000000000000001000010011-11101000000000100000000000000001
ETHMALOSA	0000000100000000000000000000001100010000100000000000000000100000000000000100C01000011011-11101000000000100000000000000001
S. APULIENSIS	0000?0????0??????0000???00000011000?0?00100000?0000???000?0????????00????00?????11011-1100100000000???0?0???0?????0
CHANOS	00001001000000000000000100000000110001001010000000100000000100?000000000010001000011011-11101000000000100000000000000000
GONORYNCHUS	10001001001000000000000000010001100010010100100001000000000100?000?00000000001000011011-11101000000100100000000000000100
PHRACTOLAEMUS	00001000010000000000000011000001001-001001010000000001000100100?0?01-1000010001000011011-11101010000000100000100000000000
GRASSEICHTHYS	00001001 0?0?00000001001--00000??1000100?010000000000000?01001001000??0000?0001000011011-111??000000000??0000000000000
CROMERIA	00001001000000000001001--00000001-00100?01000000000000?00001001000??0000?0001000011011-111??000000000110000000000000000
PARAKNERIA	0000100100000000010000010000010011000100001000000000000100001000 0?01-00000000010000110 11-11101100000000110000000000000000
KNERIA	000010010000000001000001000001001100010000100000000000010000100000 01-00000000010000110 11-11101100000000110000000000000000
C. MAROCCANUS	00001001 0000?0????000010000000011000?0?001?000000000???000?0??????????????00?????11011-11?0100000000???0?0???0?????0
S. DIASII	0000?00???00?????0000100000000011000?0?001?000000000???000?0????????00????00???0??1011-11???00000000???0?0???0?????0
C. MACROPOMA	00001001?0000?0?0?000010000000011000 00?00100000?0000???000?0??????????????00?????11011-11101000000000???0?0???0?????0
L. CHARACIFORMIS	000010010000000?0?000010000000011000 10?001000000000000?0?000?0????????00????00???001?011-111010000000000???0?0???0?????0
CATOSTOMUS	00001001?00011001000001000000000110001000010000000010000000100?000000000000001000011011-11101000000000100000000000000000
COBITIS	00001001000011001010011000000000110001000010000000000000000010000000000000000001000011011-11101000000000100000000000000000
OPSARIICHTHYS	00001001100011001010001000000000110001000010000000000000000010000000000000000001000011011-11101000000000100000000000000000
DANIO	00001001100011001000001000000000110001000010000000000000000010010000000000000001000011011-11101000100000100000000000000000
BARBUS	00001001100011001000001000000000110001000010000000000000000010000000000000000101000011011-11101000000000100000000000000000
XENOCHARAX	0000100100000000000100000000000011002000001000000000000000001000000000000000C0001000010011-11101000000000100000000100000000
DISTICHODUS	000010010000000000001000?000000001100100000100010000000000000010000000?000001000100001001 1-11101000000000100000000100000000
CITHARINUS	000010010000000000010001000000000110010000010000000100000000010000000000000010001000010011-11101000000000100000000100000000
BRYCON	00001001000000000001000100000000011000100001000000000000000001000000 0?0000000001000010011-11101000000000100000000100000000
BRACHYHYPOPOMUS	00001011?0000?00110??01??0000001100000000100000001000000000100001000000000100010000110 11-1110000000000010000000000000000
STERNOPYGUS	00001011?0000?00110??00??0000001100000000100000101000000000100001000000000000010000100 11-1110100010000100000000000000000
GYMNOTUS	00001011?0000?00110??01??0000001100000000100000101000000000100001000000000000010000100 11-1110000000000010000000000000000
DIPLOMYSTES	0000100101000000110??01??00000?01000000011-0-00100000000000100001000000000000010000100 11-1111-00000000110000000001000000
NEMATOGENYS	0000100101000000110??01??00000?01000010011-0-0010100001000100?010000110000001000010011-1111-00100000110000000001100000
TRICHOMYCTERUS	0000100101000000100??01??00000?01000010011-0-00?01000010001000010000000000000010000100 11-1111-00100000110000000001000000
CALLICHTHYS	0000100101000000100??01??00000?01000000011-0-00100000000000100001000000000C00001000010011-1111-000000000-00000000100000
CETOPSIS	0000100101000000110??01??00000?01000000011-0-00000000000000100001000000000000010000100 11-1111-00000100110000000001100000
SILURUS	0000100101000000110??01??00000?01000000011-0-00100000000000100001000000000000010000100 11-1111-00000000010000000001100000
CHRYSICHTHYS	0000100101000000110??00??00000?010?0000011-0-00100000000000100001000000000000010000100 11-1111-00000000110000000001100000
BAGRUS	0000100101000001110??01??00000?01000000011-0-00100000000000100001000000000000010000100 11-1111-00000000110000000001100000
PIMELODUS	0000100101000001110??01??00000?01001000011-0-00100000000000100001000000000000010000100 11-1111-00000000110000000001100000

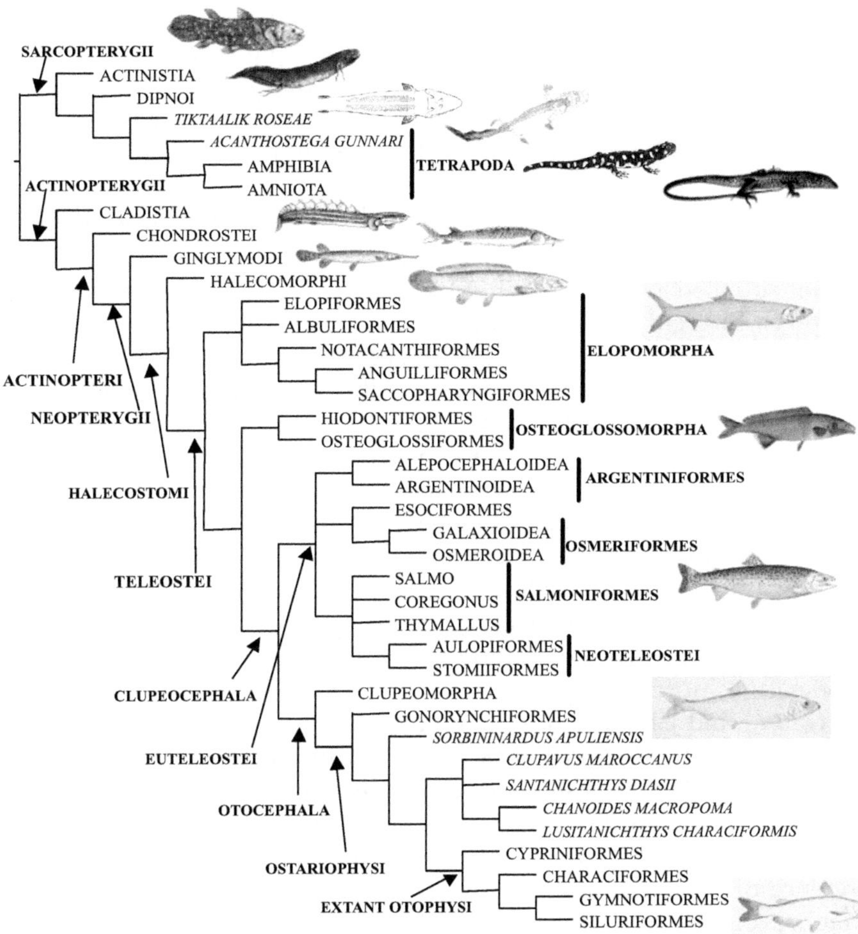

Figure 4 Relationships among the major osteichthyan groups examined, derived from the "majority fools" tree obtained in the cladistic analysis of Chapter 3 (see Fig. 3).

also provide, for most clades, a short commentary and comparison with previous hypotheses, as well as some anatomical drawings of at least a representative member of the clade. The aim is to give the reader a short and very simplified, but hopefully useful, summary of the major osteichthyan clades obtained in this work, which in a certain way resembles, but is admittedly much less extensive and detailed than, that provided in Lauder and Liem's (1983) excellent work mainly focused on actinopterygians. The numbering of the characters mentioned in the list of synapomorphies below follows that of Section 3.2. Character state changes mentioned in this list are restricted to those unambiguous character state changes occurring in the different nodes, and can be divided into two main categories: (1) state changes occurring exclusively in a certain node (in bold) and (2) state changes subsequently reversed in a more terminal node and/or independently acquired in another node (non-bold).

Clade 1 (all actinopterygians included in the cladistic analysis): [1: 0→1], [18: 0→1], [239: 0→1]

As mentioned in Chapter 1, the Osteichthyes, including bony fishes and tetrapods, is a highly diverse group of animals comprising more than 42,000 living species, which is usually divided into two major groups, the Sarcopterygii (lobefins and tetrapods) and the Actinopterygii (rayfins) (Fig. 1). In that chapter, I provided a very short, introductory summary of the scenario that is most commonly accepted in general textbooks nowadays regarding the relationships between the major osteichthyan groups, which is shown in Figs. 1 and 2. I also explained that the extant vertebrates that are usually considered to be the closest relatives of osteichthyans are the chondrichthyans (Fig. 1), but that according to most authors there is a group of fossil fishes that is more closely related to osteichthyans than are the chondrichthyans: the †Acanthodii, a group that, together with the Osteichthyes, form a clade usually named Teleostomi. Apart from the Teleostomi and Chondrichthyes, there is another group that is usually included in the gnathostomes, the †Placodermi, which is usually considered to be the sister-group of teleostomes + chondrichthyans. Osteichthyans are not the only taxa to contain bone in their skeletons, but the taxonomic term Osteichthyes ("bone" and "fish") recognizes the pervasive presence of bone, especially throughout the endoskeleton, among most members of this clade (e.g., Kardong, 2002). It is important to keep in mind that when I refer to bony fishes in the present work, I refer to non-tetrapod osteichthyans, a group that, according to most authors and according to the results of the present work, is actually non-monophyletic (Figs. 1, 3, 4).

Some examples of morphological synapomorphies that are commonly given in general textbooks to support the osteichthyan clade are the presence of dermal bones of the skull such as the maxilla and the premaxilla, the presence of endochondral bone, and the presence of dermal bones of the pectoral girdle such as the interclavicle (e.g., Pough et al. 1996; Lecointre and Le Guyader, 2001; Kardong, 2002). Another feature that is also usually referred to in such books to support this clade is the presence of air-filled sacs, although certain authors consider that this may eventually be a synapomorphy of gnathostomes; according to most researchers these sacs were originally similar to lungs, the presence of "true" swimbladders being thus a derived feature within osteichthyans (e.g., Lecointre and Le Guyader, 2001).

The results of the cladistic analysis of the present work corroborate a main division of osteichthyans into two major extant groups, the Actinopterygii (clade 1) and the Sarcopterygii (clade 64) (Figs. 3, 4). The next paragraphs will deal with the actinopterygian clades obtained in this analysis (clades 1-61). The sarcopterygian clades obtained in the analysis will be dealt with further below (clades 64-69).

The Actinopterygii is a remarkably diverse taxon, with more than 28,000 living species (Nelson, 2006) (see, e.g., Fig. 4). Some morphological features often given in general textbooks to support the monophyly of this taxon are the enlargement of the basal elements of the pectoral fin, the unique arrangement of the scales, the interlocking mechanism, and the presence of a single dorsal fin (e.g., Pough et al., 1996; Lecointre and Le Guyader, 2001; Kardong, 2002). Various other features have been proposed in more specific works as potential synapomorphies of this group (e.g., Patterson, 1982; Cloutier and Arratia, 2004; Gardiner et al., 2005). Molecular studies have also supported the monophyly of this taxon, although, as referred above, some controversial works published by Arnason and colleagues (e.g., Rasmussen and Arnason, 1999a,b; Arnason et al., 2001, 2004) have contradicted this monophyly. According to Meyer and Zardoya (2003) and other authors, the peculiar results of these controversial works were caused by noise (saturation) in the molecular data.

Within the morphological features examined in the cladistic analysis of the present work, three unambiguous synapomorphies support the grouping of all the 73 actinopterygian terminal taxa included in that analysis (Fig. 3, clade 1): the presence of a single section of the muscle intermandibularis (char. 1: 0→1, reversed in clade 5 and ultimately reacquired in a few taxa within that clade, and independently acquired in clade 66), the modification of the muscle coracomandibularis in a peculiar muscle branchiomandibularis (18: 0→1, this muscle is exclusively present in members of clade 1 but is lost, inside of it, in *Lepisosteus* and in clade 6), and the absence of separate bones corresponding to the splenials, postsplenials or splenialpostsplenials (329: 0→1, not reversed inside clade 1 but independently acquired in sarcopterygian terminal taxa such as *Lepidosiren* and *Ambystoma ordinarium*).

It should be noted that, as explained by, for example, Cloutier and Arratia (2004), although separate bones corresponding to the splenials, postsplenials or splenialpostsplenials are absent in the vast majority of extant and fossil actinopterygians, this absence is very likely not a synapomorphy of the actinopterygian clade as a whole because such bones are found in, for example, the basal actinopterygian fossil †*Dialipina*. Concerning the presence of the peculiar branchiomandibularis, it is obviously difficult to appraise whether such a feature was present or not in basal actinopterygian taxa such as †*Dialipina*, but some authors do refer to this feature as a potential actinopterygian synapomorphy (e.g., Wiley, 1979a,b). Regarding the presence of a single intermandibularis section, according to the results of the present analysis this could also constitute a potential synapomorphy of actinopterygians. However, it should be stressed that this latter feature is far more variable within osteichthyans than the presence of the branchiomandibularis, being, for example, present in non-actinopterygian taxa such as *Lepidosiren* and *Ambystoma ordinarium*, while the branchiomandibularis is exclusively found in actinopterygians.

There is a feature that exhibits an ambiguous distribution in the cladistic analysis and that, when the "fast optimization" option of Nona&Winclada is used, appears as a potential synapomorphy of clade 1: the presence of a recognizable dilatator operculi (178: 0→1). When this optimization is chosen, a recognizable dilatator operculi appears as a feature acquired in clade 1 and subsequently reversed in *Acipenser* + *Psephurus* and in teleostean taxa such as *Eurypharynx*. When a "slow optimization" is chosen, it appears as independently acquired in *Polypterus* and in clade 4, being subsequently reversed in this clade 4 and in teleosts such as *Eurypharynx*. Although these options appear equally parsimonious, in this specific case I think there are reasons to suggest that the presence of a recognizable dilatator operculi may well be a synapomorphy of a clade including at least all extant actinopterygians. In fact, the dilatator operculi of *Polypterus* has the same developmental origin (the dorsal part of the mandibular muscle plate), the same innervation (the nerve V), the same function (mainly associated with the abduction of the opercle), and the same overall configuration as the dilatator operculi of the members of clade 4, that is, of neopterygians (e.g., Edgeworth, 1935; Winterbottom, 1974; Miyake et al., 1992; this work). The absence of a dilatator operculi in adult extant acipenseriforms may well be related to the fact that these fishes peculiarly lack an opercular bone, which is precisely the structure where the dilatator operculi of most other actinopterygians is inserted. It should also be noted that extant acipenseriforms are often indicated as an example of paedomorphic taxa (see, e.g., Bemis et al., 1997; Findeis, 1997). This could help to explain why, unlike most other actinopterygians, in which the constrictor dorsalis becomes ontogenetically differentiated into two muscles, the levator arcus palatini and the dilatator operculi, adult acipenseriforms remain with a single constrictor dorsalis muscle, the protractor hyomandibulae (see Sections 3.2, 4.2, 4.3).

In summary, the presence of a recognizable dilatator operculi, of a single section of the muscle intermandibularis, and principally of a peculiar muscle branchiomandibularis, may well be synapomorphies of the clade including all extant actinopterygians. However, it is difficult to discern whether these features were also present in basal actinopterygian fossil taxa such as †*Dialipina*, and, thus, to appraise whether they constitute synapomorphies of the Actinopterygii as a whole. Some authors have, for example, suggested that there was no recognizable dilatator operculi in basal actinopterygians (e.g., Gardiner, 1984; Mallat, 1997). A detailed study on the presence/absence of these features in those basal actinopterygian fossils in which these characters can be discerned is thus needed to shed light on this issue. But even if such a study would point out that these features were not present in those fossils and were just acquired in a somewhat less inclusive actinopterygian clade (e.g., a clade excluding †*Dialipina*: see Cloutier and Arratia, 2004), this would nevertheless provide a good example to illustrate that muscular features could help supporting major osteichthyan clades (see Chapter 4, Section 4.6).

Polypterus: [76: 0→1], [85: 0→1], [136: 0→1], [215: 0→1], [275: 0→1], **[300: 0→1]**, [317: 0→1], [330: 0→1], [331: 0→1]

As mentioned above, *Polypterus* is one of the two living cladistian genera (the other genus is *Erpetoichthys*), and was thus chosen to represent the extant cladistians in the cladistic analysis of the present work. *Polypterus* and *Erpetoichthys* are both included in the family Polypteridae and in the order Polypteriformes, which, it should be noted, is not the only cladistian order recognized in the literature (e.g., Lund, 2000). The nine features listed above are thus not necessarily synapomorphies of *Polypterus*, of the Polypteridae, of the Polypteriformes, or of the Cladistia as a whole; they are rather a mixture of synapomorphies of these different taxa and/or even of more inclusive clades. One example that illustrates this is the presence of an independent, ossified dermohyal **(300: 0→1)**. Within the taxa included in the cladistic analysis, this feature is present only in *Polypterus*. But according to some authors it may well represent a synapomorphy of the Actinopterygii as a whole (e.g., Patterson, 1982; Cloutier and Arratia, 2004; Gardiner et al., 2005). It is outside the scope of the present work to list and discuss all the potential synapomorphies of taxa such as the genus *Polypterus*, the family Polypteridae, the order Polypteriformes or the Cladistia. However, it can be said, for instance, that one of the nine features listed above, the fusion of the maxilla with the infraorbitals forming a peculiar, long-toothed compound structure (215: 0→1, also found in *Lepisosteus*), has often been listed as a potential synapomorphy of the Cladistia (e.g., Lauder and Liem, 1983). Another example of character that has often been listed in the literature as a potential synapomorphy of this latter clade is the presence of peculiar "dorsal finspines" (e.g., Patterson, 1982; Lauder and Liem, 1983; Pough et al., 1996; Lecointre and Le Guyader, 2001).

Some aspects of the anatomy of *Polypterus* are illustrated in Figs. 5, 6, 7, 8, 9 and 10. An illustration of the overall shape of members of this genus is given in Fig. 5. More detailed drawings of the configuration of the cephalic and pectoral girdle osteological and myological structures are given in Figs. 6, 7, 8, 9 and 10. Figures 7 and 8 are original drawings by the author showing details of the adductor mandibulae complex, of the mandible, and of the pectoral girdle bones and muscles. These six figures will be useful for the anatomical comparisons and the discussions provided below.

Clade 2 (all actinopterans included in the cladistic analysis): [302: 0→1], [322: 0→1]

The Actinopteri, including two main extant groups, the Neopterygii and the Chondrostei, is a widely accepted actinopterygian clade (Fig. 1) that has been supported by various morphological and molecular studies (e.g., Meyer and Zardoya, 2003; Stiassny et al., 2004). It comprises more than 28,000 extant species. In fact, this can be said also for the Neopterygii and the Halecostomi, since it is the Teleostei that contributes to the vast majority of the living species of all these clades; the Halecomorphi, the Ginglymodi, the Chondrostei and the Actinistia have, in comparison with the Teleostei, a remarkably small number of extant species (Fig. 1). Examples of some morphological features often given in general textbooks to support the Actinopteri are the presence of a peculiar spiracular canal, the air-filled sacs connecting dorsally to the foregut, or the presence of fringing fulcra on fins (e.g., Pough et al., 1996; Lecointre and Le Guyader, 2001). Various other features have been proposed in more specific works as potential synapomorphies of this group (e.g., Patterson, 1982; Cloutier and Arratia, 2004; Gardiner et al., 2005).

Within the morphological features examined in the cladistic analysis, two unambiguous synapomorphies support the grouping of all the 72 actinopteran terminal taxa analyzed in a monophyletic clade (Fig. 3, clade 2): the absence of ossified interhyal in adults (302: 0→1, reversed and reacquired in certain taxa of this clade 2, and independently acquired in the sarcopterygian clade 65) and the absence of ossified gular plates (322: 0→1, reversed in certain taxa of this clade 2, such gular plates being present, for example, in *Amia*, *Elops* and *Megalops*; independently acquired in certain sarcopterygian taxa, such gular plates being lacking, for example, in *Lepidosiren*, *Neoceratodus*, †*Acanthostega gunnari*, *Ambystoma ordinarium* and *Timon*). Since both these features appear highly homoplasic and have been

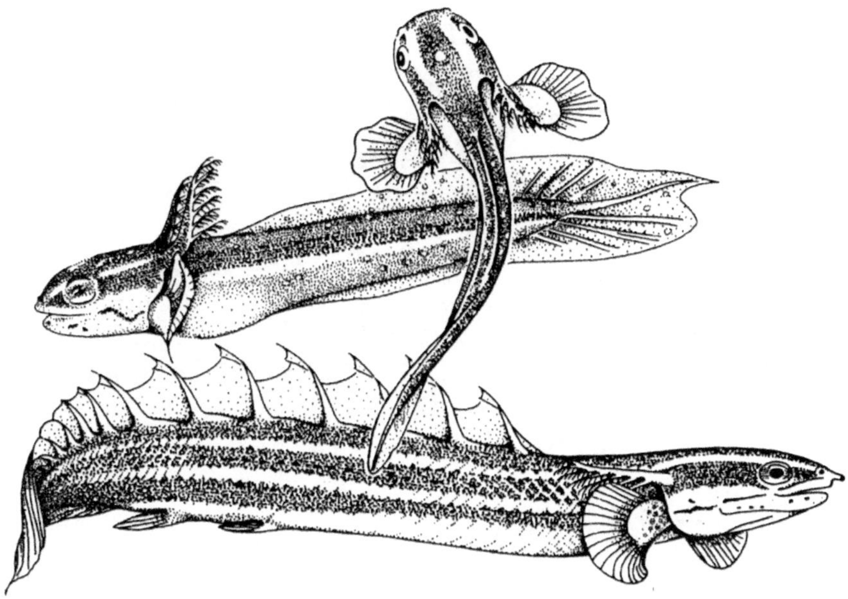

Figure 5 General aspect of members of the genus *Polypterus* (Cladistia) belonging to different developmental stages (modified from Bartsch and Gemballa, 1992).

reversed in certain taxa inside clade 2 as well as independently acquired in certain taxa outside of it, one should be cautious when considering their value as actual potential synapomorphies to support the Actinopteri.

Clade 3 (*Acipenser* + *Psephurus*): [74: 0→1], [99: 0→1], [128: 0→1], [145: 0→1], [153: 1→0], **[180: 0→1]**, [185: 0→1], **[191: 0→1]**, **[192: 1→0]**, [271: 0→1], [284: 0→1], [287: 0→1], **[316: 0→1]**, [325: 0→1]

The Chondrostei includes two extant families, the Polyodontidae, with two living genera, and the Acipenseridae, with four living genera; these two families are represented in the cladistic analysis by *Psephurus* and *Acipenser*, respectively. The Chondrostei is often defined as a group including the fossil family †Birgeriidae and the order Acipenseriformes, which includes the Acipenseridae and Polyodontidae but also other families exclusively represented by fossils (e.g., Grande and Bemis, 1991, 1996; Bemis et al., 1997). Thus, the 14 features listed above are not necessarily synapomorphies of the Chondrostei, of the Acipenseriformes, or of an eventual clade Acipenseroidei, in which some authors include the families Acipenseridae and Polyodontidae (e.g., Bemis et al., 1997). They are, instead, probably a mixture of synapomorphies of these different taxa and/or even of other

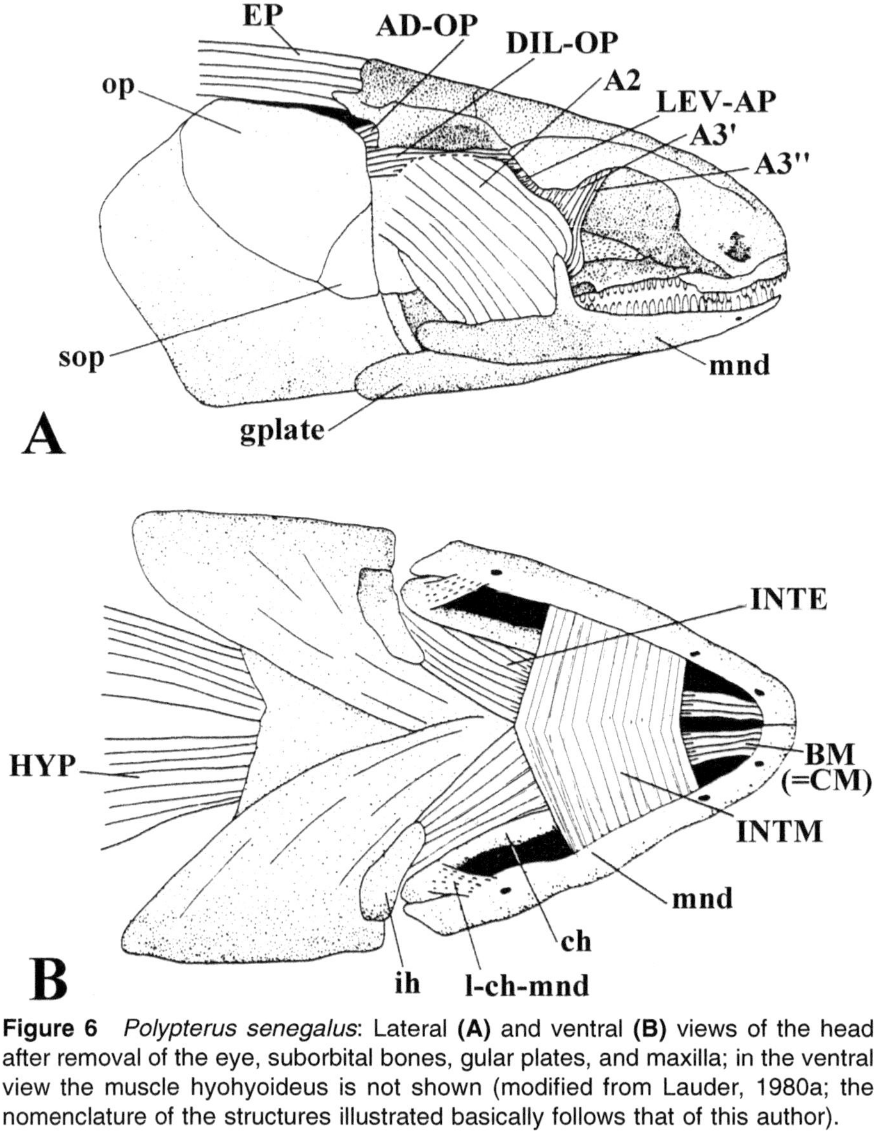

Figure 6 *Polypterus senegalus*: Lateral **(A)** and ventral **(B)** views of the head after removal of the eye, suborbital bones, gular plates, and maxilla; in the ventral view the muscle hyohyoideus is not shown (modified from Lauder, 1980a; the nomenclature of the structures illustrated basically follows that of this author).

clades (see above). As explained above, it is outside the scope of this work to list and discuss all the potential synapomorphies of clades such as the Chondrostei, the Acipenseriformes, or the Acipenseroidei. An overview of such potential synapomorphies has been given by Grande and Bemis (1996) and Bemis et al. (1997). It can be said, however, that two of the 14 features listed above, for example, are often listed as potential synapomorphies of the

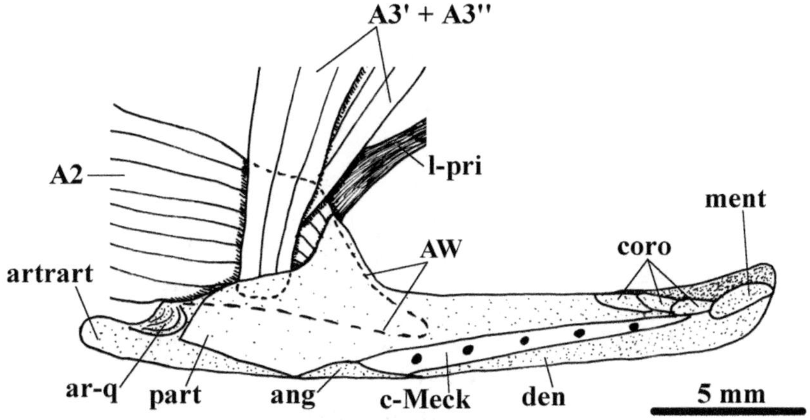

Figure 7 *Polypterus bichir*. Mesial view of adductor mandibulae and mandible; mandibular teeth are not illustrated.

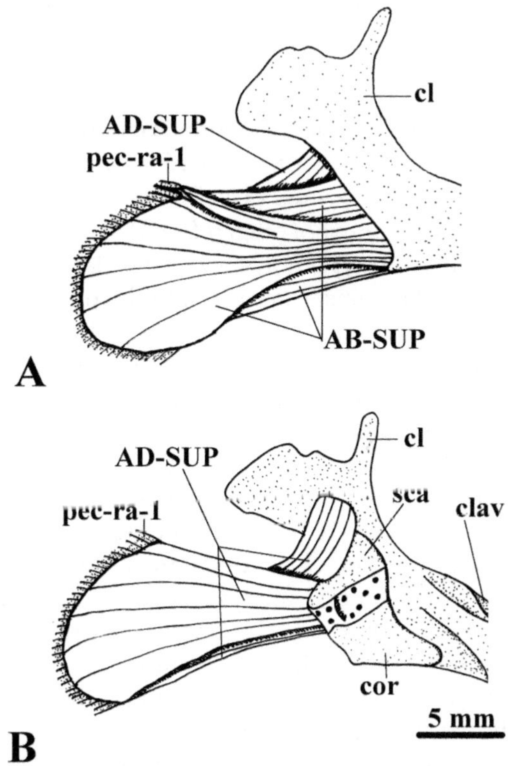

Figure 8 *Polypterus bichir*. Lateral **(A)** and mesial **(B)** views of the muscles associated with the pectoral fin; in the lateral view the adductor of the fin is also illustrated.

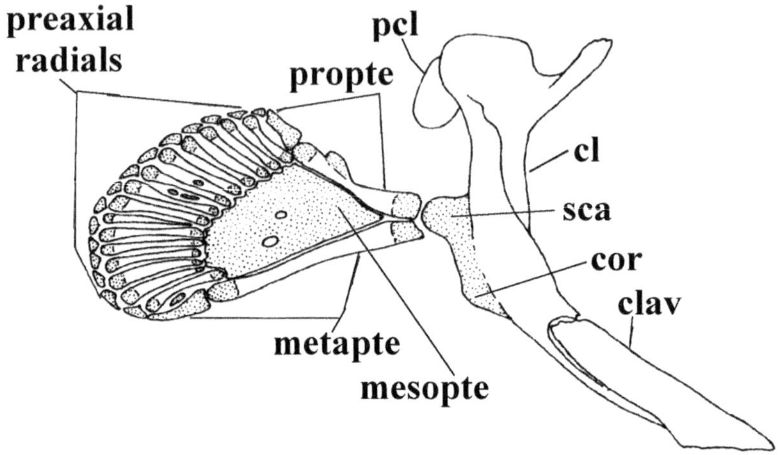

Figure 9 *Polypterus ornatipinnis*: Pectoral girdle and fin (modified from Rosen et al., 1981; the nomenclature of the structures illustrated basically follows that of these authors).

Acipenseroidei, namely the loss of opercular bone (284: 0→1) and the reduction in number and peculiar modification of the branchiostegal rays (**316: 0→1**) (e.g., Grande and Bemis, 1996; Bemis et al., 1997).

Some of these 14 features are not often referred to in the literature but may well constitute potential synapomorphies of the clades mentioned above. For example, among all the taxa examined in which this character was not coded as missing or inapplicable, together with the catfishes analyzed, *Acipenser* and *Psephurus* are the only ones that seemingly lack a ligament, either ossified or not, between the posttemporal and the posterior margin of the neurocranium (128: 0→1). Also, among all taxa examined in which this character was not coded as missing or inapplicable, *Acipenser* and *Psephurus* are the only ones in which the adductor operculi muscle inserts exclusively on the subopercle (**180: 0→1**). In addition, among all non-teleostean osteichthyan taxa examined in which this character could be discerned, that is, among all non-teleostean taxa analyzed excepting †*Tiktaalik roseae* and †*Acanthostega gunnari*, *Acipenser* and *Psephurus* are the two taxa in which there is no well-differentiated adductor mandibulae A3' (185: 0→1). They are also the only two taxa, among all groups analyzed in which this character could be discerned, to exhibit a peculiar "retractor" and a peculiar "protractor" of the hyomandibula (**191: 0→1; 192: 0→1**). It would thus be interesting to undertake a detailed study of the fossil groups usually included in the Chondrostei, in order to examine whether at least some of

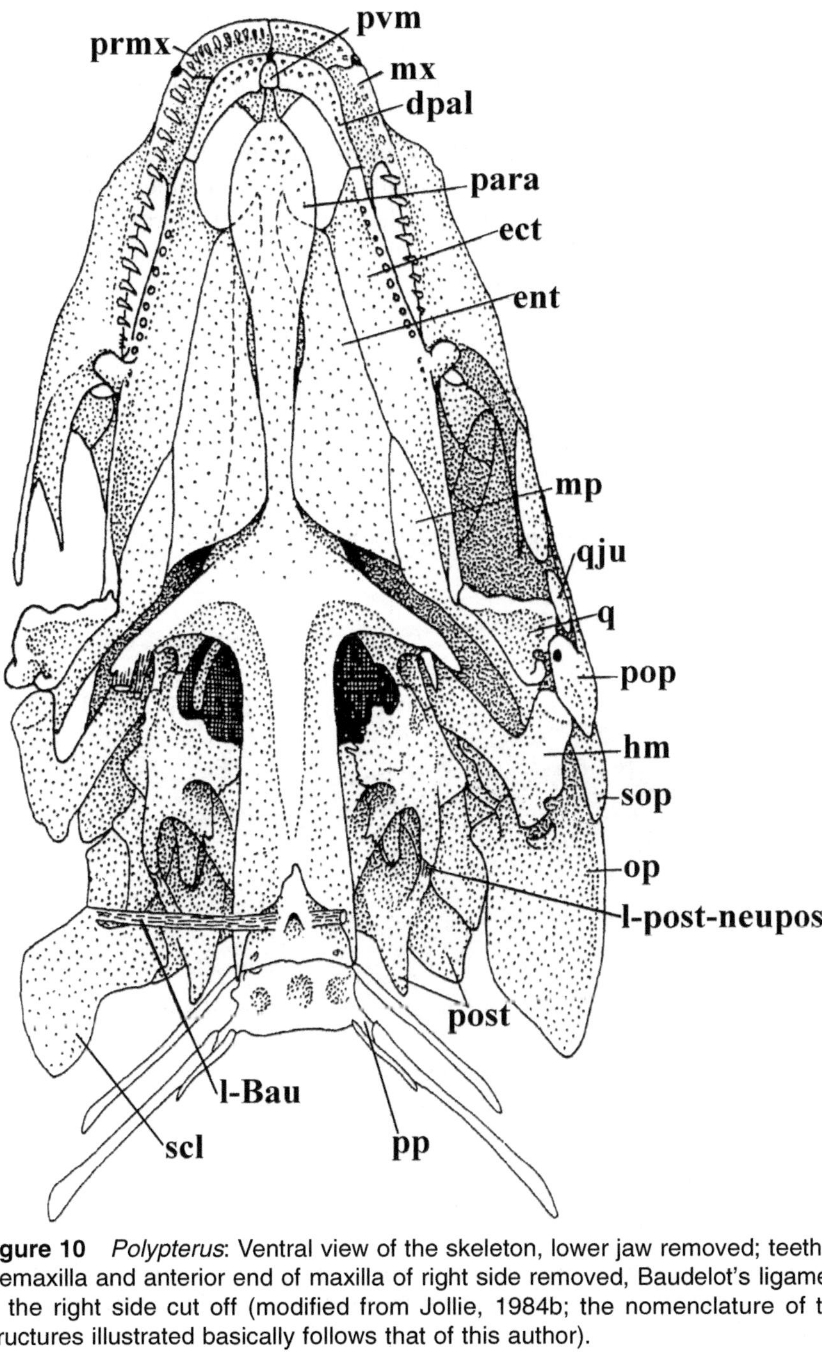

Figure 10 *Polypterus*: Ventral view of the skeleton, lower jaw removed; teeth of premaxilla and anterior end of maxilla of right side removed, Baudelot's ligament of the right side cut off (modified from Jollie, 1984b; the nomenclature of the structures illustrated basically follows that of this author).

these features can eventually be examined in those groups and to check, thus, whether they may actually constitute potential synapomorphies of the Acipenseroidei, of the Acipenseriformes, or even of the Chondrostei as a whole.

Some aspects of the anatomy of chondrosteans are illustrated in Figs. 11, 12 and 13. An illustration of the overall shape of a member of the genus *Acipenser* is given in Fig. 11. More detailed drawings of the configuration of the cephalic and pectoral girdle osteological and myological structures of fishes of the genus *Polyodon* are given in Figs. 12 and 13.

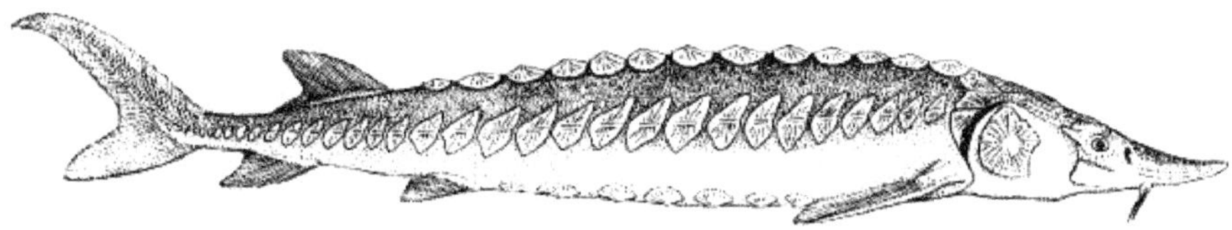

Figure 11 General aspect of a member of the genus *Acipenser* (Chondrostei) (modified from Bauchot, 1987).

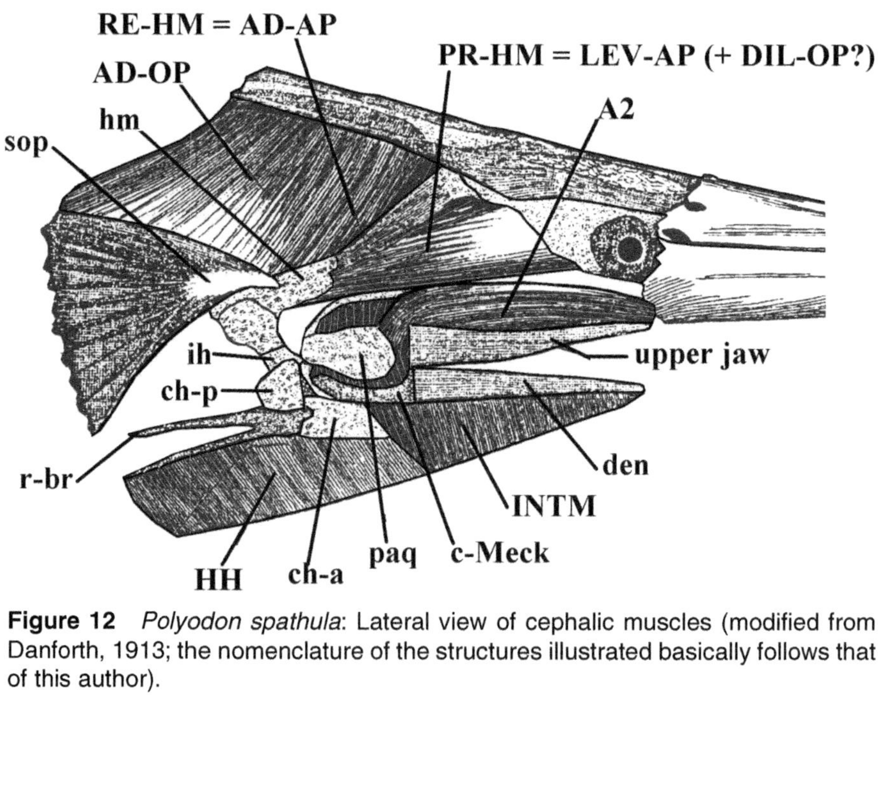

Figure 12 *Polyodon spathula*: Lateral view of cephalic muscles (modified from Danforth, 1913; the nomenclature of the structures illustrated basically follows that of this author).

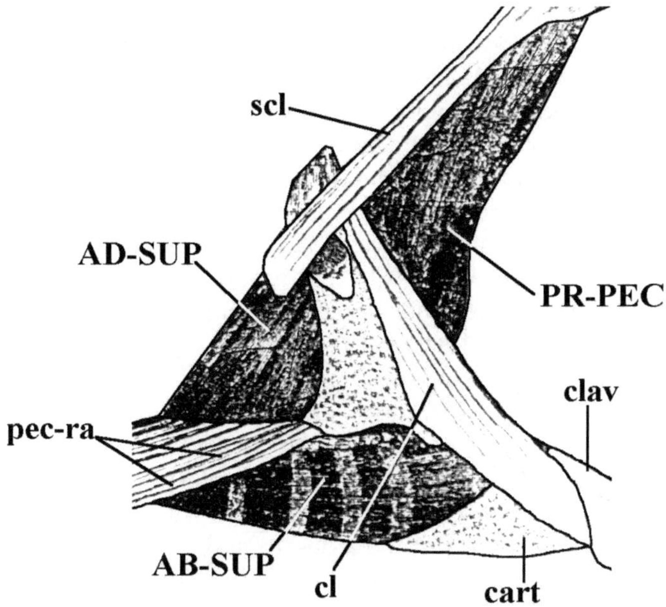

Figure 13 *Polyodon spathula*: Lateral view of the pectoral girdle muscles (modified from Danforth, 1913; the nomenclature of the structures illustrated basically follows that of this author).

Acipenser: **[2: 0→1]**, **[122: 0→1]**, **[279: 0→1]**

As explained above, these three features are not necessarily synapomorphies of the genus *Acipenser*; the supracleithrum peculiarly firmly attached to the posttemporal (**122: 0→1**) and the presence of a peculiar "cartilaginous palatal complex" (**279: 0→1**) have been considered to be potential synapomorphies of the family Acipenseridae (e.g., Bemis et al., 1997; Findeis, 1997).

Psephurus: [79: 0→1]

This feature has been proposed by Grande and Bemis (1991, 1996), Bemis et al. (1997) and others as a potential synapomorphy of the family Polyodontidae.

Clade 4 (all neopterygians included in the cladistic analysis): **[25: 0→1]**, [154: 0→1], [272: 0→1], **[299: 0→1]**, [328: 0→1]

The monophyly of the Neopterygii, a clade including three main osteichthyan extant groups, the Ginglymodi, the Halecomorphi and the Teleostei, is commonly accepted in the literature (Fig. 1). However, the support for this clade comes mostly from morphological studies, and some molecular works have in fact provided support for an alternative hypothesis in which the Ginglymodi and the Halecomorphi appear more closely related to Chondrostei than to Teleostei (e.g., Venkatesch et al., 2001; Inoue et al., 2003). But some molecular works did provide support for the monophyly of the Neopterygii, for example, Kikugawa et al. (2004). Examples of some morphological features often given in general textbooks to support the monophyly of this clade are the presence of an ossified symplectic and the rays of the dorsal and anal fins reduced to equal the number of endoskeletal supports (e.g., Pough et al., 1996; Lecointre and Le Guyader, 2001; Kardong, 2002). An updated list of other potential synapomorphies of this clade is given in the work of Gardiner et al. (2005).

Within the structures examined in the cladistic analysis, five unambiguous synapomorphies support the grouping of the 70 neopterygian taxa included in that analysis (Fig 3, clade 4): presence of a recognizable arrector dorsalis (**25: 0→1**, homoplasy-free among the groups in which this character could be discerned); the presence of "pectoral splints" (154: 0→ 1, reversed in clades 10 and 12 but not independently acquired in any terminal taxon outside of clade 4); presence of an ossified symplectic (272: 0→1, reversed in some clades inside of this clade 4; the sarcopterygian *Latimeria* has also an ossified "symplectic": see Section 3.2); presence of two independent ossified ceratohyals (**299: 0→1**, homoplasy-free among the groups in which this character was not coded as missing or inapplicable); mentomeckelian bones not present as independent ossifications (328: 0→1, independently acquired in some sarcopterygian taxa; *Acipenser* was coded as '?' for this character: see Section 3.2). This cladistic analysis thus strongly supports the hypothesis that the Ginglymodi and the Halecomorphi are more closely related to Teleostei than to Chondrostei (Figs. 3, 4).

Lepisosteus: [100: 0→1], [130: 0→1], **[162: 0→1]**, [215: 0→1], [335: 0→1]

As mentioned above, *Lepisosteus* is one of the two living genera of Ginglymodi (the other is *Atractosteus*), and was thus chosen to represent the extant members of this clade in the cladistic analysis. These genera are included in the family Lepisosteidae, and, thus, in the order Lepisosteiformes, but this is not the only order of Ginglymodi recognized in the literature (e.g., an order †Semionotiformes is also included in this clade by Grande, 2005). Thus, again, it should be kept in mind that the five characters listed above may well be a mixture of synapomorphies of clades as diverse as the genus *Lepisosteus*, the family Lepisosteidae, the order Lepisosteiformes or the Ginglymodi as a whole. For instance, the maxilla fused with infraorbitals forming a peculiar, long-toothed compound structure (215: 0→1, also found in *Polypterus*), is a feature that has often been listed in general textbooks as a potential synapomorphy of the Ginglymodi (e.g., Pough et al., 1996; Lecointre and Le Guyader, 2001). Another example of character that has often been listed in such textbooks as a potential synapomorphy of this latter clade is the presence of peculiar opisthocoelous centra (e.g., Pough et al., 1996; Lecointre and Le Guyader, 2001; Kardong, 2002). A detailed review of the Ginglymodi and of the potential synapomorphies supporting its major subgroups will be given in a voluminous work in progress by Grande and Bemis (see Grande, 2005).

An illustration of the overall shape of a member of the genus *Lepisosteus* is given in Fig. 14. More detailed drawings of the configuration of the cephalic and pectoral girdle osteological and myological structures of fishes of this genus are given in Figs. 15, 16 and 17. The two latter figures are original drawings by the author showing details of the adductor mandibulae complex, the mandible, and the pectoral girdle musculature.

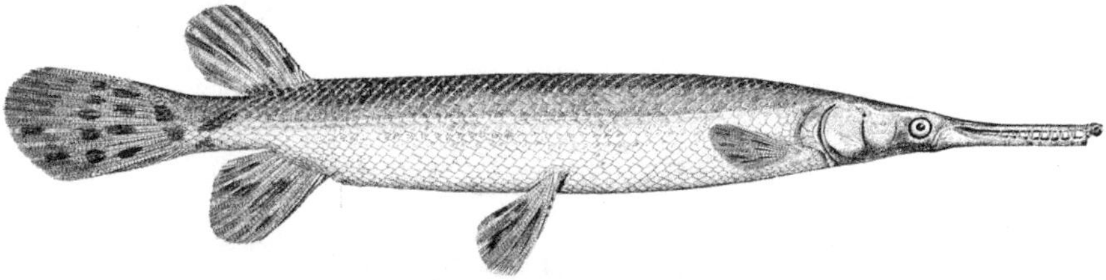

Figure 14 General aspect of a member of the genus *Lepisosteus* (Ginglymodi) (modified from Goode, 1884-1887).

Clade 5 (all halecostomes included in the cladistic analysis): [1: 1→0], [82: 1→0], [97: 0→1], [124: 0→1], **[181: 0→1]**, [209: 0→1], **[221: 0→1]**, [230: 0→1], [273: 0→1], **[282: 0→1]**, [342: 0→1]

The monophyly of the Halecostomi is commonly accepted in the literature (Fig. 1). However, as explained above, in Stiassny et al.'s (2004) overview, these authors stated that "although most workers have followed Patterson (1973) in the recognition of *Amia* as the closest living relative of the Teleostei, there remains some controversy" about the phylogenetic position of this taxon, and, thus, about the monophyly of the Halecostomi. Actually, according to Grande (2005), a study in progress by Grande and Bemis suggests that the Halecomorphi may be more closely related to the Ginglymodi than to the Teleostei. Such a view has also been supported by molecular studies such as those by Inoue et al. (2003) and Kikugawa et al. (2004). Thus, there is still much controversy concerning one fundamental question: which are the closest living relatives of teleosts, the amiids of the

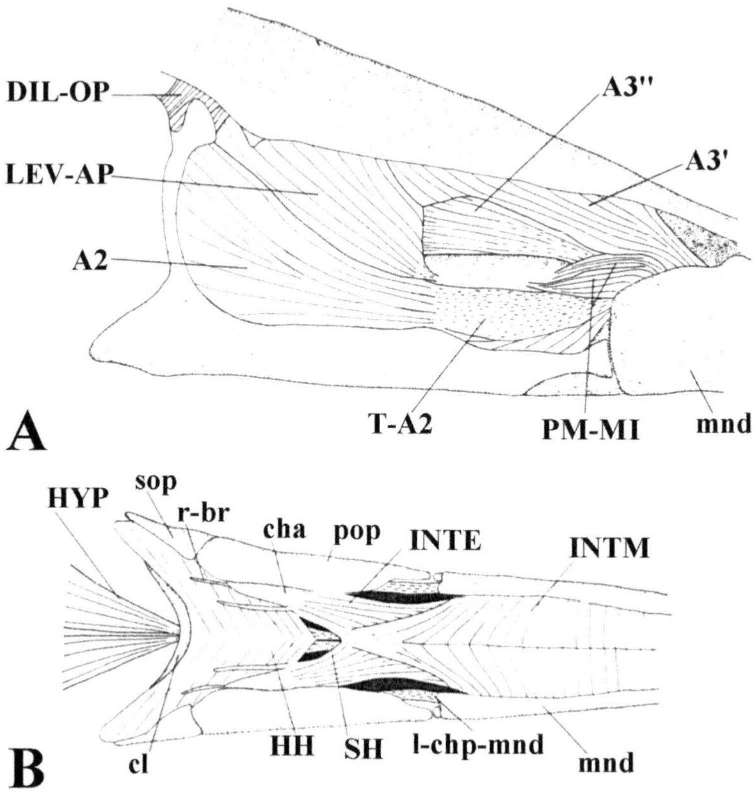

Figure 15 *Lepisosteus oculatus*: Lateral **(A)** and ventral **(B)** views of the cephalic musculature after removal of the eye (modified from Lauder, 1980a; the nomenclature of the structures illustrated basically follows that of this author).

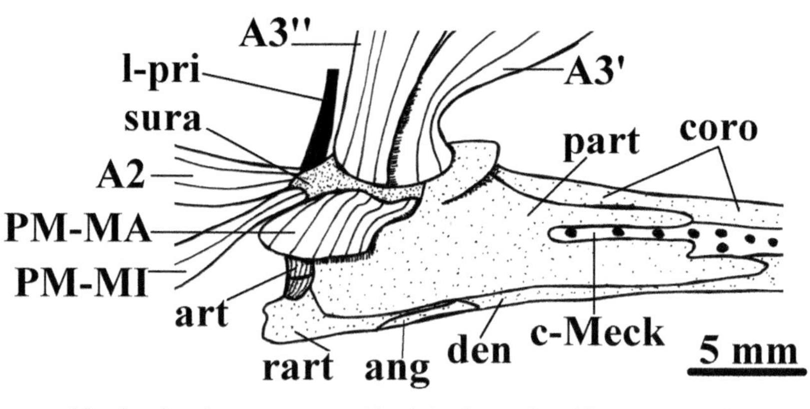

Figure 16 *Lepisosteus osseus*: Mesial view of adductor mandibulae and mandible; mandibular teeth are not illustrated.

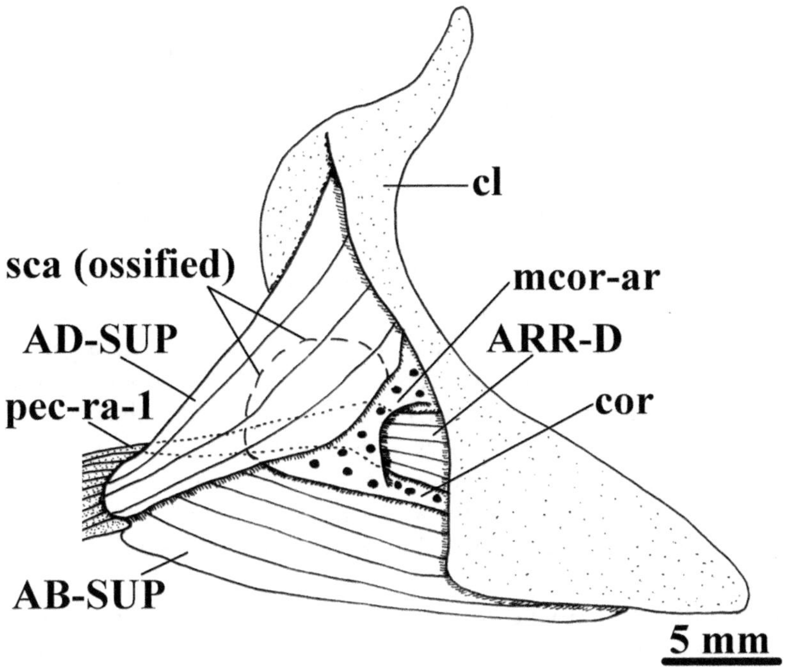

Figure 17 *Lepisosteus osseus*: Mesial view of the muscles associated with the pectoral fin.

genus *Amia*, or both these fishes and the lepisosteids of the genera *Lepisosteus* and *Atractosteus*, that is, the members of the three extant genera of an eventual clade "Holostei" including the Halecomorphi and the Ginglymodi?

The results of the cladistic analysis of the present work strongly support the first of these two hypotheses, that is, that *Amia* is the closest living relative of teleosts (Figs. 3, 4). Eleven features support this hypothesis in the analysis: presence of two sections of the intermandibularis (1: 1→0, modified in certain taxa inside clade 5 and also occurring in taxa outside of it); presence of basisphenoid as an independent ossification (82: 1→0, modified in certain taxa inside clade 5 and also occurring in taxa outside of it); presence of intercalar as an independent ossification (97: 0→1, modified in certain taxa inside clade 5 but not occurring, among the terminal taxa included in the cladistic analysis, outside of this clade); dorsomesial limb of posttemporal (or posttemporo-supracleithrum) markedly thin and mesially extended (124: 0→1, modified in certain taxa inside clade 5 but not occurring, among the terminal taxa included in the present work, outside of this clade); adductor operculi relatively well separated from adductor arcus palatini (**181: 0→1**, homoplasy-free within the terminal taxa in which this character could be discerned); presence of distinct levator operculi (209: 0→1, only reversed, inside clade 5, in *Eurypharynx*; a muscle often called "levator operculi" but seemingly not homologous with the levator operculi of the members of this clade 5 is found in *Latimeria*: see Chapter 4, Section 4.2); maxillae not markedly ankylosed with neurocranium (**221: 0→1**, homoplasy-free among the terminal taxa in which this character was not coded as missing or inapplicable); mesial surface of distal portions of maxillae/supramaxillae firmly attached to mandibles (230: 0→1, only reversed, among the terminal taxa in which this character was not coded as missing or inapplicable, in the non-diplomystid catfishes); quadratojugals not present as independent ossifications (273: 0→1, also occurring in certain taxa outside of clade 5, but not reversed inside of it); interopercle present as an independent element (**282: 0→1**, homoplasy-free within the terminal taxa included in the cladistic analysis in which this character could be discerned); presence of an independent, ossified coronomeckelian bone (342: not occurring outside clade 5 and, among the terminal taxa of this clade in which this character could be discerned, reverted in the catfish *Callichthys*).

It is worth noting that some of these features (e.g., the presence of two sections of the intermandibularis, the presence of the basisphenoid as an independent ossification and the dorsomesial limb of posttemporal (or posttemporo-supracleithrum) markedly thin and mesially extended) are rather homoplasic within the osteichthyans examined. Also, according to Grande (2005), some of the features listed above, such as the presence of intercalars and of interopercles, are seemingly present in some fossil taxa of the Ginglymodi and, thus, may well actually represent neopterygian synapomorphies. In fact, the view often defended by authors favoring a clade Ginglymodi + Halecomorphi is not that the various peculiar derived features found in the Halecomorphi and in the Teleostei were acquired independently. Instead, they often defend the view that these features were acquired in the node leading to their clade Teleostei + (Halecomorphi + Ginglymodi) and then reversed within the Ginglymodi. The presence of intercalars and interopercles in certain fossil taxa of the Ginglymodi are examples provided by these authors to support such a view. However, it should be stressed that besides the presence of intercalars and interopercles there are many other peculiar features that appear as almost, or even completely, homoplasy-free in the present cladistic analysis and that support a close relationship between *Amia* and the Teleostei, such as the adductor operculi well separated from the adductor arcus palatini, the presence of a distinct levator operculi, the maxillae not markedly ankylosed with the neurocranium, the mesial surface of the distal portions of maxillae/

supramaxillae firmly attached to mandibles, and the presence of an independent ossified coronomeckelian bone. In fact, it is important to stress that the presence of these peculiar features in *Amia* and in most teleosts is importantly related to the presence of certain characteristic feeding and breathing mechanisms in these fishes that are absent in any other osteichthyan group (e.g., Lauder, 1980a; Lauder and Liem, 1983). If one were to follow the view of the defenders of the Halecomorphi + Ginglymodi hypothesis, one would have to consider that all these peculiar features and these unique feeding and breathing mechanisms were acquired in the node leading to a clade including these two clades plus the Teleostei and that all of them were subsequently lost within the Ginglymodi, as they are all missing in at least some members of this group, such as *Lepisosteus*. Such a scenario seems, at least at first sight, rather unsound. But, of course, one cannot completely exclude the hypothesis that such a scenario did occur during evolution. However, I do think that, unless the defenders of the Halecomorphi + Ginglymodi hypothesis could explain the reasons that may have caused all these reversions of features and mechanisms that seemingly have played, and continue to play, a crucial role in the success of the numerous fishes having them, most researchers will continue to favor the grouping of Halecomorphi + Teleostei. In fact, I consider that this is a good example to illustrate the potential contribution of myological and/or functional characters for clarifying the higher-level phylogeny of major vertebrate groups: many of the characters listed above to support the monophyly of halecostomes concern myological and/or functional features (e.g., the presence of a distinct levator operculi and thus of an opercular mechanism to open the mouth [209], the high mobility of the maxilla [221] and its association with the movements of the mandible [230], and the marked division between the adductor operculi and the adductor arcus palatini [209]).

Amia: [109: 0→1], [125: 0→1], [157: 0→1], [158: 0→1], **[163: 0→1]**, [274: 0→1], [275: 0→1], [311: 0→1], [322: 0→1]

As mentioned above, *Amia*, and namely the species *Amia calva*, is the only living taxon of the Halecomorphi and thus was chosen to represent the extant members of this clade in the cladistic analysis of the present work. *Amia* is included in the family Amiidae and in the order Amiiformes. But this is not the only order of Halecomorphi recognized in the literature (e.g., an order †Ionoscopiformes is also included in this clade by Grande and Bemis, 1998). Thus, once again, the nine characters listed above may well be a mixture of synapomorphies of clades as diverse as the genus *Amia*, the Amiidae, the Amiiformes or the Halecomorphi as a whole. For instance, the presence of two articulatory points of contact between the ventral portion of the suspensorium/palatoquadrate and the mandible (274: 0→1, also occurring in certain other taxa examined such as *Latimeria*) is a feature often listed in general textbooks as a potential synapomorphy of the Halecomorphi (e.g., Pough et al., 1996; Lecointre and Le Guyader, 2001). For a review of the major subgroups of Halecomorphi and of their potential synapomorphies see, for example, Grande and Bemis (1998).

An illustration of the general shape of a member of the genus *Amia* is given in Fig. 18. More detailed drawings of the configuration of the cephalic and pectoral girdle osteological and myological structures of fishes of this genus are given in Figs. 19, 20, 21, 22 and 23. The three latter figures are original drawings by the author showing details of the adductor mandibulae complex, the mandible, and the pectoral girdle musculature.

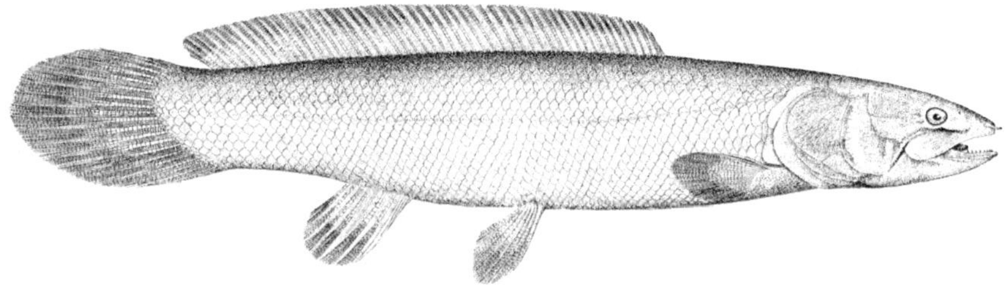

Figure 18 General aspect of a member of the genus *Amia* (Halecomorphi) (modified from Goode, 1884-1887).

Clade 6 (all teleosts included in the cladistic analysis): [3: 0→1], **[5: 0→1]**, [20: 0→1], **[21: 0→1]**, **[24: 0→1]**, [27: 0→1], [28: 0→1], [76: 0→1], [101: 0→1], [129: 0→1], [142: 0→1], **[155: 0→1]**, [186: 0→1], [225: 0→1], [231: 0→1], [248: 0→1], [317: 0→1], [325: 0→1], [326: 0→1]

The Teleostei, with about 28,000 living species, is the most speciose group of vertebrates (Nelson, 2006). The extraordinary taxonomic diversity of teleostean fishes is associated with an enormous variety of morphological forms (see, e.g., Fig. 24) and adaptations to very different freshwater, brackish, and marine habitats, from high elevation mountain springs over 5,000 meters above sea level to the ocean abyss almost 8500 meters below (e.g., Arratia, 2000, Stiassny et al., 2004). As expected, the teleostean taxa included in the cladistic analysis appear more closely related to each other than to other osteichthyans (Figs. 3, 4). Among the 19 unambiguous synapomorphies listed above supporting the clade including these teleostean taxa (Fig. 3, clade 6), features such as those concerning characters 3 (posterior intermandibularis integrated in protractor hyoidei, but also deeply associated with anterior intermandibularis), 20 (anteroventromesial

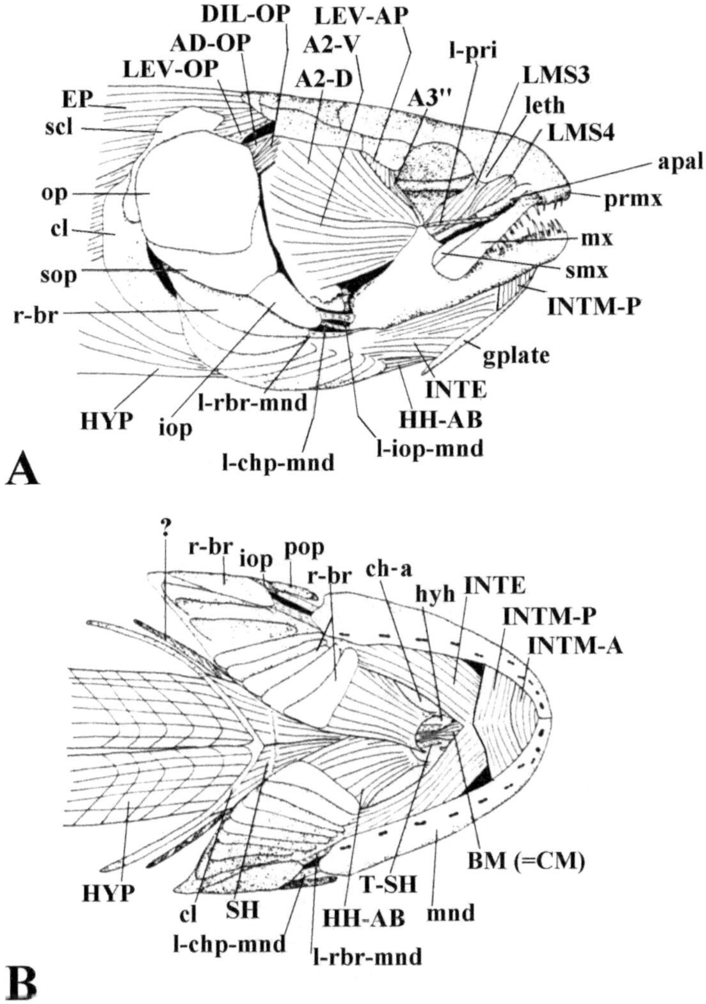

Figure 19 *Amia calva*: Lateral **(A)** and ventral **(B)** views of the cephalic musculature after removal of the eye, dorsal portion of the preopercle, and gular plate; in the ventral view the hyohyoideus inferior and the hyohyoidei adductores are not shown (modified from Lauder, 1980a; the nomenclature of the structures illustrated basically follows that of this author).

portion of hypaxialis continuous with posteroventromesial portion of sternohyoideus), 21 (sternohyoideus consolidated into a single median muscle), 24 (presence of distinct muscle arrector ventralis), 76 (prevomer unpaired), 101 (ossification of chondral supraoccipital), 142 (mesocoracoid arch ossified), 155 (first pectoral ray articulating directly with scapula

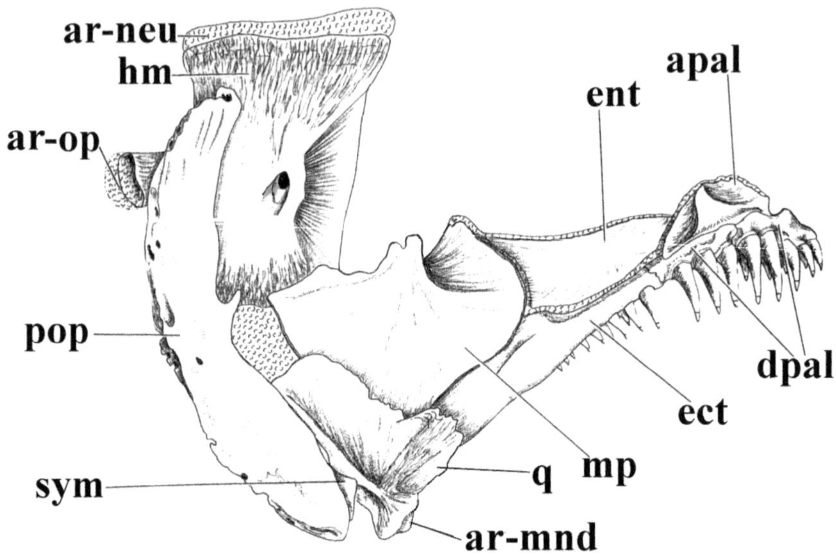

Figure 20 *Amia calva*: Lateral view of the preopercle and suspensorium (modified from Arratia and Schultze, 1991; the nomenclature of the structures illustrated basically follows that of these authors).

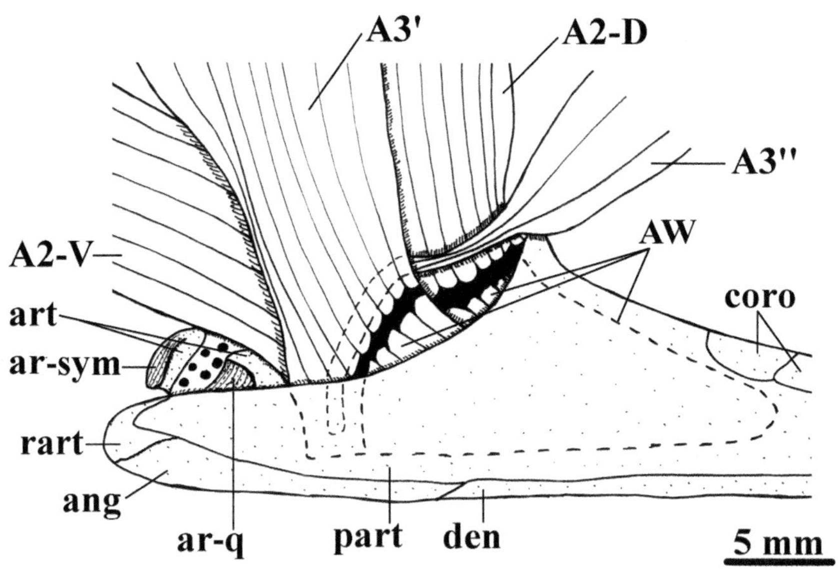

Figure 21 *Amia calva*: Mesial view of mandible and of adductor mandibulae sections A2, A3', A3'' and Aw; the levator maxillae superioris 3 and 4 and the mandibular teeth are not illustrated.

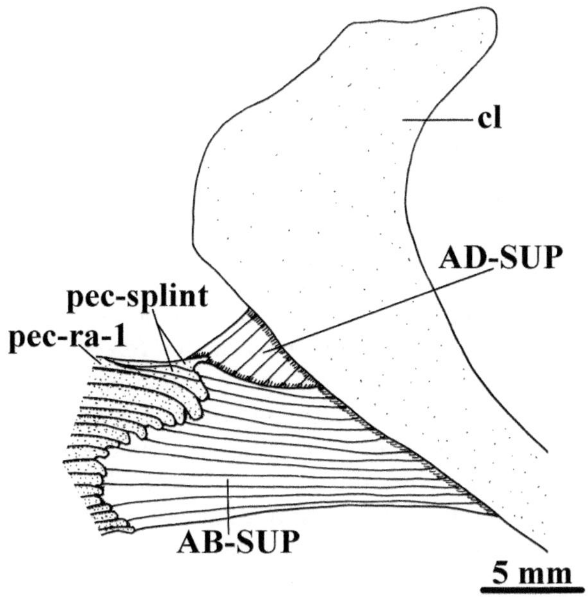

Figure 22 *Amia calva*: Lateral view of the muscles associated with the pectoral fin.

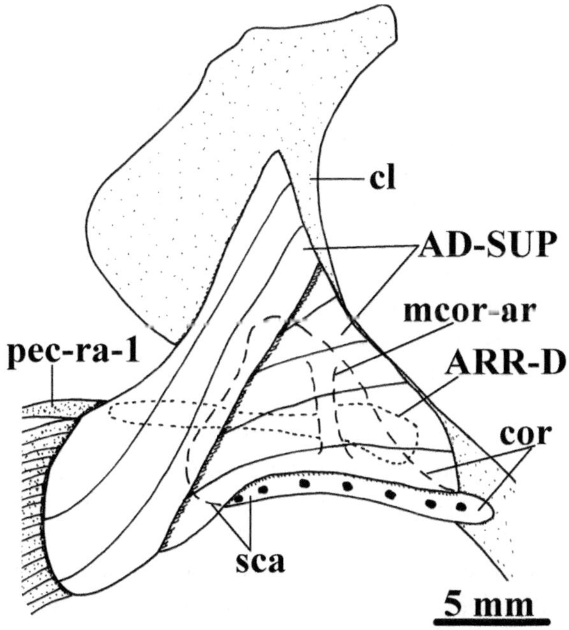

Figure 23 *Amia calva*: Mesial view of the muscles associated with the pectoral fin; the abductor of the fin was removed.

and/or eventually coracoid), 225 (premaxillae not markedly ankylosed with neurocranium), 248 (mobile articulation, either direct or indirect, between autopalatine/dermopalatine and maxilla), 317 (presence of ossified urohyal/"urohyal"/parahyoid/"tendon urohyal"), 325 (coronoid bones absent as independent ossifications) and 326 (prearticulars absent as independent ossifications) have been previously proposed as potential teleostean synapomorphies (e.g., Schaeffer and Rosen, 1961; Lauder and Liem, 1983; Jollie, 1986; De Pinna, 1996; Arratia and Schultze, 1990; Arratia, 1997, 1999). However, some other unambiguous synapomorphies supporting this clade 6 concern features that are not often referred to in the literature as potential synapomorphies of the Teleostei or of the clade including the various extant teleostean groups: e.g., those concerning characters 5 (significant part of interhyoideus associated with mandible), 27 (arrector dorsalis subdivided into two well-developed sections), 28 (arrector dorsalis attaching on both the first and second pectoral rays), 129 (Baudelot's ligament attaching proximally on anterior free vertebrae, and eventually also with neurocranium), 186 (absence of distinct adductor mandibulae A3'), and 231 (primordial ligament attaching posteriorly on posterolateral surface of mandible).

Apart from the 19 unambiguous synapomorphies listed above, there are some features that according to the results of the cladistic analysis have an ambiguous distribution but that may eventually be interpreted as potential synapomorphies of clade 6. For instance, the loss of the muscle branchiomandibularis (= coracomandibularis) (17: 0→ 1): according to the results of the cladistic analysis, this feature may have been acquired in neopterygians and then reversed in *Amia* or may have been independently acquired in *Lepisosteus* and in teleosts. Although the two hypotheses appear equally parsimonious, in my opinion it seems rather unsound that *Amia* has independently acquired a muscle that is strikingly similar to, and has precisely the same developmental origin and the same innervation as, the characteristic muscle branchiomandibularis of other actinopterygians (e.g., Lauder, 1980a; Wilga et al., 2000; see also Chapter 4, Section 4.2). Thus, this might eventually be a further synapomorphy to support clade 6. The loss of protractor pectoralis (29: 0→1) is also a feature that may have occurred in neopterygians and then reversed in *Amia* or that may have occurred independently in *Lepisosteus* and in teleosts. For the reasons explained just above, I would be inclined to favor the latter hypothesis; therefore, this might eventually also represent a synapomorphy to support clade 6. Other features are, for example, the presence of distinct, strong ligaments connecting the anterior surface/anterior cartilage of autopalatines and/or dermopalatines and the maxilla and/or premaxillae (222: 0→1) and the presence of an ossified interhyal (302: 0→1). According to the results of the cladistic analysis, these two latter features might represent synapomorphies of the Clupeocephala and of the Elopomorpha, or, instead, might represent synapomorphies of the clade including all teleostean taxa included in the analysis that were subsequently lost in the Osteoglossomorpha (as well as in other, more derived taxa: see below). These two features could thus eventually be interpreted as synapomorphies of the Elopomorpha + Clupeocephala, if the Osteoglossomorpha were considered to be the most basal extant teleostean clade, as is done by various authors (see, e.g., Fig. 2). However, as can be seen in Fig. 3, all the most parsimonious trees (100%) obtained in the cladistic analysis of the present work support the Elopomorpha, and not the Osteoglossomorpha, as the most basal teleostean group included in the analysis (see clade 12 below).

Clade 7 (all elopomorphs included in the cladistic analysis): [66: 0→1], [331: 0→1], **[353: 0→1]**

As explained above, the monophyly of a clade Elopomorpha including elopiforms, albuliforms, notacanthiforms, anguilliforms, and the peculiar saccopharyngiforms has been questioned by some authors (e.g., Filleul, 2000; Filleul and Lavoué, 2001). As stressed by these authors, no published morphological cladistic analysis has in fact included representatives of all these taxa and supported their grouping in a monophyletic clade. Regarding molecular works, some have supported the inclusion of these taxa in a monophyletic group (e.g., Wang et al., 2003; Inoue et al., 2004), but others have contradicted this view (e.g., Obermiller and Pfeiler, 2003).

In the cladistic analysis of the present work the elopiform, albuliform, notacanthiform, saccopharyngiform and anguilliform fishes included in that analysis do appear grouped in a monophyletic clade (Figs. 3, 4). This is thus the first published morphological cladistic analysis supporting the monophyly of a clade including all these fishes. Although there are only three unambiguous synapomorphies supporting this monophyly, there are other features with an ambiguous distribution on the tree that might eventually also represent potential elopomorph synapomorphies. It should be emphasized that that all the equally parsimonious trees obtained in the analysis (100%) support the elopomorph clade (Fig. 3). Curiously, in 50% of these trees the albuliforms appear as the sister-group of the clade including the Notacanthiformes, Anguilliformes and Saccopharyngiformes, as is commonly accepted in the literature (see Fig. 2), but in the other 50% they appear more closely related to the elopiforms. Consequently, in the tree of Fig. 3 the Albuliformes, the Elopiformes and the clade comprising Notacanthiformes, Anguilliformes and Saccopharyngiformes appear in an unresolved trichotomy.

The first feature listed above supporting the monophyly of elopomorphs concerns the placement of the anterior margin of the prevomer/vomer well posteriorly to the anterior margin of the mesethmoid (66: 0→1). It is found in the specimens examined of the genera *Elops*, *Albula* and *Notacanthus* and cannot be discerned in the anguilliform fishes analyzed (due to a complete fusion between the prevomer and the mesethmoid) and saccopharyngiform fishes analyzed (such a fusion might also occur, but this is not clear); it was independently acquired in a few other teleostean taxa (see below). The second feature listed above concerns the absence of the retroarticular as an independent ossification (331: 0→1). Within the actinopterygian taxa included in the cladistic analysis, this is a rather rare feature, being found in elopomorphs, *Polypterus*, catfishes and *Mormyrus* (*Hiodon* was coded as '?': see Section 3.2). However, it should be noted that this feature is also found in some teleostean fishes not included in the analysis (see, e.g., Nelson, 1973). The third feature listed above, which is homoplasy-free within the taxa included in the analysis in which this feature could be discerned, concerns the presence of a "leptocephalus larva" (**353: 0→1**).

Two features with ambiguous distributions that may eventually be interpreted as potential synapomorphies of the elopomorphs included in the analysis are the presence of distinct, strong ligaments connecting the anterior surface/anterior cartilage of autopalatines and/or dermopalatines and the maxillae and/or premaxillae (222: 0→1) and the presence of an ossified interhyal (302: 0→1) (see comments to clade 6 above).

Some aspects of the anatomy of elopomorphs are illustrated in Figs. 24, 25, 26, 27, 28, 29 and 30. Illustrations of the general shape of an elopiform fish of the genus *Elops* and of an anguilliform fish of the genus *Gymnothorax* are given in Fig. 24. More detailed drawings of the configuration of the cephalic and pectoral girdle osteological and myological structures of fishes of the genus *Elops* are given in Figs. 25, 26, 27, 28, 29 and 30. This latter figure is an original drawing by the author showing details of the pectoral girdle muscles and bones of these fishes.

Albula: [1: 0→1], [3: 1→0], [85: 0→1], [145: 0→1], [157: 0→1], [161: 0→1], [216: 0→1], [228: 0→1], [258: 0→1], **[265: 0→1]**, [307: 0→1]

Clade 8 (*Elops* + *Megalops*): [311: 0→1], [322: 1→0]

A close relationship between *Elops* and *Megalops* has been defended by Greenwood et al. (1966), Nelson (1973), Forey et al. (1996) and other authors and has received further support in some molecular studies (e.g., Obermiller and Pfeiler, 2003; Wang et al., 2003; Inoue et al., 2004). However, in some studies *Megalops* was placed together with *Elops* and the other elopomorphs

Figure 24 Diversity of teleosts: general shape of members of the genera *Elops* (right, top), *Ictalurus* (right, middle), *Gymnothorax* (right, below), and *Hippocampus* (left) (modified from Goode, 1884-1887).

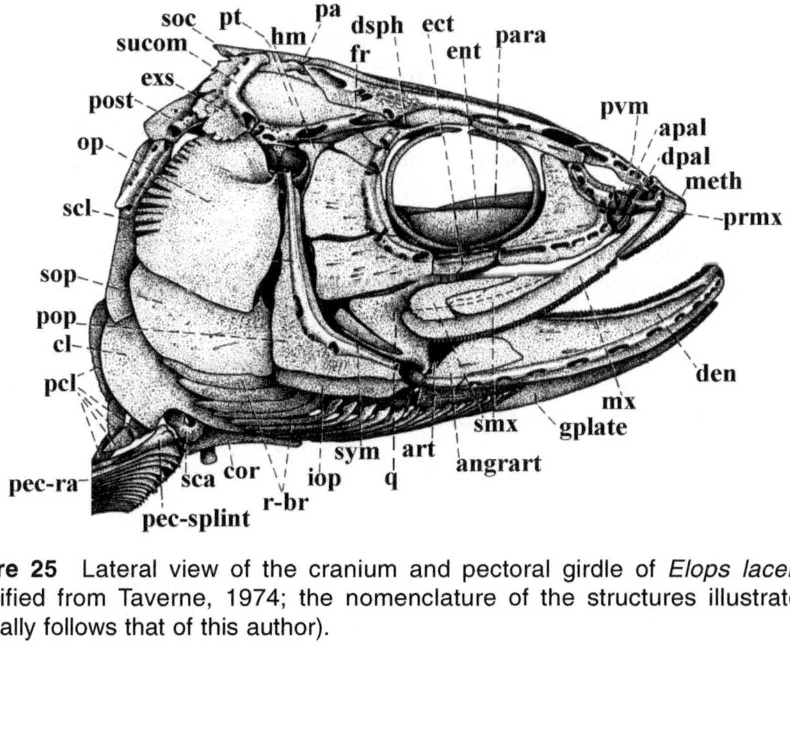

Figure 25 Lateral view of the cranium and pectoral girdle of *Elops lacerta* (modified from Taverne, 1974; the nomenclature of the structures illustrated basically follows that of this author).

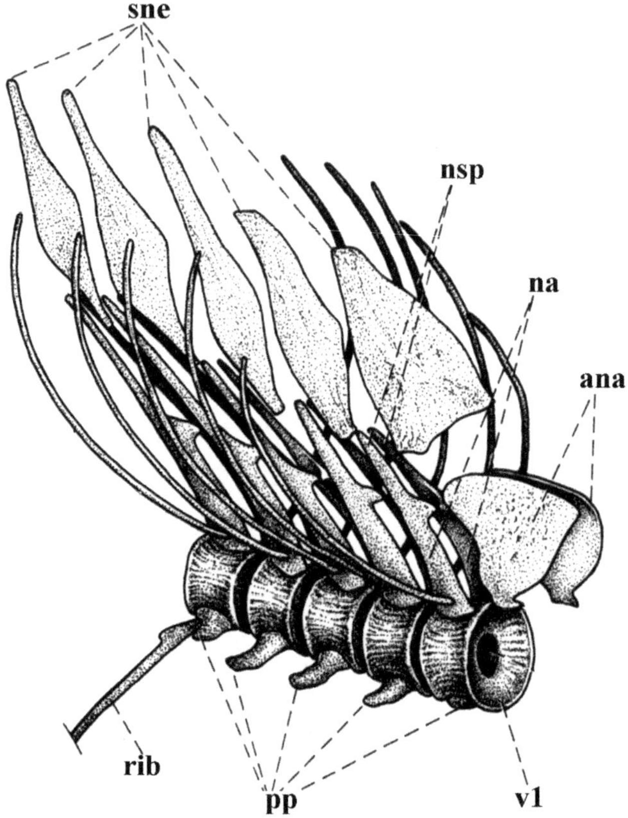

Figure 26 Dorsolateral view of anterior vertebrae of *Elops lacerta* (modified from Taverne, 1974; the nomenclature of the structures illustrated basically follows that of this author).

in an unresolved trichotomy (e.g., Patterson and Rosen, 1977) or was placed as the sister-group of a clade including *Elops* and the remaining elopomorphs (e.g., Forey, 1973b). In the cladistic analysis of the present work the two features listed above support a closer relationship between the elopiforms *Elops* and *Megalops* than between any of these two taxa and the other elopomorph taxa included in the analysis.

Elops: [139: 0→1], [185: 0→1], [277: 0→1]

Megalops: [66: 1→0] , [138: 0→1]

Clade 9 (*Notacanthus* + *Eurypharynx* + *Conger* + *Anguilla*): [3: 1→2], [81: 0→1], [97: 1→0], [99: 0→1], [120: 0→1], [141: 0→1], [187: 1→0], [263: 0→1], [280: 0→1]

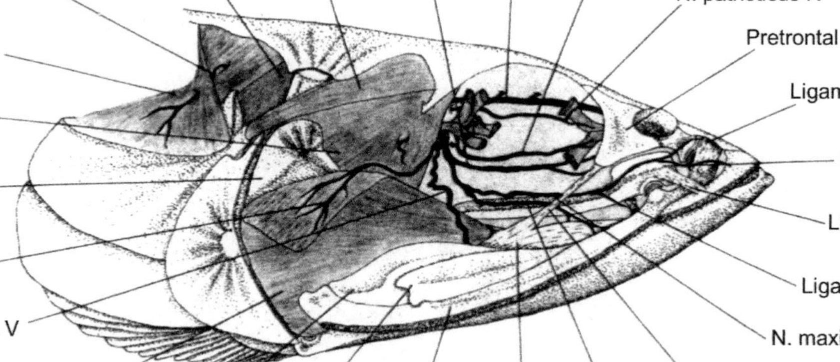

Figure 27 Lateral view of cephalic muscles of *Elops saurus*; the dorsal edge of the opercular bone has been folded over to expose the muscle adductor operculi (modified from Vrba, 1968; the nomenclature of the structures illustrated basically follows that of this author).

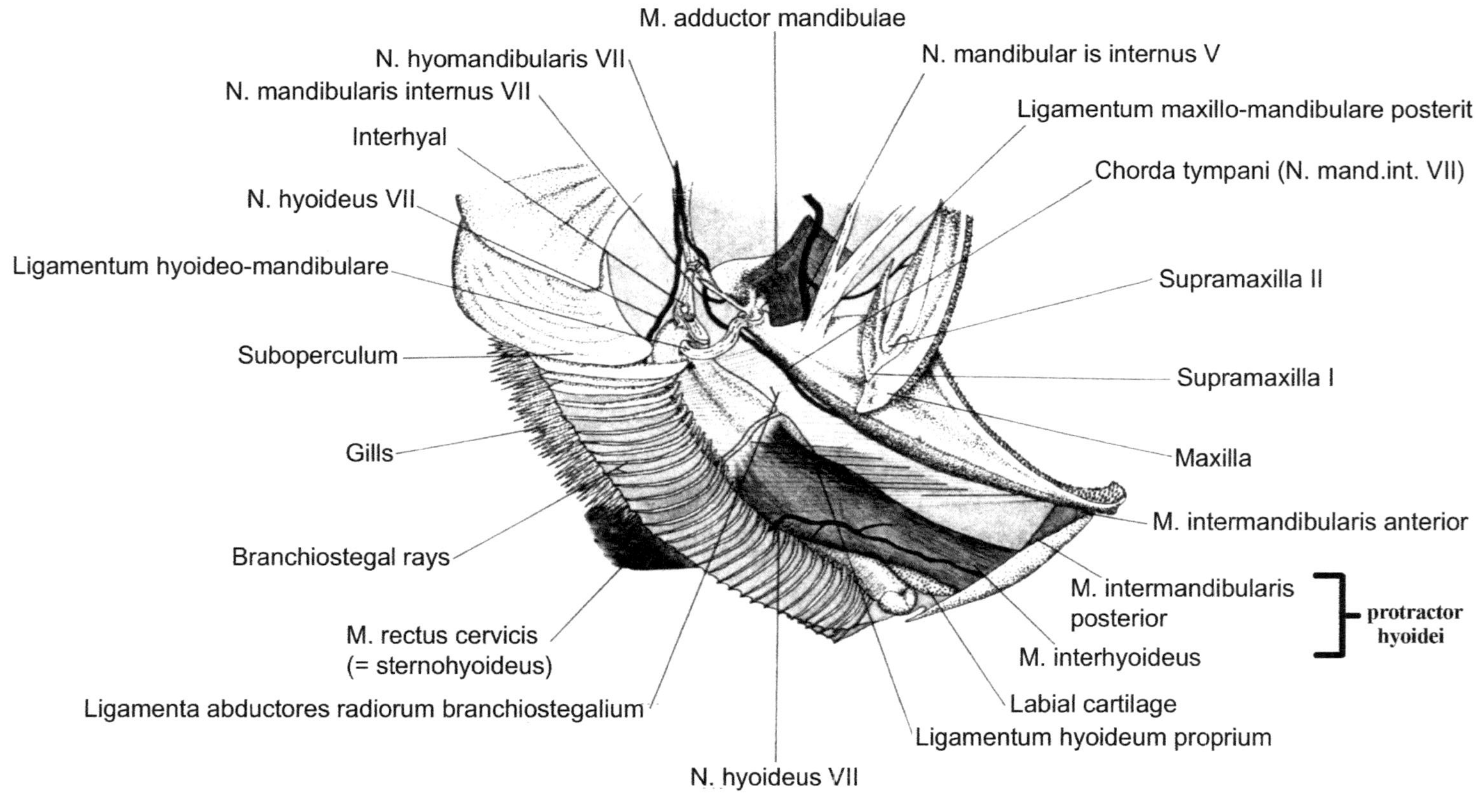

Figure 28 Ventral head region of *Elops saurus*; right lateral view of fully extended jaw apparatus, the interopercle has been removed (modified from Vrba, 1968; the nomenclature of the structures illustrated basically follows that of this author).

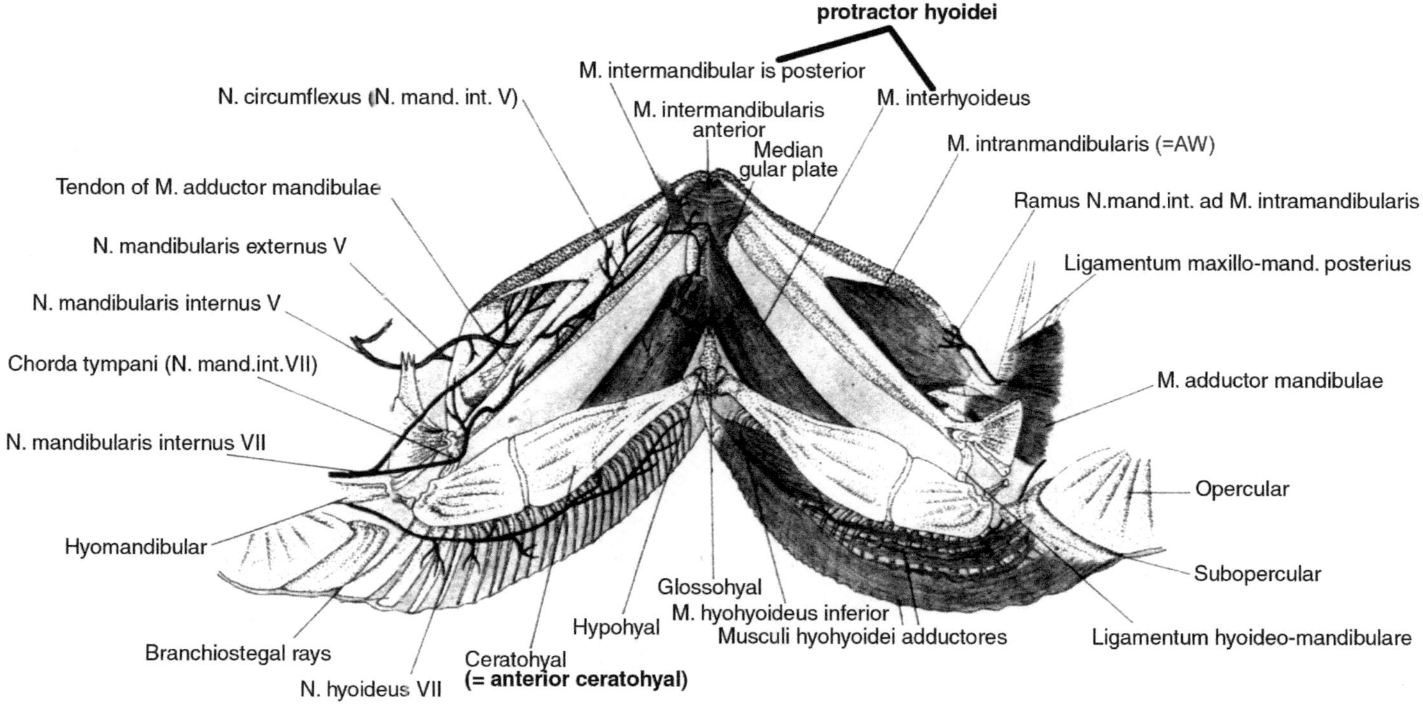

Figure 29 Dorsal view of the nerves and muscles of the lower jaw of *Elops saurus*; on the left side the adductor mandibulae, intermandibularis and hyohyoidei have been removed and the posterior gular plate insertion of the intermandibularis posterior and anterior gular plate origin of the interhyoideus have been turned up (modified from Vrba, 1968; the nomenclature of the structures illustrated basically follows that of this author).

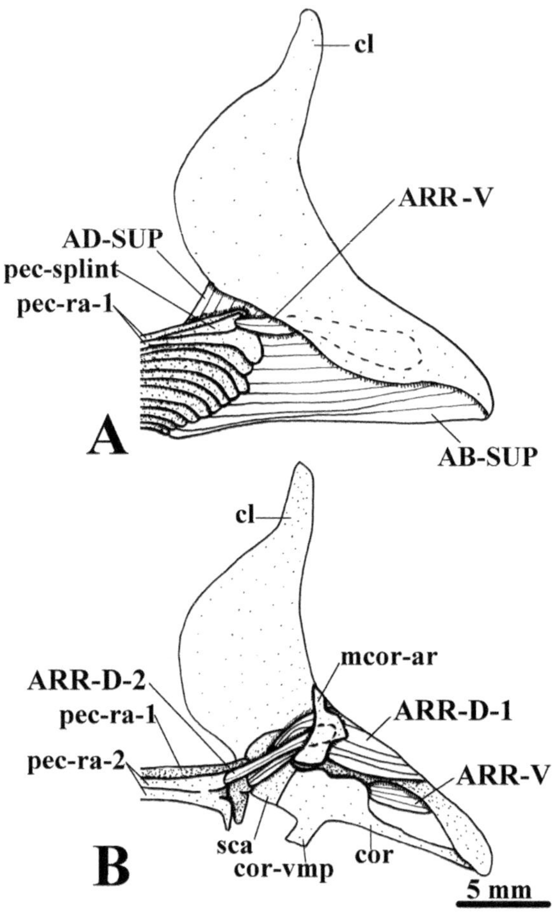

Figure 30 Lateral (**A**) and mesial (**B**) views of the pectoral girdle musculature of *Elops saurus*; in the mesial view the adductor superficialis and the abductor superficialis are not shown.

The grouping of Notacanthiformes, Anguilliformes and Saccopharyngiformes in a monophyletic clade is strongly supported by these nine features. To my knowledge, the posterior intermandibularis forming the protractor hyoidei and not being deeply mixed with the anterior intermandibularis (3: 1→2) and the absence of adductor mandibulae Aw (187: 1→0) have not been previously proposed in the literature as synapomorphies of this group clade.

Notacanthus: [78: 0→1], [139: 0→1], [157: 0→1], [161: 0→1], [182: 0→1], [203: 0→1], [216: 0→1], [258: 0→1], [289: 0→1]

Clade 10 (*Eurypharynx + Conger + Anguilla*): [12: 1→0], [110: 0→1], [154: 1→0], [271: 0→1], [272: 1→0], [302: 0→1], [320: 0→1], [330: 0→1]

The grouping of anguilliform and saccopharyngiform fishes is expected (see Fig. 2) and well corroborated.

***Eurypharynx*:** **[4: 0→1]**, **[19: 0→1]**, [101: 1→0], [121: 0→1], [127: 0→1], [134: 0→1], **[168: 0→1]**, [178: 1→0], [179: 0→1], [185: 0→1], [209: 1→0], [217: 0→1], [281: 0→1], [284: 0→1], [287: 1→0], [315: 0→1], [317: 1→0]

Clade 11 (*Conger + Anguilla*): [177: 0→1], [277: 0→1]

The order Anguilliformes is usually considered a monophyletic group, and in that respect the grouping of the anguilliform genera *Conger* and *Anguilla* is expected (see, e.g., Greenwood et al., 1966; Forey, 1973b; Nelson, 1973; Patterson and Rosen, 1977; Obermiller and Pfeiler, 2003). However, some authors have defended the position that certain anguilliforms (e.g., congroids or, alternatively, anguilloids) are more closely related to saccopharyngiforms than to other anguilliforms (Forey et al., 1996; Belouze, 2002; Wang et al., 2003; Inoue et al., 2004). This latter view is not supported by the results of the present cladistic analysis. However, it is evident that only a study including numerous anguilliform and saccopharyngiform taxa, as well as numerous other elopomorph and even non-elopomorph fishes, can help to address this question in a more conclusive way. The general aspect of an anguilliform fish, namely of a member of the genus *Gymnothorax*, is shown in Fig. 24.

***Anguilla*:** [307: 0→1]

***Conger*:** [87: 0→1; 187: 1→0]

Clade 12 (all non-elopomorph teleosts included in the cladistic analysis): [138: 0→1], [154: 1→0], [185: 0→1]

As mentioned above (see comments to clade 7), all the most parsimonious trees obtained place the Elopomorpha, and not the Osteoglossomorpha, as the most basal teleostean group examined (Figs. 3, 4). Three unambiguous synapomorphies support the clade including the osteoglossomorph and remaining non-elopomorph teleosts included in the analysis: mesial limb of coracoids (or scapulo-coracoids) broad and anteroposteriorly elongated (138: 0→1, subsequently reversed in some taxa of clade 12 and independently occurring, within the teleostean taxa included in the analysis, in the elopomorph *Megalops*); absence of pectoral splints (154: 1→ 0, such pectoral splints were not reacquired in any taxa of this clade 12 and were only independently lost, within the teleostean taxa included in the analysis and not belonging to this clade 12, in the elopomorph anguilliforms + saccopharyngiforms); and absence of separate section A3' of adductor mandibulae (185: 0→1, subsequently reversed in some taxa of clade 12 and independently occurring, within the teleosts included in the analysis, in the elopomorph *Elops*; the situation in *Eurypharynx* is not clear).

Apart from these three features, there are some features with ambiguous distributions that may eventually provide support for this clade, such as the fusion of at least some parapophyses of the first two free vertebrae to their respective centra (110: 0→1). Such a feature is absent, within the halecostome taxa included in the analysis, in *Amia*, *Elops*, *Megalops*, *Albula* and *Notacanthus*, as well as in some more derived teleosts; it is present in *Eurypharynx*, *Anguilla*, *Conger* and the vast majority of the members of clade 12. If the plesiomorphic condition found in the first halecostomes were to lack this feature, the most parsimonious option would be to consider that it was independently acquired in the elopomorph node leading to Anguilliformes + Saccopharyngiformes and in the node leading to clade 12. Thus, this would constitute a further potential synapomorphy to support this clade 12. However, the inclusion of other actinopterygians in the cladistic analysis did not help to clarify whether such a feature was missing or not in the node leading to halecostomes. This was because the observation of the adult *Lepisosteus* specimens analyzed in the present work did not allow us to confirm, with total confidence, the descriptions of Regan (1923) and others, according to which such a feature (i.e., the presence of fusion) is found in the adult members of this genus. Thus, this genus was coded as '?' in the cladistic analysis. The chondrosteans examined do not have ossified, independent vertebrae and, thus, were also coded as '?'. However, it should be noted that even if one considers that such a feature was present in the node leading to halecostomes, this feature would still possibly constitute, under a "fast optimization", a potential synapomorphy of clade 12. In fact, under such an optimization, one would have this feature being lost in the first halecostomes and then independently acquired in the elopomorph node leading to Anguilliformes + Saccopharyngiformes and in clade 12 (three steps in total). This option would be as parsimonious as that suggested by a "slow optimization", that is, that the feature was independently lost in Halecomorphi such as *Amia* and in the Elopomorpha, and then reacquired in the elopomorph node leading to Anguilliformes + Saccopharyngiformes (three steps in total). Thus, two of the three scenarios described just above would actually support the monophyly of clade 12. Nevertheless, due to the uncertainties regarding this issue, I consider that until more conclusive data is available one cannot refer to it as strong evidence to support this clade 12.

One other feature with an ambiguous distribution that may eventually constitute a potential synapomorphy of clade 12 is that concerning the articular mainly fused with angular (and/or retroarticular) (330: 0→1). Under both a "slow" optimization and a "fast" optimization this feature appears as plesiomorphically absent in halecostomes as well as in teleosts. However, under a "slow" optimization one would have the feature being independently acquired in the elopomorph node leading to Anguilliformes + Saccopharyngiformes, in the osteoglossomorph node leading to *Xenomystus*, *Pantodon* and *Mormyrus*, and in the node leading to clupeocephalans (see Fig. 3) (three steps in total; note that the feature is subsequently lost in some derived clupeocephalans). Under a "fast" optimization one would have this feature being independently acquired in the node leading to Anguilliformes + Saccopharyngiformes and in the node leading to the clade 12, being then subsequently lost in the osteoglossomorph *Hiodon* (three steps in total). Therefore, under this latter scenario, this feature would provide support for clade 12.

As mentioned in Chapter 1, many authors consider that osteoglossomorphs occupy a more basal position within the Teleostei than the elopomorphs (e.g., Patterson, 1977a; Patterson and Rosen, 1977; Lauder and Liem, 1983; Ishiguro et al., 2003; Obermiller and Pleifer, 2003; Inoue et al., 2003, 2004; Wang et al., 2003; see Fig. 2). However, it is worth noting that the hypothesis advanced in the present work, that is, that elopomorphs are the most basal extant teleosts, has also been defended by various authors, such as Greenwood et al. (1966), Li (1996), Shen (1996), and Arratia (1997, 1999).

Clade 13: [109: 0→1], [133: 0→1], [247: 0→1], **[346: 0→1]**

As stated by Lavoué and Sullivan (2004), although the Osteoglossomorpha is widely accepted as a monophyletic unit, the only cladistic analyses that have tested the monophyly of this group, by including representatives of all its four families (sensu Hilton, 2003) and an appropriate sample of other teleostean taxa, are essentially molecular ones. In the present cladistic analysis, the representatives of these four osteoglossomorph families (*Hiodon*: Hiodontidae; *Pantodon*: Osteoglossidae; *Xenomystus*: Notopteridae; *Mormyrus*: Mormyridae) do appear grouped in a monophyletic clade (Figs. 3, 4), which is supported by the four synapomorphies listed above. The first concerns the absence of distinct parapophyses of the first two free vertebrae (109: 0→1, reversed in *Pantodon* and also occurring in a few non-elopomorph groups examined such as *Amia* and the clade 52). The second concerns the peculiar anteroventrolateral bifurcation of the posttemporal into a shorter, lateral arm carrying a sensorial canal and a longer, mesial arm corresponding to the ossified "ligament between the posttemporal and the posterior margin of the neurocranium" sensu this work (133: 0→1). This feature is found in all the osteoglossomorphs examined except *Pantodon*, in which the latter ligament is not ossified; it is missing in all other taxa analyzed. The third synapomorphy concerns the "autopalatine missing or being almost completely, or completely, unossified" (247: 0→1), a feature that is also only found in a few taxa examined in the present work such as gymnotiforms and the members of clade 65. The fourth synapomorphy concerns the presence of a "tongue-bite mechanism" with dorsal teeth on parasphenoid, a feature that is homoplasy-free within the taxa included in the cladistic analysis (346: 0→1).

Contrary to the second synapomorphy listed above, the other three synapomorphies have been proposed as potential characters supporting osteoglossomorph monophyly in previous works (e.g., Lauder and Liem, 1983; Li and Wilson, 1996; Arratia, 1997, 1999; Hilton, 2003). As mentioned above, there are two features with ambiguous distributions that might eventually constitute synapomorphies of the clade including the osteoglossomorphs examined: the absence of distinct, strong ligaments connecting the anterior surface/anterior cartilage of the autopalatines and/or dermopalatines and the maxillae and/or premaxillae (222: 1→0) and the absence of an ossified interhyal (302: 0→1) (see comments to clade 6).

Some aspects of the anatomy of osteoglossomorphs are illustrated in Figs. 31, 32, 33, 34, 35 and 36. An illustration of the general shape of an osteoglossomorph fish of the genus *Mormyrus* is given in Fig. 31. More detailed drawings of the configuration of the osteological and myological structures of the cephalic and pectoral girdle of fishes of this genus and of the genus *Hiodon* are given in Figs. 32, 33, 34, 35 and 36. This last figure is an original drawing by the author showing details of the posttemporal bone and its associated ligaments in a member of the genus *Mormyrus*.

***Hiodon*:** [79: 0→1], [197: 0→1], [204: 0→1], [277: 0→1]

Clade 14 (*Xenomystus* + *Pantodon* + *Mormyrus*): [176: 1→0]

Although a detailed discussion of the relationships between the four osteoglossomorph families (see above) is clearly beyond the main scope of the present work, it is worthy to note that in the majority (75%) of the most parsimonious trees obtained *Hiodon* appears as the sister-group of a clade including the other osteoglossomorphs included in the analysis, as expected (see Fig. 2). However, there is only one unambiguous synapomorphy supporting this clade 14 in the "majority fools" tree shown in Fig. 3 (176: 1→0). This clade can, however, be eventually supported by

Figure 31 General shape of a fish of the genus *Mormyrus* (Teleostei: Osteoglossomorpha) (modified from Poll, 1967).

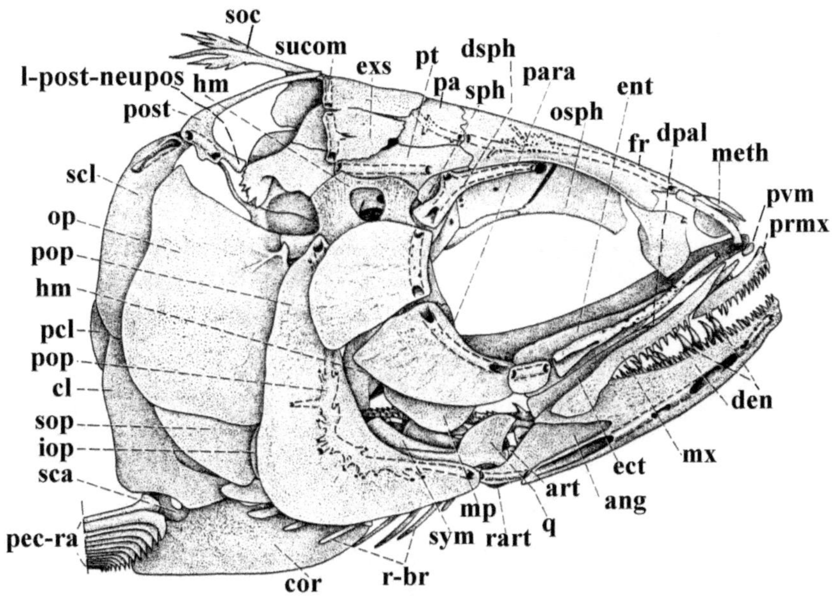

Figure 32 Lateral view of the cranium and pectoral girdle of *Hiodon tergisus* (modified from Taverne, 1977b; the nomenclature of the structures illustrated basically follows that of this author).

other features if a "fast optimization" is chosen (e.g., 12: 1→0; 26: 0→1; 68: 0→1; 281: 0→1) or, alternatively, if a "slow optimization" is selected (e.g., 330: 0→1). It should be emphasized that in the 25% most parsimonious trees in which this clade 14 does not appear, *Hiodon* is grouped with *Pantodon* + *Mormyrus*, with features 78 (0→1), 143 (1→0) and 204 (0→1), and eventually 68 (1→0), supporting this grouping in these trees but only if a "fast optimization" is chosen. It can thus be said that this latter hypothesis is not completely contradicted by the results of the present work, although it is important to keep in mind that, overall, the majority of the equally

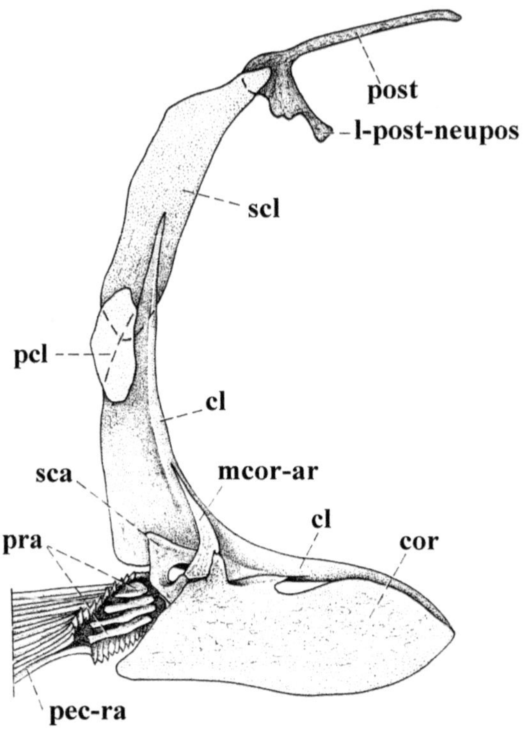

Figure 33 Mesial view of the pectoral girdle of *Hiodon tergisus* (modified from Taverne, 1977b; the nomenclature of the structures illustrated basically follows that of this author).

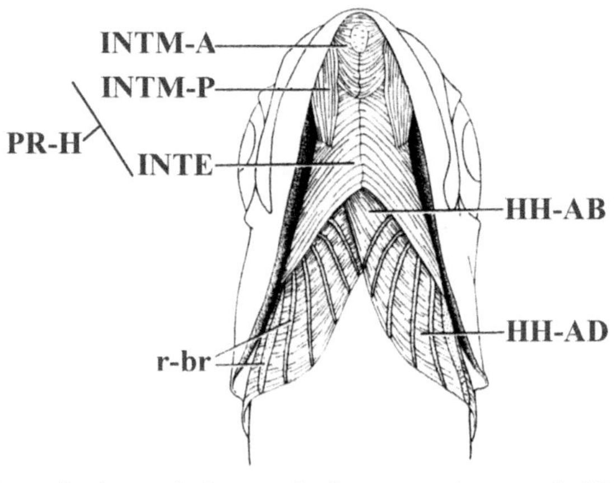

Figure 34 Ventral view of the cephalic musculature of *Hiodon alosoides* (modified from Greenwood, 1971; the nomenclature of the structures illustrated basically follows that of this author).

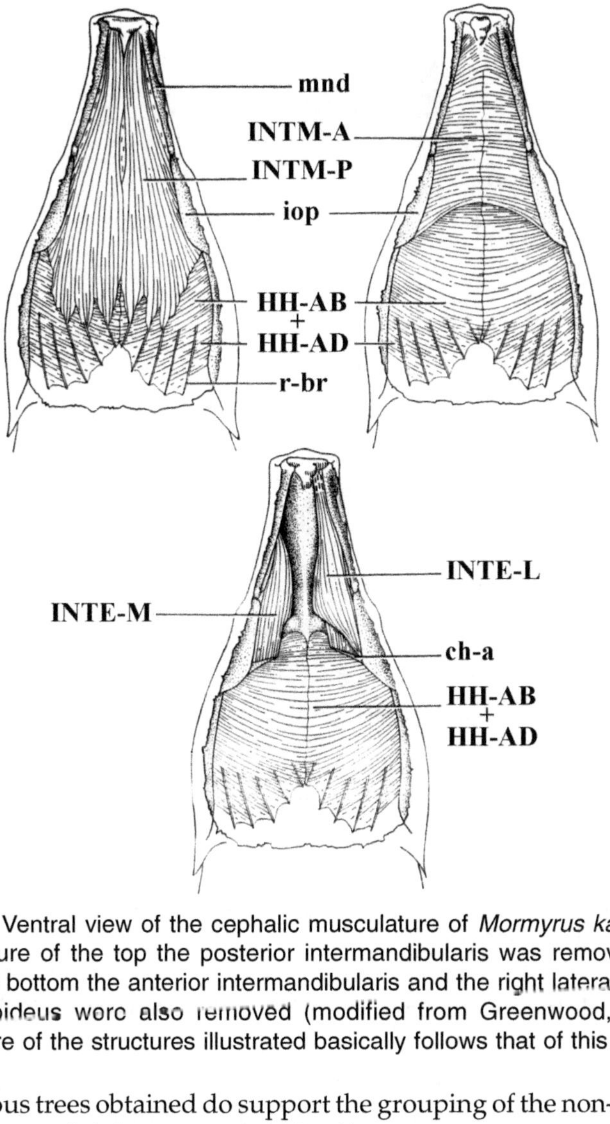

Figure 35 Ventral view of the cephalic musculature of *Mormyrus kannume*; on the right figure of the top the posterior intermandibularis was removed; on the figure of the bottom the anterior intermandibularis and the right lateral division of the interhyoideus were also removed (modified from Greenwood, 1971; the nomenclature of the structures illustrated basically follows that of this author).

parsimonious trees obtained do support the grouping of the non-hiodontid osteoglossomorph fishes examined (Fig. 3).

Xenomystus: [20: 1→0], **[32: 0→1]**, [228: 0→1], [246: 0→1], [307: 0→1]

Clade 15 (*Pantodon* + *Mormyrus*): [82: 0→1], [100: 0→1], [136: 0→1], [157: 0→1], [187: 0→1], [224: 0→1], [272: 1→0]

Although a detailed discussion of the relationships between the four osteoglossomorph families is beyond the main scope of this work (see

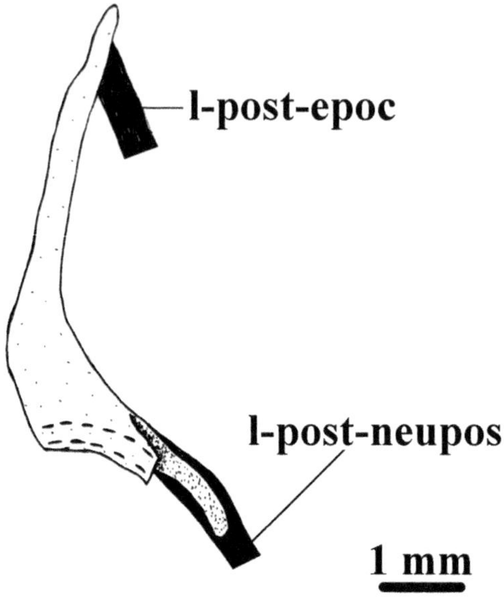

Figure 36 Lateral view of the posttemporal and associated ligaments in *Mormyrus tapirus.*

above), it is worthy to note that the grouping of the mormyrid *Mormyrus* and the osteoglossid *Pantodon* (sensu Hilton, 2003) is strongly supported in the cladistic analysis.

Pantodon: [3: 1→2], [109: 1→0], [132: 0→1], [133: 1→0], [139: 0→1], **[144: 0→1]**, [225: 1→0], [302: 1→0]

Mormyrus: [3: 1→0], **[14: 0→1]**, [66: 0→1], [78: 0→1], [97: 1→0], **[170: 0→1]**, [204: 0→1], [216: 0→1], [245: 0→1], [258: 0→1], **[261: 0→1]**, [263: 0→1], **[312: 0→1]**, [313: 0→1], [331: 0→1]

Clade 16 (all clupeocephalans included in the cladistic analysis): [3: 1→2], **[22: 0→1]**, [96: 0→1], [99: 0→1]

The assembly of the non-elopomorph and non-osteoglossomorph teleosts examined in clade 16 is expected (Figs. 1, 2: Clupeocephala) and is well supported by four unambiguous synapomorphies: posterior intermandibularis included in protractor hyoidei and not deeply mixed with anterior intermandibularis (3: 1→2, not reversed in fishes of the clade 16 and independently occurring in *Pantodon* and in Notacanthiformes + Saccopharyngiformes + Anguilliformes); presence of distinct muscle "arrector 3" (22: 0→1, homoplasy-free within the taxa examined); main bodies of parietals (or of parieto-extrascapulars) widely separated from each other in dorsal view (96: 0→1, not independently acquired within the other actinopterygian taxa examined, but subsequently reversed in some more derived groups of clade 16: see below); absence of parasphenoid teeth (99: 0→1, not subsequently reversed, within the taxa included in the analysis, in members of clade 16, and only occurring independently in those taxa of clades 3, 9 and 65). Unlike the two latter features, the two former ones have not been previously proposed as clupeocephalan synapomorphies in the literature. Some other features with ambiguous distributions may be interpreted as synapomorphies of this clade 16 if a "fast optimization" is chosen (13: 0→1; 216: 0→1; 263: 0→1) or, alternatively, if a "slow optimization" is selected (143: 0→1; 222: 0→1; 302: 1→0; 330: 0→1; 322: 0→1) (see, e.g., comments to clade 6 above).

Clade 17 (all euteleosteans included in the cladistic analysis): [123: 0→1], [280: 0→1], [307: 0→1]

The grouping of the euteleostean fishes included in the analysis in clade 12 is expected (see Figs. 1, 2) and is supported by these three features: main body of posttemporal (or posttemporo-supracleithrum) lying considerably far from neurocranium, with almost no contact between these two structures (123: 0→1, not reversed within the taxa examined, and only occurring independently in a few taxa outside the clade 17, such as *Gonorynchus*); well-developed articulatory facet on the posterolateral margin of suspensorium for mesial surface of preopercle (280: 0→1, not reversed within the taxa examined of clade 17, and only occurring independently in a few taxa outside the clade such as some derived elopomorphs and some gonorynchiforms); and mandibulohyoid and mandibulointeropercular ligaments not well separated from each other (307: 0→1, not reversed within the taxa examined of clade 17, and only occurring independently in a few taxa outside the clade such as *Anguilla*, *Albula*, *Denticeps* and *Xenomystus*). None of these three features is commonly listed in the literature as an example of euteleostean synapomorphy. There are two features with ambiguous distributions that may eventually be interpreted as synapomorphies of this clade if a "slow optimization" is chosen (13: 0→1; 263: 0→1).

Three main euteleostean clades appear in an unresolved trichotomy in the "majority fools" tree obtained in the present work (Figs. 3, 4). In fact, 50% of the most parsimonious trees obtained have favored the grouping of argentiniforms and the clade salmoniforms + neoteleosts; examples of features with unambiguous or ambiguous distributions supporting such a scenario in these trees are those concerning characters 67 (0→1), 220 (0→1), 232 (0→1), and 263 (0→1). The other 50% most parsimonious trees obtained have favored the grouping of the clade osmeriforms + esociforms and the clade salmoniforms + neoteleosts; examples of features with unambiguous or ambiguous distributions supporting such a scenario in these trees are those regarding characters 68 (0→1) and 176 (0→1).

The relationships between basal euteleosts have been a subject of much controversy (e.g., Greenwood et al., 1966; Gosline, 1969; Rosen, 1973, 1974, 1985; Fink and Weitzman, 1982; Lauder and Liem, 1983; Fink, 1984; Begle, 1991, 1992; Johnson and Patterson, 1996; Sanford, 2000). For instance, Ishiguro et al. (2003) consider that the osmeriforms are closely related to the Argentinoidea, and that the esociforms are closely related to the salmoniforms. Fink and Weitzman (1982), Lauder and Liem (1983), and Begle (1991, 1992) consider that the osmeriforms are closely related to the argentiniforms, and that the esociforms are not closely related to the salmoniforms. Researchers such as Johnson and Patterson (1996), in turn, defend a close relationship between osmeriforms and salmoniforms, and between the clade formed by these two groups and the argentiniforms. A brief, updated summary of these and other hypotheses concerning basal euteleostean relationships has been provided by Ishiguro et al. (2003). In short, it can be said that, together with the phylogenetic hypothesis shown in Figs. 3 and 4 of the present work, almost all possible combinations between the major euteleostean groups mentioned above have already been proposed in the literature. The results of the cladistic analysis of the present work support the monophyly of a clade Alepocephaloidea + Argentinoidea, of a clade Esociformes + (Galaxioidea + Osmeroidea), and of a clade Salmoniformes + Neoteleostei (see below), but do not allow us to confidently point out which two of these three major euteleostean clades obtained may be more closely related.

Some aspects of the anatomy of the Euteleostei are illustrated in Figs. 37, 38, 39, 40, 41, 42, 43 and 44. An illustration of the general shape of a fish of the genus *Salmo* is given in Fig. 37. More detailed drawings of the configuration of the osteological and myological structures of the cephalic region and pectoral girdle of fishes of the genera *Alepocephalus*, *Xenodermichthys* and *Aulopus* are given in Figs. 38, 39, 40, 41, 42, 43 and 44, which are original drawings by the author.

Clade 18 (all argentiniforms included in the cladistic analysis): [68: 0→1], **[84: 0→1]**, [176: 1→0], [214: 0→1], [231: 1→0], [255: 0→1], **[349: 0→1]**

The grouping of the argentinoid and alepocephaloid fishes examined is strongly supported by seven synapomorphies: posterodorsal portion of the mesethmoid (or rostrodermethmoid and/or supraethmoid) appearing markedly compressed transversally when seen in dorsal view (68: 0→1, not

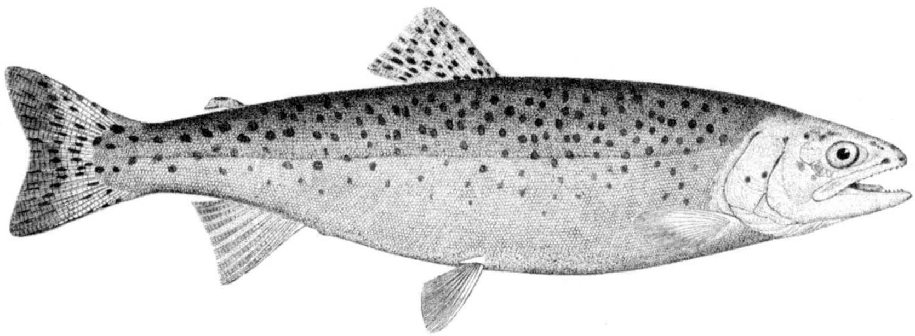

Figure 37 General aspect of a member of the genus *Salmo* (Teleostei: Euteleostei) (modified from Goode, 1884-1887).

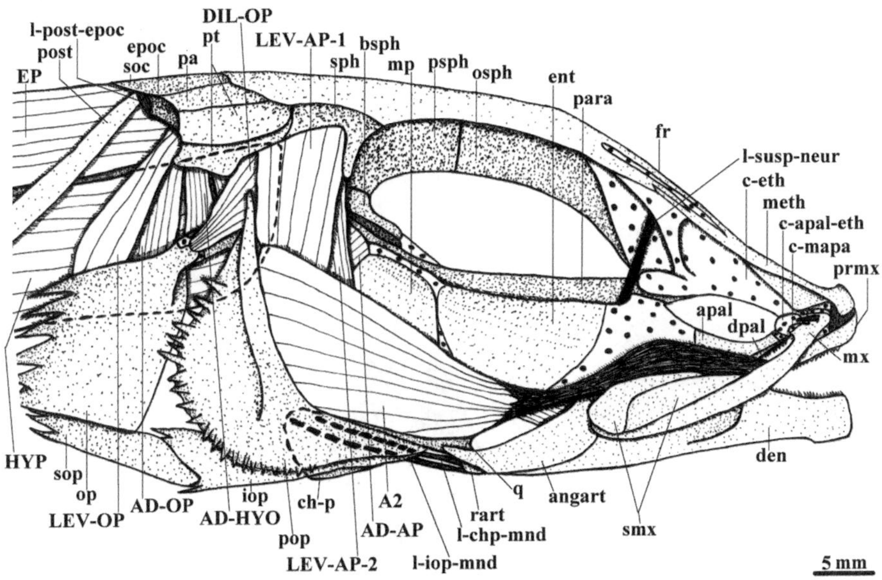

Figure 38 Lateral view of the cephalic musculature of *Alepocephalus rostratus*. The pectoral girdle muscles are not illustrated; most elements of the pectoral girdle, as well as the nasals and infraorbitals, were removed.

reversed within the taxa examined of clade 18, but independently occurring in some teleostean groups outside this clade such as clupeiforms, characiforms, gymnotiforms and siluriforms); both autopterotic and dermopterotic bones present as distinct, independent ossifications (84: 0→1, homoplasy-free within the taxa examined in which this character could be discerned); mainly undivided A2 not attaching on mesial surface of mandible by means of two well-distinguished tendons (176: 1→0, not modified in the taxa examined of clade 18 but also occurring in *Engraulis*,

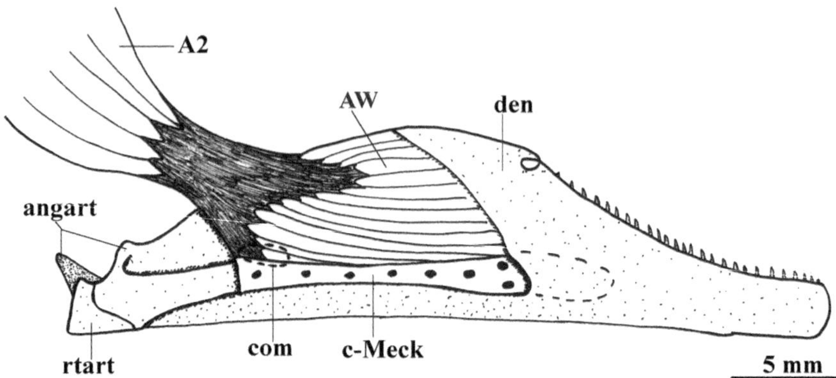

Figure 39 Mesial view of the left mandible and adductor mandibulae of *Alepocephalus rostratus*.

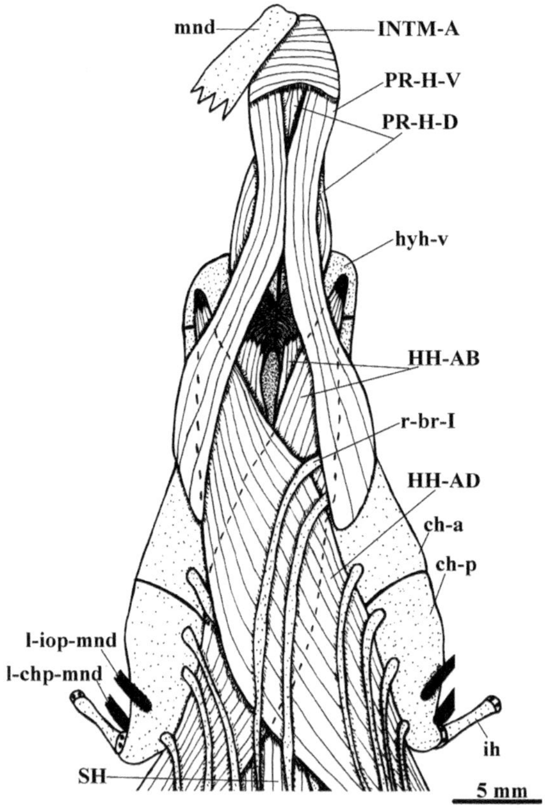

Figure 40 Ventral view of the cephalic musculature of *Alepocephalus rostratus*. On the left side, the mandible was cut; on the right side, the mandible was removed.

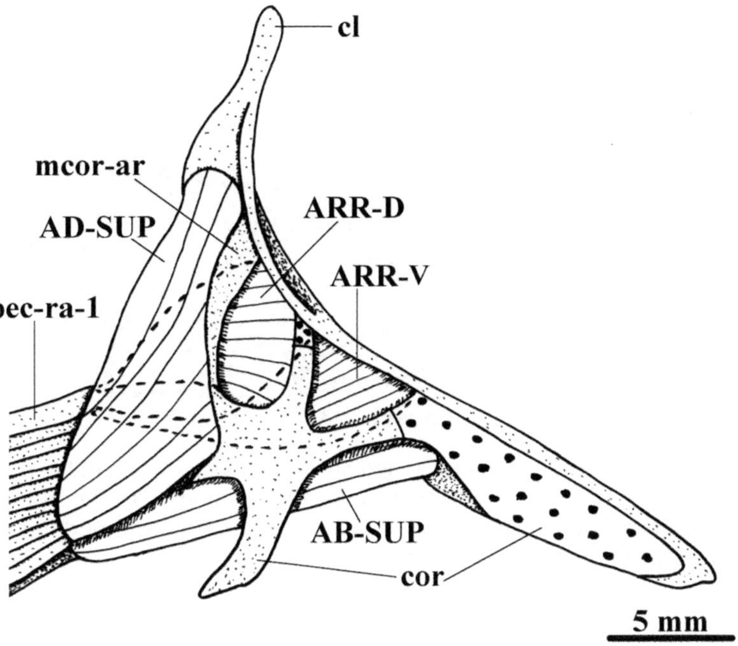

Figure 41 Mesial view of the pectoral girdle muscles of *Alepocephalus rostratus*.

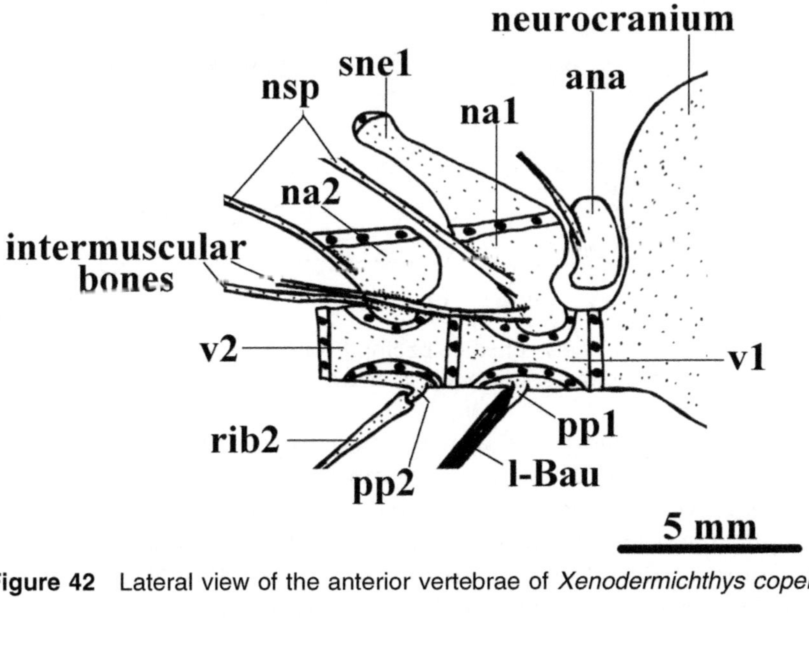

Figure 42 Lateral view of the anterior vertebrae of *Xenodermichthys copei*.

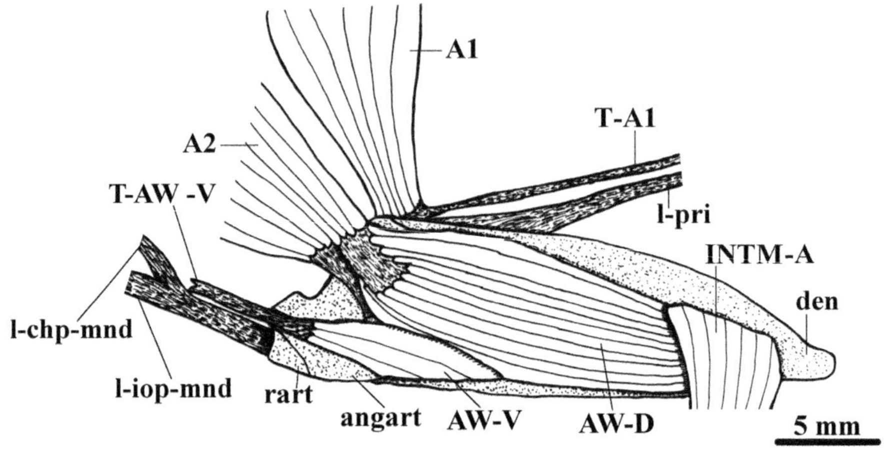

Figure 43 Mesial view of the left mandible and adductor mandibulae of *Aulopus filamentosus*, the anterior intermandibularis and the primordial ligament, as well as the ligaments between the mandible, posterior ceratohyal and interopercle, are also shown; mandibular teeth were removed.

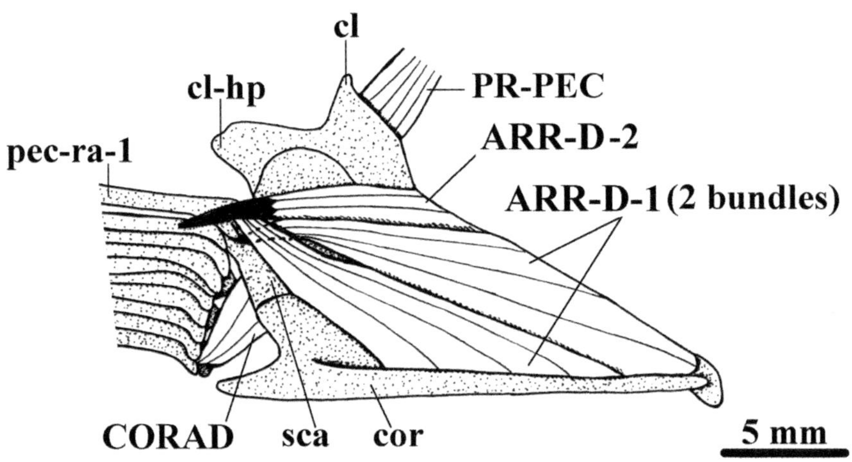

Figure 44 Mesial view of the pectoral girdle musculature of *Aulopus filamentosus*; the protractor pectoralis is shown.

some osteoglossomorphs and most ostariophysans); fibers of hypaxialis and/or epaxialis peculiarly covering much of the neurocranial floor (214: 0→1, not reversed inside clade 18 and only independently occurring in the aulopiforms + stomiiforms examined); primordial ligament attaching posteriorly on dorsal surface of coronoid process (231: 1→0, found in some teleostean groups outside the clade 18 but seemingly not reversed inside of it; *Bathylagus*, *Xenodermichthys* and *Searsia* were coded as "Inapplicable" for this character as they seemingly do not have a distinct primordial ligament); peculiar dorsoventral enlargement of posterior portion of autopalatine (255: 0→1, not reversed inside clade 18 and only occurring outside of it in the osmeroids examined); presence of peculiar "accessory cartilage of the fifth ceratobranchial" (**349: 0→1**, homoplasy-free within the taxa examined). Many features exhibiting an ambiguous distribution in the analysis may eventually be interpreted as potential synapomorphies of this clade 18 if a "fast optimization" is chosen (20: 1→0; 139: 0→1; 157: 0→1; 289: 0→1).

It should be stressed that, contrary to what is commonly accepted (e.g., Greenwood and Rosen, 1971; Rosen, 1973, 1974, 1985; Fink and Weitzman, 1982; Lauder and Liem, 1983; Fink, 1984; Begle, 1991, 1992; Nelson, 1994, 2006; Johnson and Patterson, 1996; Sanford, 2000; this work), some molecular cladistic analyses (e.g., Ishiguro et al., 2003; Lavoué et al., 2005) have contradicted the monophyly of Argentiniformes (sensu this work: see Figs. 2, 4). According to these molecular analyses, the Alepocephaloidea is the sister-group of either the Clupeomorpha or the Ostariophysi, but not of the Argentinoidea. This latter point makes me particularly reticent about the conclusions of these molecular analyses. That a clade Alepocephaloidea + Argentinoidea is eventually placed closer to certain Otocephala taxa than to certain euteleostean groups would eventually not seem too unsound. But that the Alepocephaloidea is placed inside the Otocephala and the Argentinoidea is not does seem rather unsound in light of the large amount of data (provided by numerous authors and by various types of morphological characters) available to support a sister-group relationship between the Alepocephaloidea and the Argentinoidea (e.g., Greenwood and Rosen, 1971; Rosen, 1973, 1974; Begle, 1991, 1992; Johnson and Patterson, 1996; Sanford, 2000; this study).

Clade 19 (*Argentina* + *Bathylagus*): [197: 0→1], [203: 0→1], [228: 0→1], [232: 0→1], [238: 0→1], [258: 0→1], **[305: 0→1]**, [313: 0→1]

The grouping of the two argentinoid taxa examined is expected (see Fig. 2) and well corroborated.

Argentina: [96: 1→0], [138: 1→0], [177: 0→1], [323: 0→1], **[344: 0→1]**

Bathylagus: [30: 0→1], [141: 0→1], [187: 0→1]

Clade 20 (*Alepocephalus* + *Xenodermichthys* + *Searsia*): [12: 1→0], [27: 1→0], [28: 1→0], [67: 0→1], [220: 0→1]

The grouping of the alepocephaloid fishes examined is expected (see Fig. 1) and well corroborated.

Alepocephalus: [203: 0→1]

Xenodermichthys: [138: 1→0], **[297: 0→1]**

Searsia: [100: 1→0], **[234: 0→1]**, [263: 1→0], **[345: 0→1]**

Clade 21 (all salmoniform and neoteleostean taxa included in the cladistic analysis): [67: 0→1], [110: 1→0], [220: 0→1], [223: 0→1], [232: 0→1], [252: 0→1]

The assembly of the salmoniform and neoteleostean fishes included in the cladistic analysis in this clade 21 is supported by six unambiguous synapomorphies: presence of anterolateral processes of mesethmoid supporting and/or articulating with premaxillae (67: 0→1, also occurring in certain groups outside of this clade 21 and reversed inside of it in the aulopiforms examined; the members of the genus *Coregonus* might be either CS0 or CS1 for this character: see Section 3.2); parapophyses of two first free vertebrae not fused to centra (110: 1→ 0, occurring in certain groups outside the clade 21 and modified, inside of it, in *Aulopus* and *Astronesthes*); supramaxillae present as independent ossifications (220: 0→1, occurring in various groups outside the clade 18 but not reversed inside of it, although it should be noted that *Astronesthes* and *Stomias* were coded as '?' for this character); presence of well-developed "rostral" cartilaginous or cartilaginous-like structures associated with the posterior surface of well-developed premaxillary dorsomedial processes attached to/articulating with ethmoid region (223: 0→ 1, outside clade 18 it is only found in *Osmerus*); presence of strong, well-defined ligament between premaxilla and proximal surface of maxilla (232: 0→ 1, occurring in a few groups outside of clade 21, but not reversed inside of it); anterior portion and/or anterior cartilage of autopalatine forming peculiar "broad hook" covering a great portion of proximal portion of maxilla in lateral view (252: 0→1, found exclusively in the taxa of clade 18, and only reversed, within the taxa examined, in *Coregonus* and *Stomias*). One feature exhibiting an ambiguous distribution in the analysis may eventually be interpreted as a potential synapomorphy of this clade if a "fast optimization" is chosen (216: 1→0).

It should be stressed that, although many authors nowadays consider the Esociformes as the probable sister-group of the Neoteleostei (see Figs. 1, 2), some previous studies have defended, as does the present work, a sister-group relationship between salmoniforms and neoteleosteans (e.g., Lauder and Liem, 1983; Fink, 1984). It should also be noted that, curiously, the three salmoniform taxa included in the present work do not appear grouped in a monophyletic clade in the "majority fools" tree obtained in the cladistic analysis (see Figs. 3, 4; but see Chapter 4, Section 4.6). The monophyly of the Salmoniformes has, however, been strongly supported in the extensive work of Sanford (2000).

Salmo: No unambiguous features

Coregonus: [13: 1→0], [78: 0→1], [245: 0→1], [252: 1→0]

Thymallus: [81: 0→1]

Clade 22 (*Aulopus* + *Chlorophthalmus* + *Stomias* + *Astronesthes*): [119: 0→1], [157: 0→1], [214: 0→1], **[347: 0→1]**

The grouping of the neoteleostean taxa included in the analysis in clade 22 is expected (see Figs. 1, 2) and is well supported by four unambiguous synapomorphies: peculiarly large "precervical gap" filled mainly with connective tissue between first free vertebra and neurocranium (**119: 0→1**, homoplasy-free within the taxa examined in which this character could be discerned); adductor mandibulae attaching not only on mandible and/or primordial ligament, near its mandibular insertion, but also on other structures (157: 0→1, not reversed in the fishes examined of the clade 22 but occurring in other fishes outside of it); fibers of hypaxialis and/or epaxialis peculiarly covering great part of neurocranial floor (214: 0→1, not reversed in taxa examined of clade 22 and only occurring also in the argentiniform fishes analyzed); presence of peculiar muscle retractor dorsalis (262: 0→1, homoplasy-free within the taxa examined in which this character could be discerned). There are some features exhibiting an ambiguous distribution in the analysis that may eventually be interpreted as potential synapomorphies of this clade 22 if a "fast optimization" is chosen (27: 1→2; 81: 1→0; 96: 1→0; 141: 0→1; 197: 0→1; 228: 0→1). It is worth noting that although certain features listed above might prove to be eventual potential synapomorphies of the Neoteleostei, this can obviously be examined appropriately only in a study including numerous other representative taxa of this group.

Clade 23 (*Aulopus* + *Chlorophthalmus*): [23: 0→1], [29: 1→0], [67: 1→0], [68: 0→1], [138: 1→0], **[160: 0→1]**, **[188: 0→1]**

The grouping of the two aulopiform taxa included in the analysis is expected (see Figs. 1, 2) and well supported by these seven features. As explained in Chapters 1 and 2, a detailed discussion on the synapomorphies of derived osteichthyan taxa such as the Eurypterygii (Aulopiformes + Ctenosquamata: see Fig. 1) is clearly beyond the main scope of the present work; the aulopiforms are in fact the only eurypterygian taxa included in this work. Nevertheless, it is important to clarify that some of these nine synapomorphies may be synapomorphies of the Eurypterygii as a whole, and not only of Aulopiformes or of certain specific subgroups of this order. For instance, the consistent presence of the coracoradialis (**23: 0→1**), of the protractor pectoralis (29: 1→0) and of the adductor mandibulae A1

(**160: 0→1**) have been proposed by some authors as potential eurypterygian synapomorphies (e.g., Winterbottom, 1974; Greenwood and Lauder, 1981; Lauder and Liem, 1983; Gosline, 1986; Wu and Shen, 2004). Another example concerns the attachment of the subdivision of the Aw on the suspensorium and/or opercular series (**188: 0→1**), a feature that, as will be explained in Section 3.2, is found in many non-aulopiform eurypterygian fishes and may thus eventually constitute a potential eurypterygian synapomorphy. However, the taxonomic distribution of these features can only be investigated appropriately in a study including numerous other representative eurypterygian taxa.

Aulopus: [110: 0→1], [263: 1→0], [277: 0→1]

Chlorophthalmus: [204: 0→1]

Clade 24 (*Stomias* + *Astronesthes*): [130: 0→1], [161: 0→1], **[319: 0→1]**, **[321: 0→1]**, **[355: 1→0]**

The grouping of the two stomiiform taxa examined is expected (see Figs. 1, 2) and well supported by these five features. As explained above, it should be emphasized that these features are not necessarily synapomorphies of the Stomiiformes.

Astronesthes: [110: 0→1]

Stomias: [132: 0→1], [252: 1→0], **[303: 0→1]**

Clade 25 (all esociform and osmeriform taxa included in the cladistic analysis): [81: 0→1], [266: 0→1]

The assembly of the esociform and osmeriform fishes examined in clade 25 is supported by these two unambiguous synapomorphies: absence of ossified orbitosphenoid (81: 0→1, occurring in certain groups outside this clade but not reversed inside of it); presence of peculiar, prominent hyomandibular lateral spur at or below the level of the opercular process (266: 0→1, occurring exclusively in the taxa of the clade and only reversed, within the taxa included in the analysis, in *Stokellia* + *Retropinna*). Some features with ambiguous distributions may eventually be interpreted as potential synapomorphies of this clade if a "fast optimization" is chosen (82: 0→1; 266: 0→1) or, alternatively, if a "slow optimization" is selected (216: 0→1). Although various authors (e.g., Rosen, 1973, 1974; Johnson and Patterson, 1996; Springer and Johnson, 2004) place the Salmoniformes as the probable sister-group of the Osmeriformes (see Fig. 2), some previous studies have partly supported a closer relationship between Esociformes and Osmeriformes than between this latter order and the Salmoniformes, as suggested in the present work (see, e.g., Waters et al., 2000: figs. 4, 5).

Clade 26 (*Esox* + *Umbra*): [138: 1→0], [185: 1→0]

The grouping of the two esociform taxa examined is expected (see Figs. 1, 2) and supported by these two features.

***Esox*:** [220: 0→1], [258: 0→1]

***Umbra*:** [67: 0→1], [197: 0→1], [231: 1→0]

Clade 27 (all osmeriforms included in the cladistic analysis): [228: 0→1], [263: 1→0]

This clade is expected (see Fig. 2) and is supported, in the "majority fools" tree obtained (Fig. 3), by these two unambiguous features. Interestingly, although this expected clade is supported by the majority (75%) of the most parsimonious trees obtained, in 25% of these trees the galaxioids examined appear grouped with esociforms, and not with the osmeroids analyzed. One unambiguous feature supports the grouping of galaxioids and esociforms in these latter trees (141: 0→1).

Clade 28 (*Osmerus* + *Plecoglossus*): [12: 1→0], [255: 0→1], [290: 0→1]

The grouping of the two osmeroid taxa examined is expected (see Fig. 2) and supported by these three features.

***Osmerus*:** [216: 1→0], [220: 0→1], [223: 0→1]

***Plecoglossus*:** [78: 0→1], [187: 0→1], **[235: 0→1], [236: 0→1], [237: 0→1]**, [245: 0→1], [323: 0→1], [334: 0→1]

Clade 29 (*Galaxias* + *Retropinna* + *Stokellia*): [132: 0→1]

The grouping of the three galaxioid taxa examined is expected (see Fig. 2) and supported by this unambiguous feature. It should, however, be noted that although the monophyly of Galaxioidea is accepted by most researchers (particularly strong evidence to support this taxon is provided by Johnson and Patterson, 1996), in a molecular analysis some galaxioid fishes appear more closely related to certain osmeroids than to other galaxioids (see, e.g. López et al., 2004; fig. 2).

***Galaxias*:** [78: 0→1], [205: 0→1], [206: 0→1], [245: 0→1], [260: 0→1]

Clade 30: [109: 0→1], [129: 1→0], [246: 0→1], [266: 1→0]

The grouping of the retropinnid galaxioids examined is expected (see, e.g., Patterson and Johnson, 1995) and supported by these four unambiguous features.

Retropinna: [216: 1→0]

Stokellia: [20: 1→0]

Clade 31 (all otocephalans included in the cladistic analysis): [90: 0→1], [116: 0→1], [117: 0→1], [277: 0→1]

The clade including the clupeomorph and ostariophysan fishes examined is strongly supported by these four unambiguous synapomorphies: position of saccular and lagenar otoliths more posterior and principally nearer to midline (**90: 0→1,** homoplasy-free within the taxa examined in which this character could be discerned); swimbladder with a silvery peritoneal tunic covering at least part of its anterior portion (**116: 0→1,** homoplasy-free within the taxa examined in which this character could be discerned); swimbladder markedly divided into peculiar anterior and posterior chambers (**117: 0→1,** homoplasy-free within the taxa examined in which this character could be discerned); hyomandibula exhibiting two articulatory heads for neurocranium (277: 0→1, also occurring in some taxa outside this clade 31 and reversed in some taxa inside of it: see below). The three first features were proposed by Rosen and Greenwood (1970) and others to be potential synapomorphies of the Ostariophysi. However, Grande and De Pinna (2004) have maintained that these features are also found in many clupeomorphs and that they may well constitute synapomorphies of the Otocephala as a whole. This latter view is supported by the present work. Some features with ambiguous distributions may eventually be interpreted as potential synapomorphies of this clade 31 if a "fast optimization" is chosen (78: 0→1; 238: 0→1; 323: 0→1).

As explained in Chapter 1, the otocephalan clade is nowadays accepted by most researchers (e.g., Lecointre, 1995; Johnson and Patterson, 1996; Arratia, 1997, 1999; Filleul and Lavoué, 2001; Inoue et al., 2001; Elmerot et al., 2002; Wang et al., 2003; Zaragüeta-Bagils et al., 2002; Stiassny et al., 2004) (see Figs. 1, 2). However, some authors have argued, on the basis of molecular cladistic analyses, that the Otocephala should be enlarged in order to include the Alepocephaloidea (Ishiguro et al., 2003; e.g., Lavoué et al., 2005). This subject is discussed above (see clade 18). These authors have argued that the results of their molecular studies do not directly contradict the results of most morphological cladistic analyses, since, in fact, these latter analyses almost never included in a single matrix representatives of the Clupeomorpha, Ostariophysi, and Alepocephaloidea, and of other teleostean taxa to which these three groups should be compared. I fully agree with this point. However, it should be noted that, with the present work, there are now already two extensive morphological cladistic analyses that have included these three groups together with many other teleostean taxa in a single matrix and that have contradicted the inclusion of the Alepocephaloidea inside the otocephalan clade (Patterson and Johnson, 1996; this work).

Clade 32 (all clupeomorphs included in the cladistic analysis): [68: 0→1], **[89: 0→1], [93: 0→1], [356: 0→1]**

The grouping of the clupeomorph fishes included in the analysis is expected (see Figs. 1, 2) and is well supported by these four features, three of which are homoplasy-free within the taxa included in the cladistic analysis in which these features could be discerned. The Clupeomorpha is one of the main teleostean groups. The extant members of this group are mainly marine. This group comprises some basal groups exclusively represented by fossil taxa, such as the †Ellimmichthyiformes, as well as the order Clupeiformes (herrings, anchovies and relatives). This latter order thus includes all extant clupeomorph species, which are among the most economically important of all fish species: heavily exploited by man for food, the immensity of their numbers also makes them an important food source for larger food fishes, as well as for a host of other marine life (e.g., Nelson, 2006). Clupeiformes are usually subdivided into Clupeoidei and Denticipitoidei (see Fig. 2). Within Clupeoidei, three major extant groups are often recognized: Clupeoidea, Engrauloidea and Pristigasteroidea (Fig. 2); within Denticipitoidei, one single living species is recognized, *Denticeps clupeoides* (e.g., Grande, 1985a; Gouréne and Teugels, 1994; Di Dario, 2002, 2004). *Denticeps* is thus often considered to be the most basal extant member of order Clupeiformes (see Fig. 2) and, consequently, of the superorder Clupeomorpha (e.g., Grande, 1985a; Di Dario, 2002, 2004). The phylogenetic results of the present work support such a basal position of *Denticeps* (Fig. 3).

Some aspects of the anatomy of clupeomorphs are illustrated in Figs. 45, 46, 47, 48, 49, 50, 51 and 52. An illustration of the general shape of a fish of the genus *Clupea* is given in Fig. 45. More detailed drawings of the configuration of the osteological and myological structures of the cephalic region and pectoral girdle of fishes of the genus *Denticeps* are given in Figs. 46, 47, 48, 49, 50, 51 and 52, all original drawings by the author.

Denticeps: [124: 1→0], [197: 0→1], **[207: 0→1]**, [222: 1→0], **[291: 0→1]**, [307: 0→1]

Clade 33 (*Ethmalosa* + *Ilisha* + *Engraulis* + *Thryssa*): [203: 0→1], [220: 1→0], [228: 0→1]

As expected (see Fig. 2), the four taxa of Clupeoidei included in the cladistic analysis of the present work were grouped together (Fig. 3).

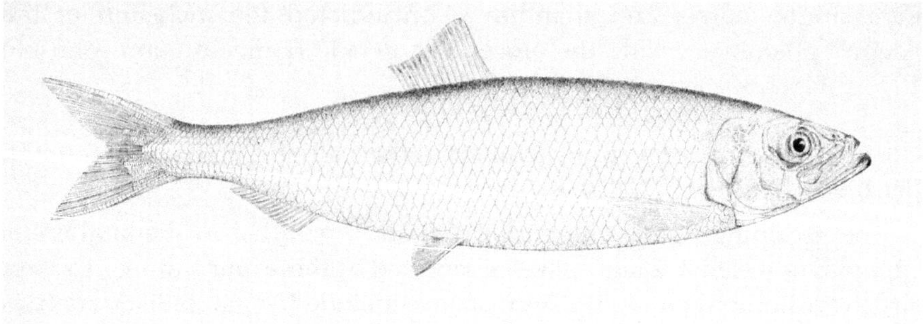

Figure 45 General aspect of a member of the genus *Clupea* (Teleostei: Clupeomorpha) (modified from Goode, 1884-1887).

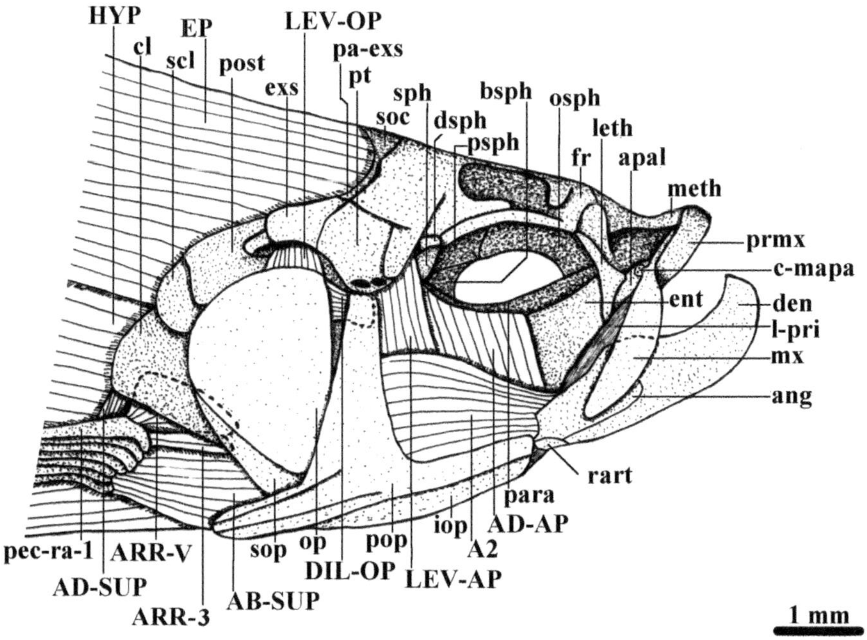

Figure 46 Lateral view of the cephalic musculature of *Denticeps clupeoides*. All muscles are exposed; the teeth of the jaws, as well as the onodontes, nasals, infraorbitals and postcleithra, were removed.

Ethmalosa: [29: 1→0], [67: 0→1], [205: 0→1], [313: 0→1]

Clade 34 (*Ilisha* + *Engraulis* + *Thryssa*): [112: 0→1]

As seen in Fig. 2, many authors consider the relationships between the Clupeoidea, the Engrauloidea and the Pristigasteroidea still unresolved. This subject was revised by Di Dario (2002), who defended a sister-group

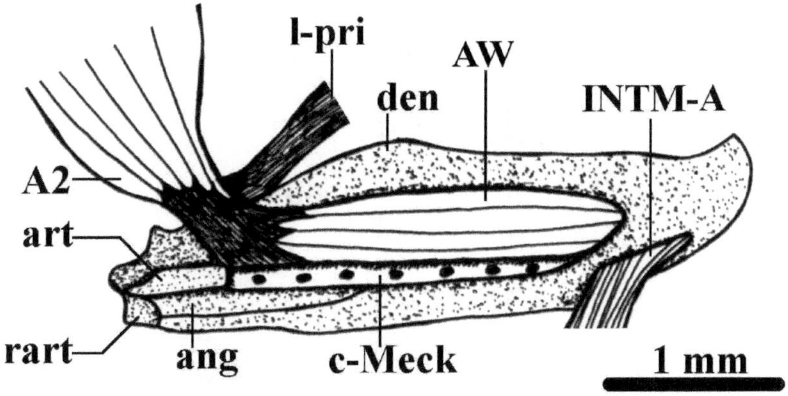

Figure 47 Mesial view of the left mandible and the adductor mandibulae of *Denticeps clupeoides*.

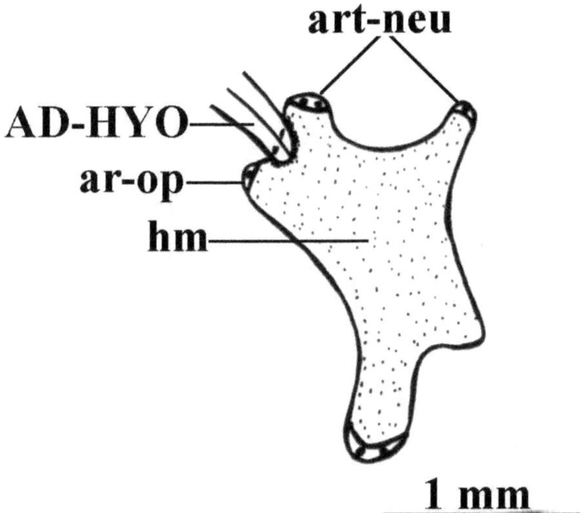

Figure 48 Mesial view of the left hyomandibula and the adductor hyomandibulae of *Denticeps clupeoides*.

relationship between the Clupeoidea and the Engrauloidea. The present work does not support this view, since the engrauloid and pristigasteroid fishes included in the cladistic analysis are grouped together in this clade 34. Apart from the synapomorphy listed above (112: 0→1), this clade 34 may eventually be supported by three other features if a "fast optimization" is chosen (216: 1→0; 238: 1→0; 323: 1→0). However, it should be recognized that the evidence provided in the present work to support this clade is rather weak, and, of course, the relationships between the Clupeoidea, Engrauloidea

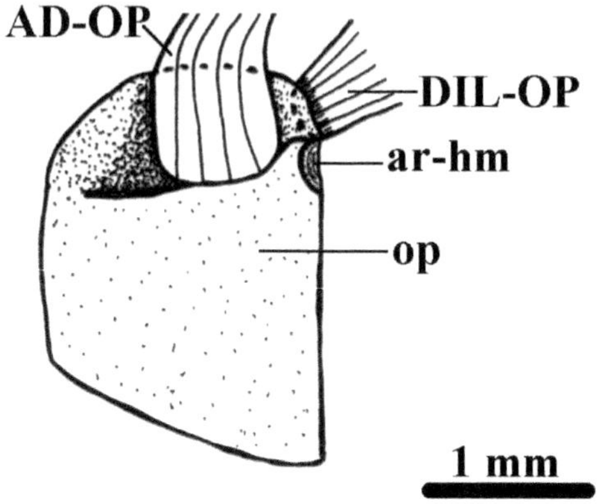

Figure 49 Mesial view of the left opercle and the dilatator operculi and adductor operculi of *Denticeps clupeoides*.

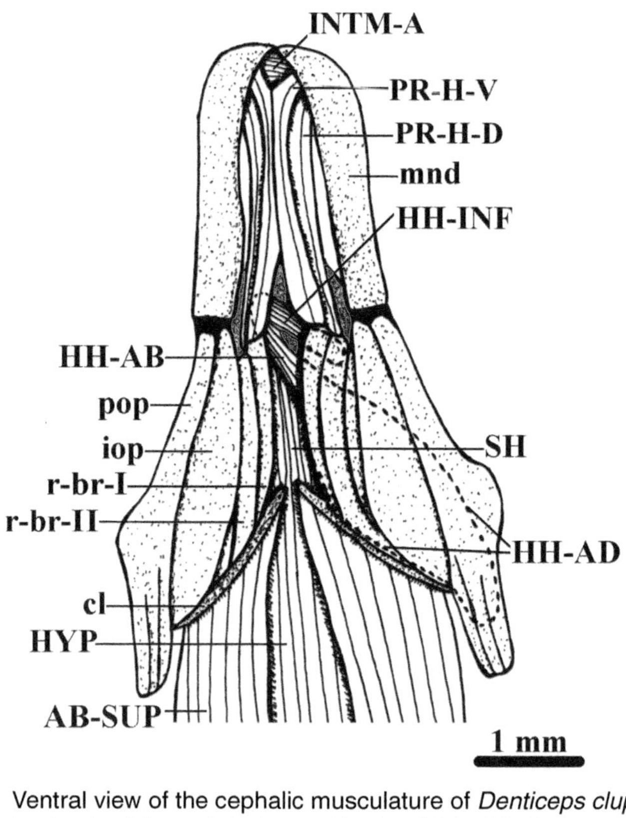

Figure 50 Ventral view of the cephalic musculature of *Denticeps clupeoides*. On the left side the hyohyoideus abductor and hyohyoidei adductores were removed.

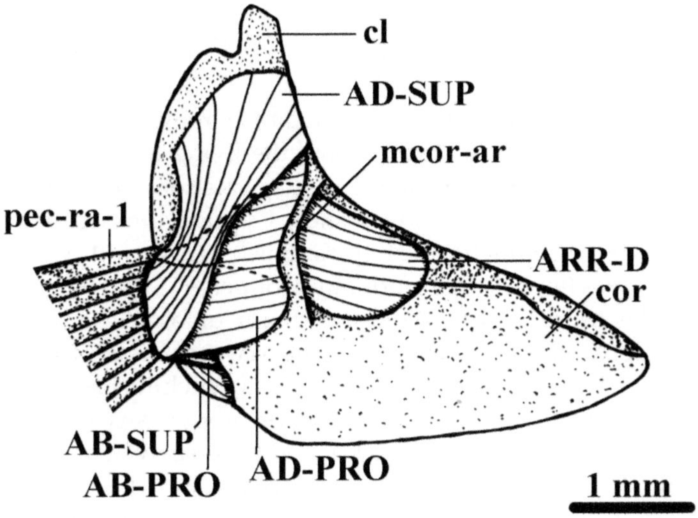

Figure 51 Mesial view of the pectoral girdle musculature of *Denticeps clupeoides*, the lateral muscles abductor superficialis and abductor profundus are also shown.

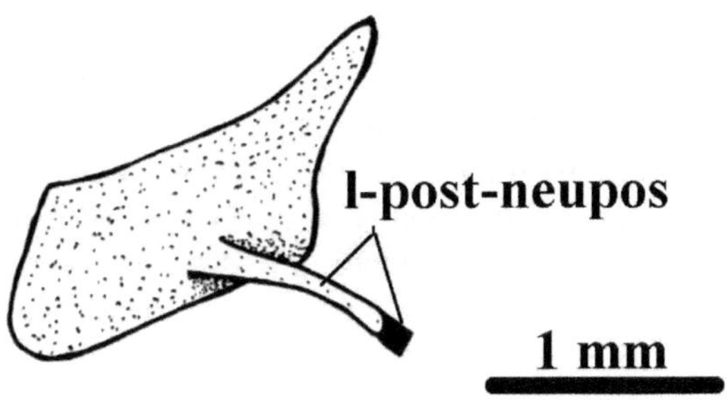

Figure 52 Mesial view of posttemporal of *Denticeps clupeoides*; the anterior portion of the bone is shown on the right side of the figure.

and Pristigasteroidea can only be appropriately examined in a cladistic analysis including many other representatives of these three groups.

Ilisha: [204: 0→1], [222: 1→0], [288: 0→2]

Clade 35 (*Engraulis* + *Thryssa*): [66: 0→1], [157: 0→1], **[283: 0→1]**

As expected (see Fig. 2), the two engrauloid taxa examined are grouped together.

Engraulis: [205: 0→1]

Thryssa: [124: 1→0], [228: 1→0]

Clade 36 (all ostariophysans included in the cladistic analysis): [29: 1→0], [82: 0→1], [112: 0→1], [157: 0→1], **[159: 0→1]**, [176: 1→0], [231: 1→0], [245: 0→1]

The Ostariophysi are usually divided into Otophysi and Anatophysi (see Figs. 1, 2), including more than 25% of the living teleostean species and about 80% of all extant freshwater fishes (e.g., Nelson, 2006). The otophysan clade includes the Siluriformes (catfishes), Cypriniformes (carps, minnows, and relatives), Characiformes (piranhas, tetras, and relatives) and Gymnotiformes (electric eels), formally grouped by Sagemehl (1885) on the basis of the presence of a complex chain of ossicles connecting the swimbladder to the inner ear—the Weberian apparatus (see Chapter 4, Section 4.5). Anatophysi includes the Gonorynchiformes (milkfish and relatives), considered the sister-group of otophysans mainly after the publication of Greenwood et al. (1966) and Rosen and Greenwood (1970).

Therefore, the assembly of the ostariophysan fishes included in the cladistic analysis in a monophyletic clade is expected. Eight unambiguous synapomorphies strongly support, in this analysis, the monophyly of clade 36: presence of protractor pectoralis (29: 1→0); absence of basisphenoid as an independent ossification (82: 0→1); ribs/parapophyses of third free vertebra highly modified (112: 0→1); adductor mandibulae attaching not only on mandible and/or primordial ligament, near its mandibular insertion, but also on other structures (157: 0→1); presence of adductor mandibulae A1-OST (**159: 0→1**); mainly undivided A2 not attaching on mesial surface of mandible by means of two well-distinguished tendons (176: 1→0); primordial ligament attaching posteriorly on dorsal surface of coronoid process (231: 1→0); absence of toothed dermopalatine (245: 0→1).

It is important to note that the scoring of the first feature as an ostariophysan synapomorphy might well be an artificial result related to the use of the specific clupeomorph taxa included in the present cladistic analysis, since, in fact, apart from *Ethmalosa* (the only clupeomorph taxon analyzed with CS0) many other clupeomorphs do have a protractor pectoralis (e.g., Greenwood and Lauder, 1981). One should also be prudent about the feature concerning the attachment of the primordial ligament on the dorsal surface of the coronoid process. This feature was considered by Fink and Fink (1981, 1996) as a synapomorphy of gymnotiforms + siluriforms. However, the cypriniforms examined in the present work also exhibit a primordial ligament attaching posteriorly on the dorsal surface of the coronoid process (except *Danio*, which was coded as '?' because the observation of the adult specimens of this genus examined in this work did not allow us to discern this character appropriately). Since the condition of gonorynchiforms and of the fossil ostariophysan taxa included in the cladistic analysis is not clear, either because it was difficult to discern this character or because a distinct primordial ligament is seemingly missing, such an attachment of this ligament on the dorsal surface of the coronoid process was consequently scored in the "majority fools" tree obtained as a potential ostariophysan synapomorphy. However, precisely because the situation in these taxa is not clear, one should be reticent regarding the acceptance of this feature as a potential synapomorphy of the Ostariophysi. The other feature listed above and not considered by Fink and Fink (1981, 1996) as an ostariophysan synapomorphy concerns the presence of an A1-OST. This feature is exclusively, and consistently, found in the Ostariophysi and does seems to provide strong support for the monophyly of a clade including at least the extant ostariophysans (e.g., Gosline, 1989; Diogo and Chardon, 2000; Diogo, 2004a; this work). The grouping of living ostariophysans is therefore strongly supported, thus corroborating the results of some molecular studies (e.g., Lavoué et al., 2005) and contradicting the results of others, in which some or all gonorynchiforms were placed as the sister-group of some or all clupeiforms included in those studies (e.g., Ishiguro et al., 2003; Saitoh et al., 2003; Wang et al., 2003; Inoue et al., 2004).

Some aspects of the anatomy of ostariophysans are illustrated in Fig. 24 and in Figs. 53 to 83. An illustration of the general shape of a fish of the catfish genus *Ictalurus* is given in Fig. 24. More detailed drawings of the configuration of the osteological and myological structures of the cephalic region and pectoral girdle of fishes of various ostariophysan taxa are given in Figs. 53 to 83. Most of these figures are original drawings by the author. It is worth noting that only a few anatomical illustrations of siluriforms are included in the present work (e.g., Figs. 69, 71D), since numerous illustrations of members of various families of this order were recently provided (Diogo, 2004a).

Clade 37 (all gonorynchiforms included in the cladistic analysis): [81: 0→1], [228: 0→1]

The grouping of the Gonorynchiformes examined is expected (see Figs. 1, 2) and is supported, within the characters examined in the cladistic analysis of the present work, by these two features.

Clade 38 (*Chanos* + *Gonorynchus*): [138: 1→0], [290: 0→1]

In an extensive morphological cladistic analysis of the Gonorynchiformes (Grande and Poyato-Ariza, 1999) the family Gonorynchidae, in which *Gonorynchus* is included, was placed as the sister-group of the Kneriidae. However, in a molecular cladistic analysis of this order (Lavoué et al., 2005) *Gonorynchus* appears as the sister-group a clade

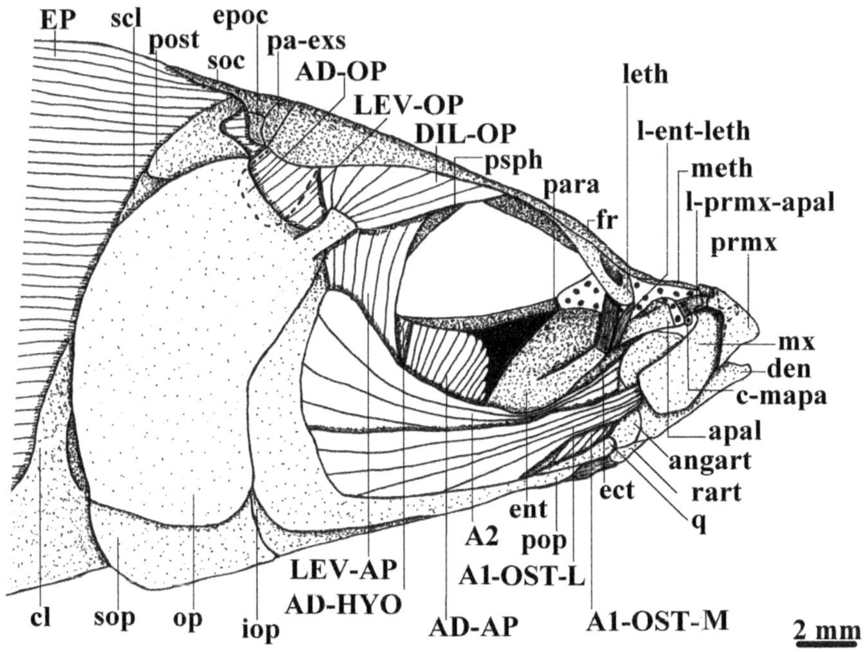

Figure 53 Lateral view of the cephalic musculature of *Chanos chanos*. The pectoral girdle muscles are not illustrated; the nasals and infraorbitals were removed.

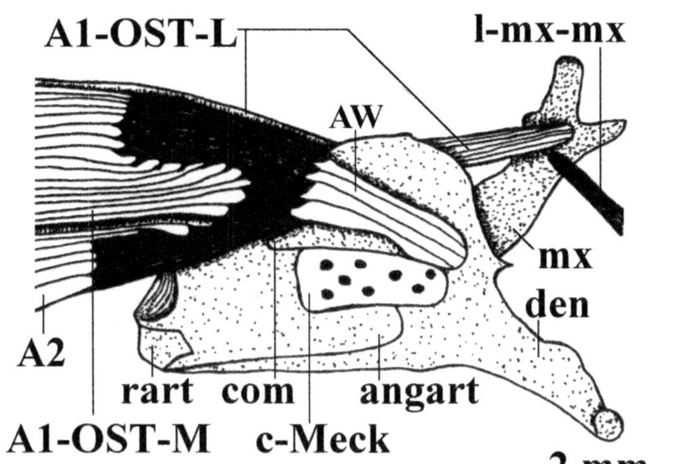

Figure 54 Mesial view of the left mandible, adductor mandibulae and maxilla of *Chanos chanos*.

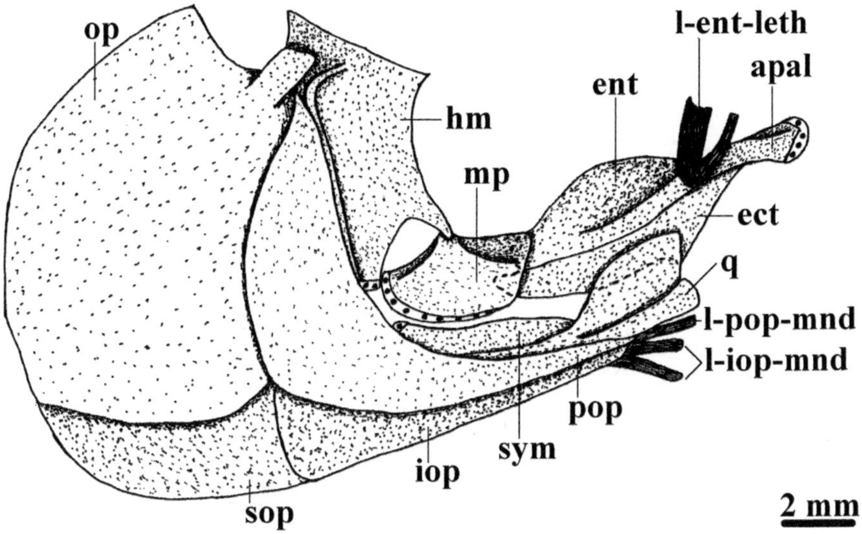

Figure 55 Lateral view of the suspensorium and opercular series of *Chanos chanos*.

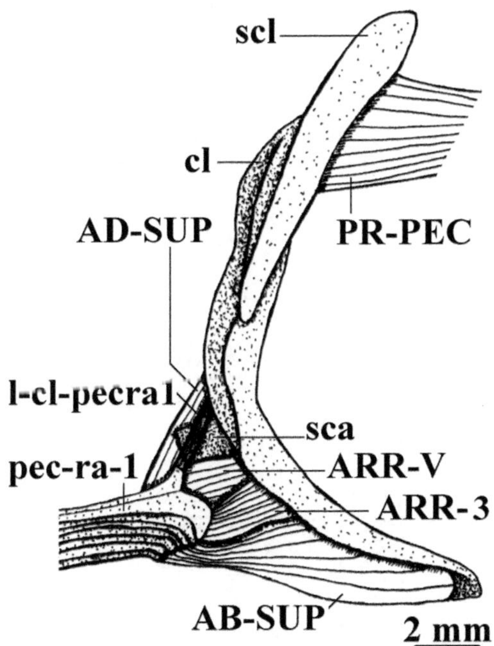

Figure 56 Lateral view of the pectoral girdle musculature of *Chanos chanos*; the protractor pectoralis is shown.

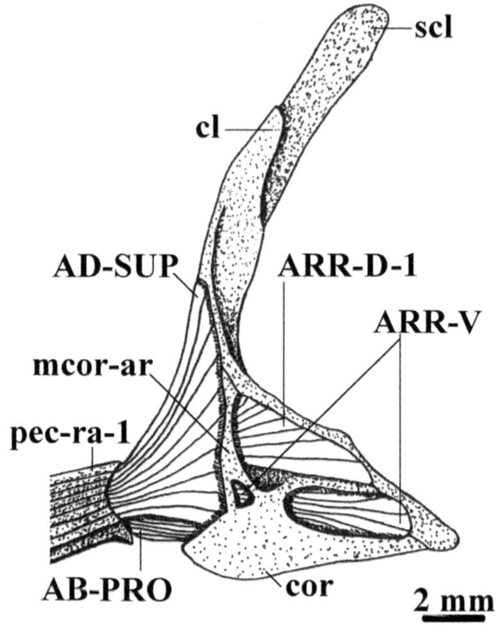

Figure 57 Mesial view of the pectoral girdle musculature of *Chanos chanos*.

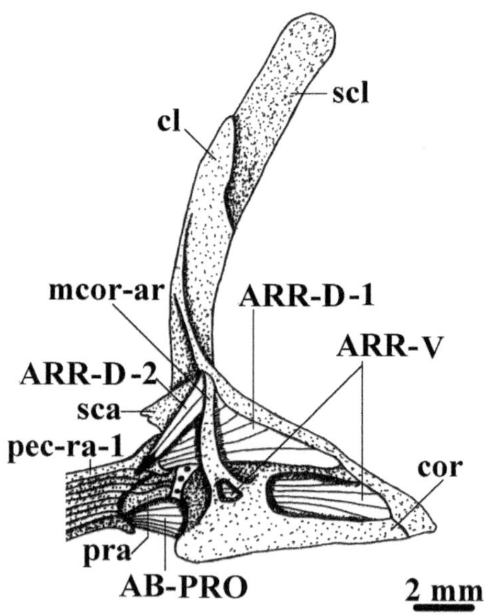

Figure 58 Mesial view of the pectoral girdle musculature of *Chanos chanos*; the adductor superficialis was removed.

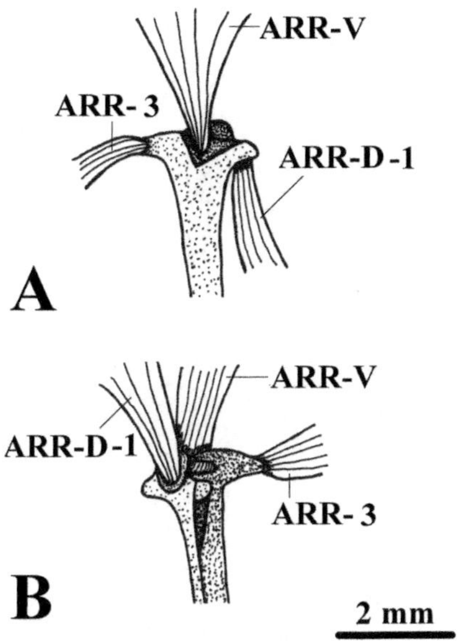

Figure 59 Lateral (**A**) and mesial (**B**) views of the anterior portion of the first pectoral ray and the insertions of the arrector 3, of the arrector ventralis, and of the section 1 of the arrector dorsalis in *Chanos chanos*.

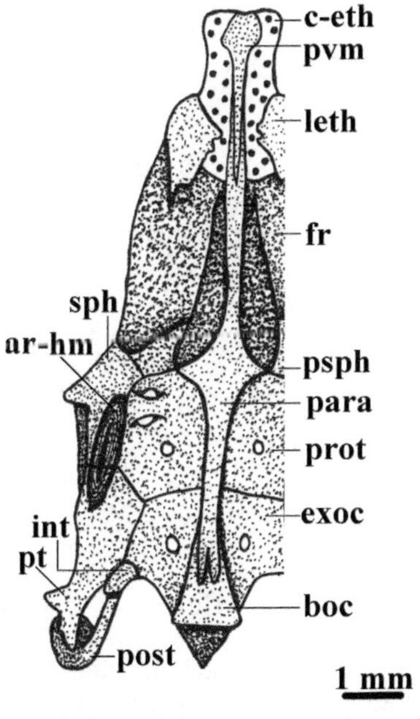

Figure 60 Ventral view of the neurocranium of *Chanos chanos*.

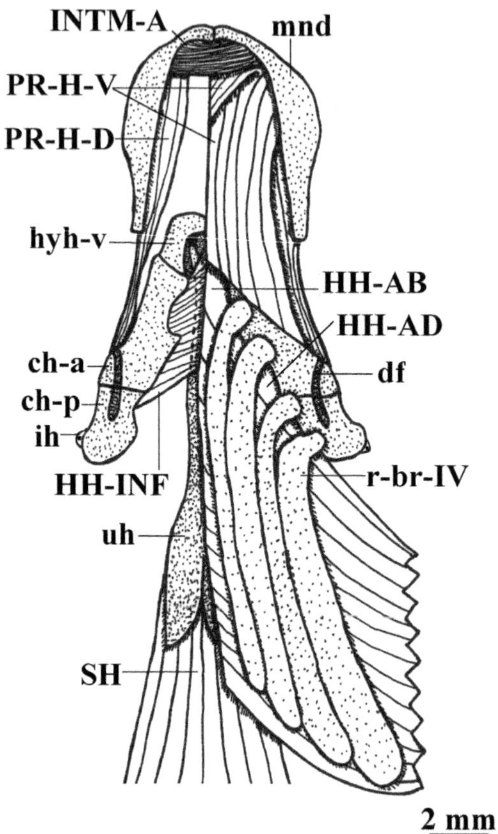

Figure 61 Ventral view of the cephalic musculature of *Chanos chanos*. On the right side all the hyoid muscles are exposed; on the left side the dorsal section of the protractor hyoidei, the hyohyoideus abductor and the hyohyoidei adductores were removed.

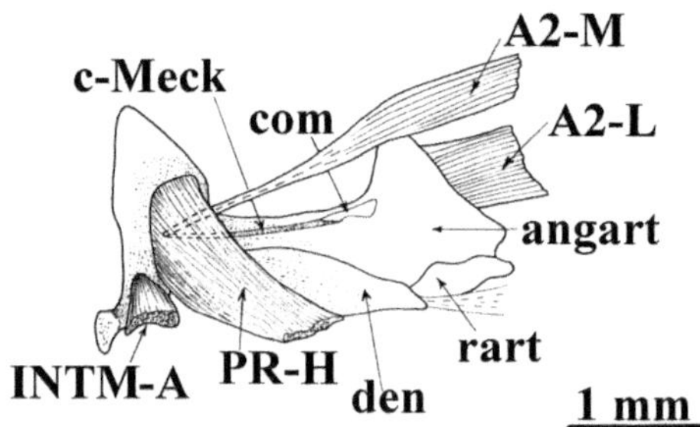

Figure 62 Mesial view of the left mandible, adductor mandibulae, protractor hyoidei and intermandibularis of *Gonorynchus gonorynchus* (modified from Howes, 1985a).

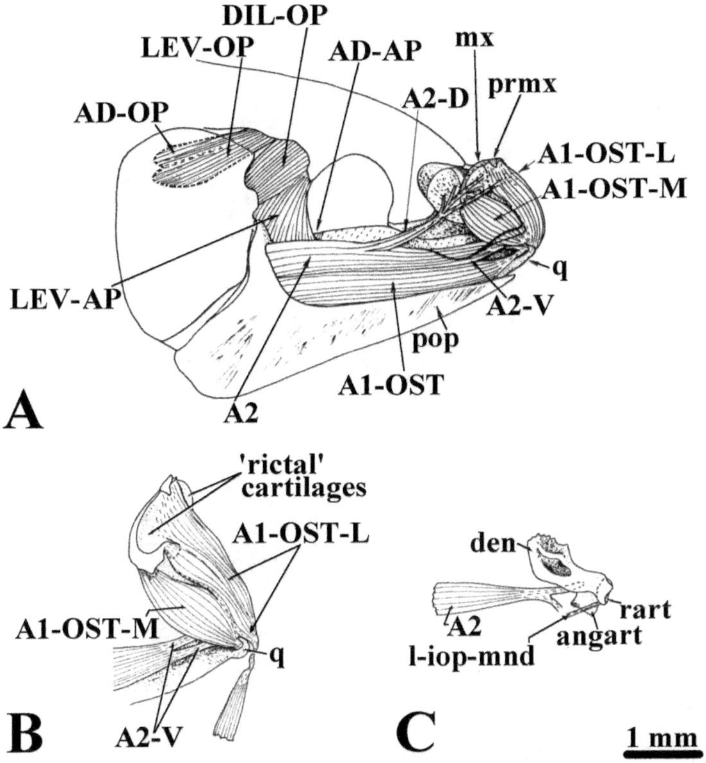

Figure 63 Cephalic musculature of *Phractolaemus ansorgei* (modified from Howes, 1985a). **A)** Lateral view of the cephalic muscles. **B)** Dorsolateral view of the upper and lower jaws, quadrate and adductor mandibulae; the A2-D has been removed and the A1-OST-L has been cut posteriorly and moved lateral to the quadrate to expose the lower portion of the A1-OST-M. **C)** Lateral view of the lower jaw and the section A2 of the adductor mandibulae.

including *Chanos* plus the remaining gonorynchiforms. In all the most parsimonious trees obtained in the present work (Figs. 3, 4) *Gonorynchus* appears as the sister-group of *Chanos*, but the evidence supporting this sister-group relationship is not particularly strong (138: 1→0, 290: 0→1). Thus, it can be said that, as stated about 20 years ago by Howes (1985a), the position of *Gonorynchus* within gonorynchiforms continues to be a particularly problematic issue. It is hoped that a future work together with T. Grande and F. Poyato-Ariza, including all known fossils of this order and, at the same time, all osteological and myological characters available for its extant taxa, will help to clarify this question.

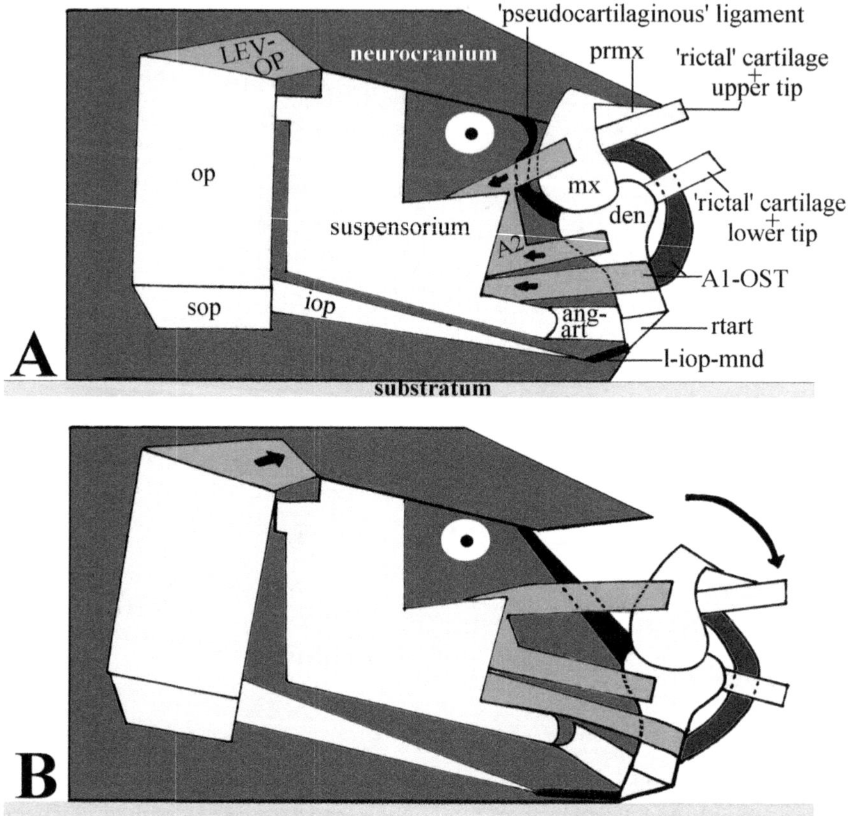

Figure 64 Illustration of the author's hypothesis concerning the mechanisms of mouth closure by the action of the adductor mandibulae and mouth opening by the action of the levator operculi in *Phractolaemus ansorgei* (note that only some cephalic structures are illustrated and that the ventral cephalic muscles and the mechanisms of mouth opening and mouth closure associated with these muscles are not represented; the movements shown are exaggerated, in order to facilitate the understanding of the illustration). **A)** The mouth is closed because of contraction of the adductor mandibulae. **B)** The mouth is opened and projected anteroventrally because of contraction of the levator operculi.

Chanos: [94: 0→1], [110: 1→0], [313: 0→1]

Gonorynchus: **[8: 0→1]**, [87: 0→1], [123: 0→1], [132: 0→1], [158: 0→1], [177: 0→1], **[196: 0→1]**, [197: 0→1], [217: 0→1], **[233: 0→1]**, **[241: 0→1]**, **[251: 0→1]**, [263: 1→0], **[268: 0→1]**, **[285: 0→1]**, **[339: 0→1]**, **[354: 0→1]**

Clade 39 (all kneriids included in the cladistic analysis): [70: 0→1], [175: 0→1], [306: 0→1]

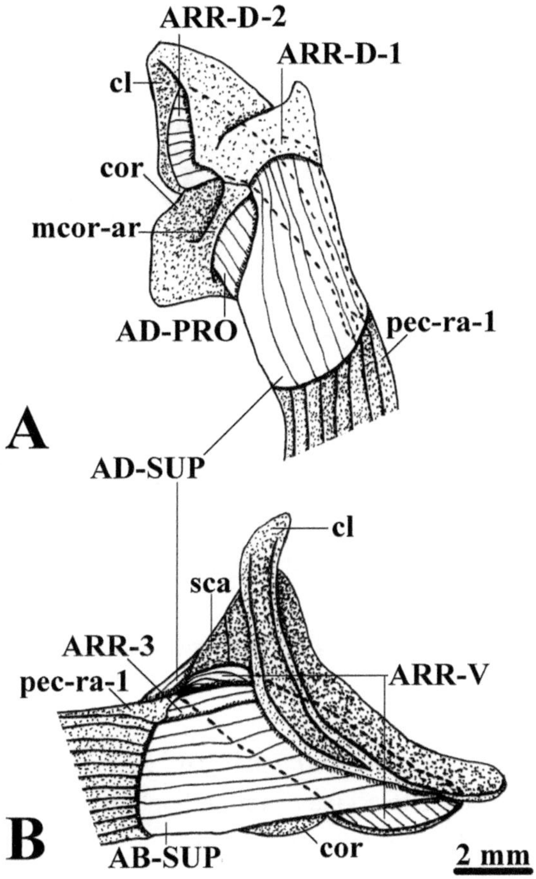

Figure 65 Dorsal (**A**) and lateral (**B**) views of the pectoral girdle musculature of *Phractolaemus ansorgei*.

The grouping of the kneriid taxa examined is expected (e.g., Grande and Poyato-Ariza, 1999) and is well supported.

***Phractolaemus*:** **[6: 0→1]**, **[69: 0→1]**, **[77: 0→1]**, [121: 0→1], [124: 1→0], **[135: 0→1]**, [158: 0→1], **[171: 0→1]**, **[172: 0→1]**, [217: 0→1], **[218: 0→1]**, [222: 1→0], **[239: 0→1]**, **[240: 0→1]**, [248: 1→0], [250: 0→1], **[264: 0→1]**, **[270: 0→1]**, **[292: 0→1]**, [296: 0→1], **[308: 0→1]**, [313: 0→1], [334: 0→1], **[348: 0→1]**

Clade 40 (*Cromeria* + *Grasseichthys* + *Kneria* + *Parakneria*): [97: 1→0], **[226: 0→1]**, **[343: 0→1]**

The grouping of these four taxa is expected (e.g., Grande and Poyato-Ariza, 1999) and is well supported.

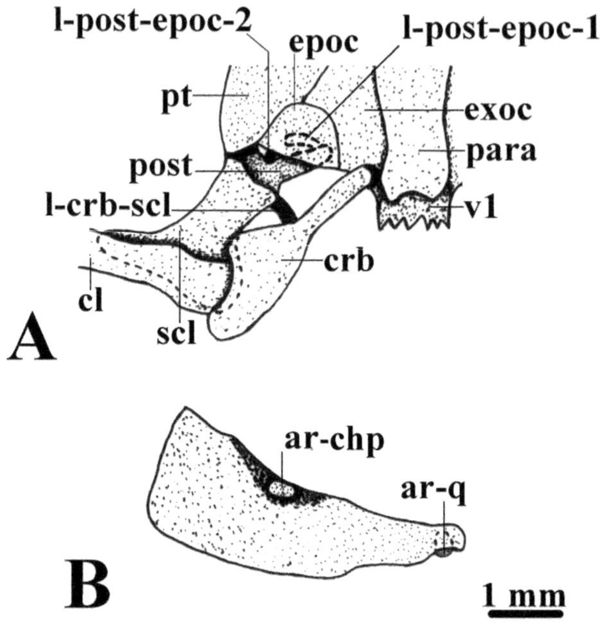

Figure 66 *Parakneria abbreviata*. **A)** Ventral view of the posterior region of the neurocranium and dorsal elements of the pectoral girdle. **B)** Mesial view of the interopercle.

Clade 41 (*Cromeria* + *Grasseichthys*): [83: 0→1], [260: 1→0] , [302: 1→0]

The grouping of the two taxa examined belonging to the Cromeriini is expected (e.g., Grande and Poyato-Ariza, 1999).

Grasseichthys: [296: 0→1]

Cromeria: No unambiguous features

Clade 42 (*Kneria* + *Parakneria*): [145: 0→1], [257: 0→1], **[269: 0→1], [333: 0→1]**

The grouping of the two taxa examined belonging to the Kneriini is expected (e.g., Grande and Poyato-Ariza, 1999).

Parakneria: No unambiguous features

Kneria: [86: 0→1], [206: 0→1]

Clade 43 (all non-gonorynchiform ostariophysans included in the cladistic analysis): [66: 0→1]

In Taverne's (1999) paper about †*Sorbininardus apuliensis*, that author considered it probable (on the basis of a hand-made tree) that this taxon is an

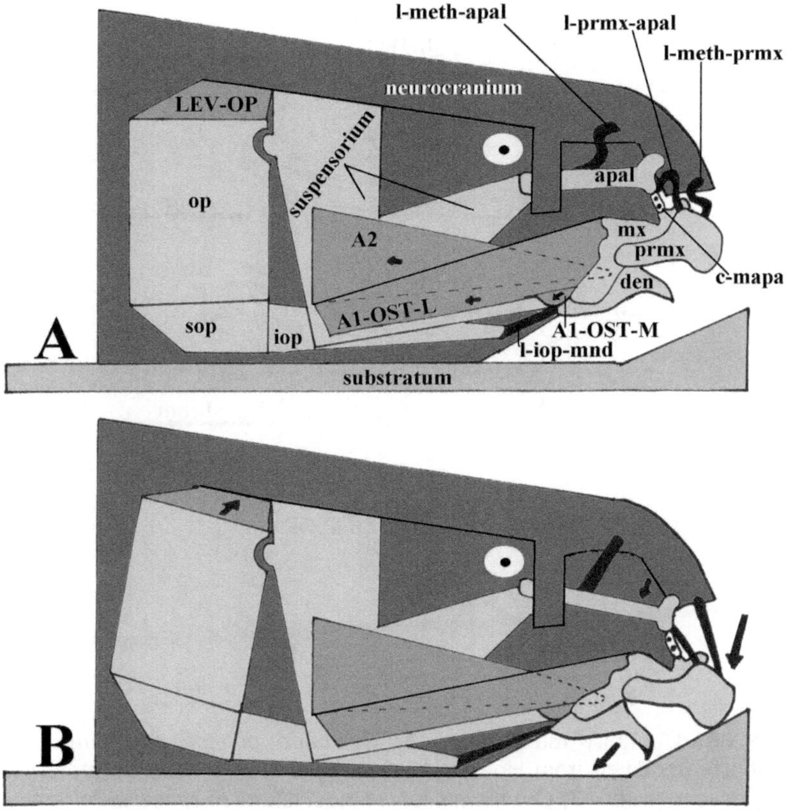

Figure 67 Illustration of the author's hypothesis concerning the mechanisms of mouth closure by the action of the adductor mandibulae and mouth opening by the action of the levator operculi in *Parakneria abbreviata* (note that only some cephalic structures are illustrated and that the ventral cephalic muscles and the mechanisms of mouth opening and mouth closure associated with these muscles are not represented; the movements shown are exaggerated in order to facilitate the understanding of the illustration). **A)** The premaxilla is retracted and the mandible is raised because of contraction of the adductor mandibulae. **B)** The premaxilla is protracted and the mandible is lowered because of contraction of the levator operculi.

ostariophysan, namely the sister-group of Gonorynchiformes. The results of the present study support the first hypothesis, but not the second: within the ostariophysans examined, †*Sorbininardus apuliensis* appears more closely related to non-gonorynchiforms than to gonorynchiforms (Figs. 3, 4). This scenario is supported by the following feature: anteroventral margin of prevomer situates well posteriorly to anteroventral margin of mesethmoid (66: 0→1, only occurring in a few taxa outside this clade 43 and only

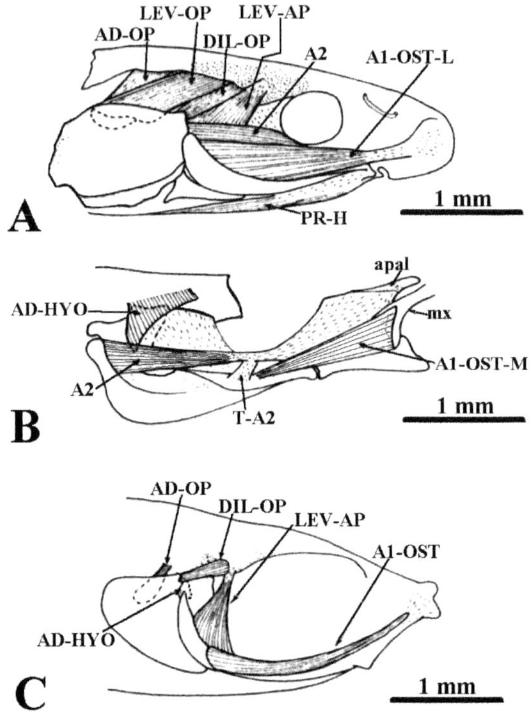

Figure 68 Cephalic musculature of *Cromeria nilotica* and *Grasseichthys gabonensis* (modified from Howes, 1985a). **A)** *Cromeria nilotica*, lateral view of the cephalic muscles. **B)** *Cromeria nilotica*, most lateral muscles were removed to show details of the adductor hyomandibulae and of the sections A2 and A1-OST-M of the adductor mandibulae. **C)** *Grasseichthys gabonensis*, lateral view of the cephalic muscles.

reversed, inside of it, in cypriniforms; the condition in †*Santanichthys diasii* and in †*Clupavus maroccanus* is not clear). Fink and Fink (1981, 1996) proposed a similar feature as a synapomorphy of the clade including siluriforms + gymnotiforms + characiforms. Since this feature, as defined in the present work, is seemingly also found in †*Sorbininardus apuliensis* (e.g., Taverne, 1999: fig. 3), in †*Chanoides macropoma* (e.g., Patterson, 1984: fig. 6B) and in †*Lusitanichthys characiformis* (e.g., Gayet, 1985: fig. 19; p. 114), it was consequently scored in the tree of Fig. 3 as a synapomorphy of this clade 43 (as explained above, the condition of †*Santanichthys diasii* and of †*Clupavus maroccanus* is not clear). One should, however, keep in mind that this is the only synapomorphy supporting this clade 43 in the analysis and, thus, that the evidence provided in the present work to support this clade is not strong, although it is stronger than that supporting a close relationship between

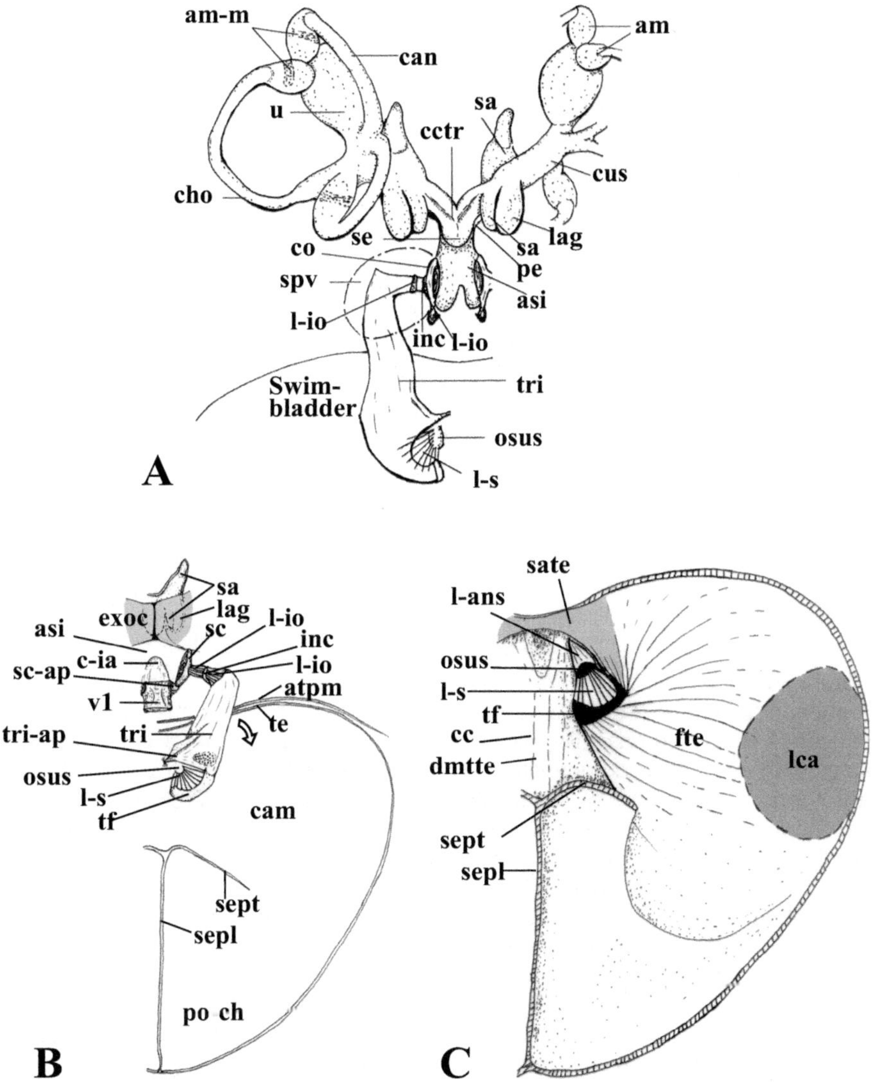

Figure 69 *Ictalurus nebulosus* (modified from Chardon et al., 2003; the nomenclature of the structures illustrated basically follows that of these authors). **A)** Dorsal view of labyrinths, sinus impar and ossicles. The right labyrinth was pushed laterally in order to show the canalis utriculo-saccularis. **B)** Dorsal half of the swimbladder with ossicles and hind part of labyrinths; note the exoccipital bridge (in light gray) covering the saccules and lagenas. The ossicles were pulled laterally (arrow) as if the swimbladder was compressed. **C)** Ventral half of the swimbladder showing the thin median area and its anterior arch and the lateral cutaneous area (both in light gray).

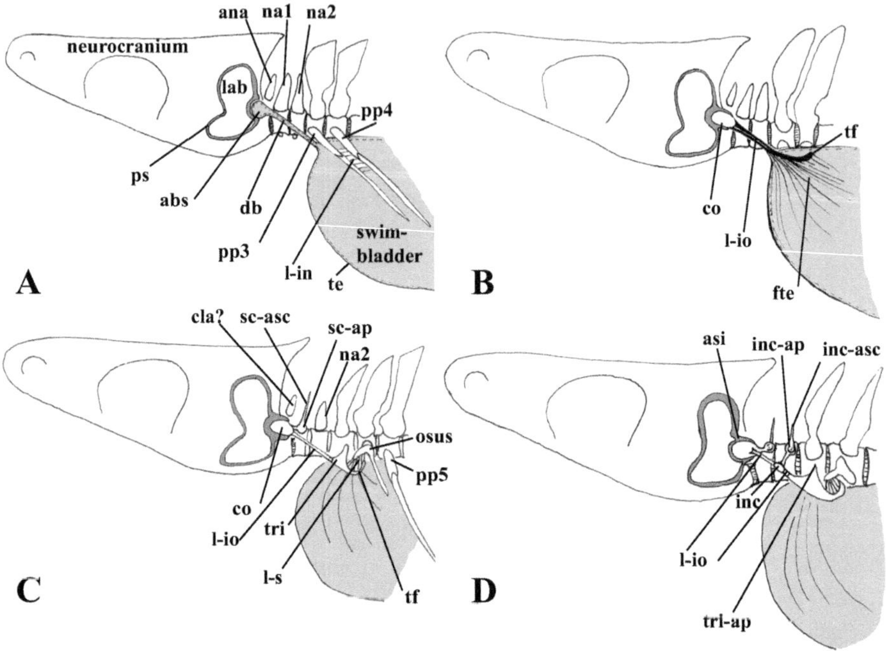

Figure 70 Illustration of Chardon et al.'s (2003) hypothesis about the origin of the Weberian apparatus (modified from Chardon et al., 2003; the nomenclature of the structures illustrated basically follows that of these authors). **A)** "Clupeiform-like" stage with an otophysic connection. The anterior duct between the swimbladder and the bulla is coated by the tunica externa; the labyrinth is surrounded by a perilymphatic space. The first two haemal arches are reduced. Parapophyses and ribs 3 and 4 are joined by an intercostal ligament. **B**) Fully hypothetical stage. The haemal arches are not figured. The endodermic epithelium and splanchnopleura have disappeared in the duct and bulla. The duct, reduced to the fibrous tunica externa, becomes the interossicular ligament; ossification of the wall of the bulla produces the concha scaphii; the wall of the swimbladder ossifies into the transformator of the tripus where the fibers of the tunica externa exert a traction. The parapophyses of free vertebrae 3 and 4 are not shown. **C)** †*Chanoides macropoma* and/or †*Lusitanichthys characiformis*-like stage. The anterior neural arch (or a paired supraneural?) becomes the claustrum. The first neural arch transforms into the scaphium (articular and ascendens processes) and fuses with the concha. The third haemal and possibly the third rib fuse with the transformator and become the tripus. The fourth haemal arch transforms into the os suspensorium, which remains attached to the third haemal arch by an intercostal ligament; this latter becomes the suspensor ligament. An osseous nodule probably appears in the interossicular ligament: the manobrium of the future intercalarium (not shown). **D)** Stage of extant otophysans (for more details, see text).

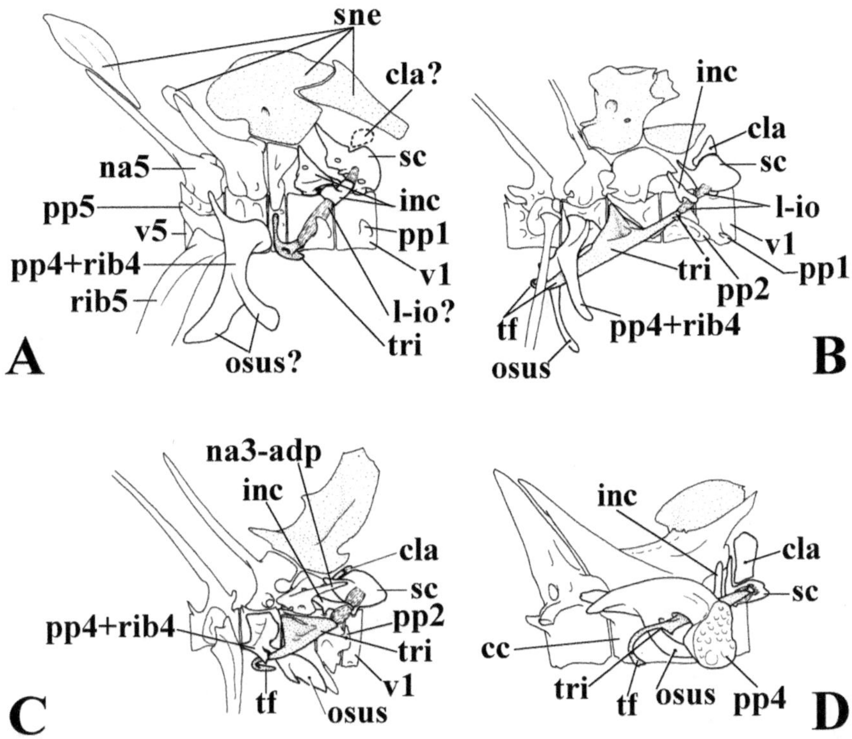

Figure 71 Outline drawings of the Weberian ossicles in the fossil †*Chanoides macropoma* **(A)**, the cypriniform *Opsariichthys uncirostris* **(B)**, the characiform *Xenocharax spilurus* **(C)**, and the siluriform *Diplomystes papillosus* **(D)** (modified from Patterson, 1984; the nomenclature of the structures illustrated basically follows that of this author).

†*Sorbininardus apuliensis* and gonorynchiforms. Some features with ambiguous distributions may be interpreted as potential synapomorphies of this clade 43 if a "fast optimization" is chosen (e.g., 101: 0→1, mesial limb of coracoids or scapulo-coracoids broad and anteroposteriorly elongated) or, alternatively, if a "slow optimization" is selected (e.g., 69: 0→1, presence of "rudimentary scaphium").

†*Sorbininardus apuliensis:* [85: 0→1], [138: 1→0], [330: 1→0]

Clade 44 (†*Clupavus maroccanus* + †*Santanichthys diasii* + †*Chanoides macropoma* + †*Lusitanichthys characiformis* + all extant otophysans included in the cladistic analysis): [96: 1→0], **[111: 0→1], [113: 0→1]**

Three unambiguous synapomorphies support this clade, which in principle can be named Otophysi (although this places †*Sorbininardus*

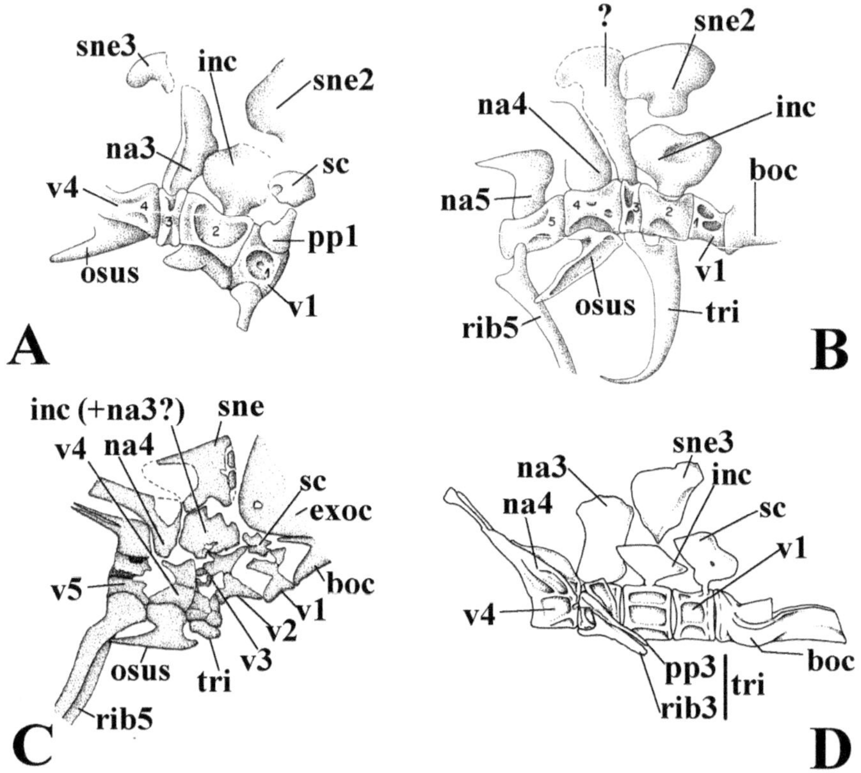

Figure 72 Outline drawings of the Weberian ossicles in the fossils †*Lusitanichthys characiformis* **(A, B)**, †*Clupavus maroccanus* **(C)**, and †*Santanichthys diasii* **(D)** (modified from Gayet, 1985; Taverne, 1995; Filleul and Maisey, 2004: the nomenclature of the structures illustrated basically follows that used by the authors of the original drawings).

apuliensis in a kind of limbo, since it is not an Otophysi but, according to the scenario proposed in Figs. 3 and 4, neither can it be considered an Anatophysi, as this will rend the Anatophysi paraphyletic): main bodies of parietals (or of parieto-extrascapulars) not widely separated from each other in dorsal view (96: 1→0); presence of a "rudimentary tripus" (**111: 0→1**, the situation in †*Clupavus maroccanus* is not clear); presence of "rudimentary os suspensorium" (**113: 0→1**, the situation in †*Santanichthys diasii* is not clear). Some features with ambiguous distributions may eventually be interpreted as synapomorphies of this clade if a "slow optimization" is chosen (104: 0→1, presence of "rudimentary scaphium"; 107: 0→1, presence of "rudimentary intercalarium": these two features cannot be appropriately discerned in †*Sorbininardus apuliensis*).

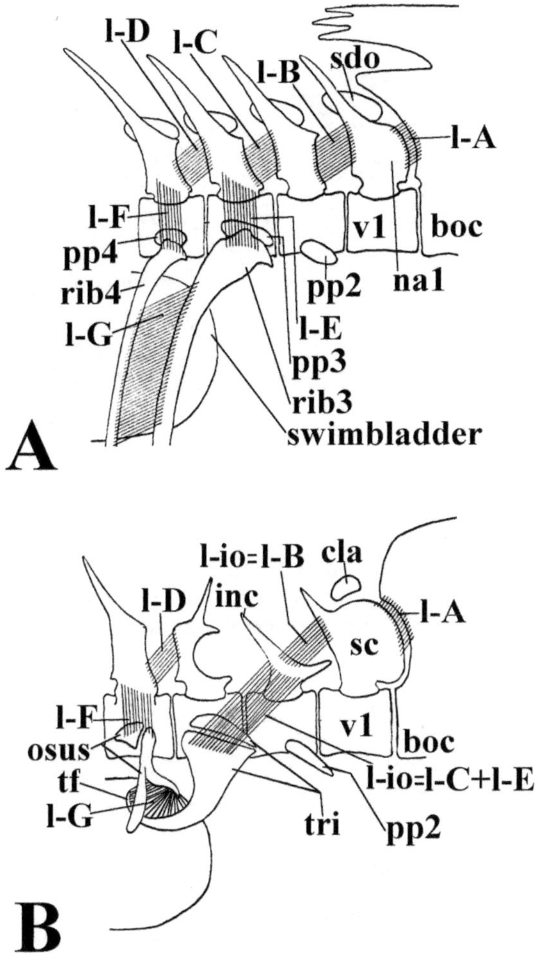

Figure 73 Diagram to explain Rosen and Greenwood's (1970) hypothesis concerning the homologies between the structures of the occipitocervical region of non-otophysans such as the gonorynchiform *Chanos* **(A)** and otophysans such as the characiform *Brycon* **(B)** (modified from Rosen and Greenwood 1970; the nomenclature of the structures illustrated basically follows that of these authors).

Clade 45 (†*Clupavus maroccanus* + †*Santanichthys diasii* + †*Chanoides macropoma* + †*Lusitanichthys characiformis*): [220: 0→1]

The single unambiguous synapomorphy supporting this clade concerns the presence of supramaxillae (220: 0→1, the four fossil taxa included in this clade are the only ostariophysans included in the cladistic analysis with supramaxillae; some extant ostariophysans not included in the analysis do

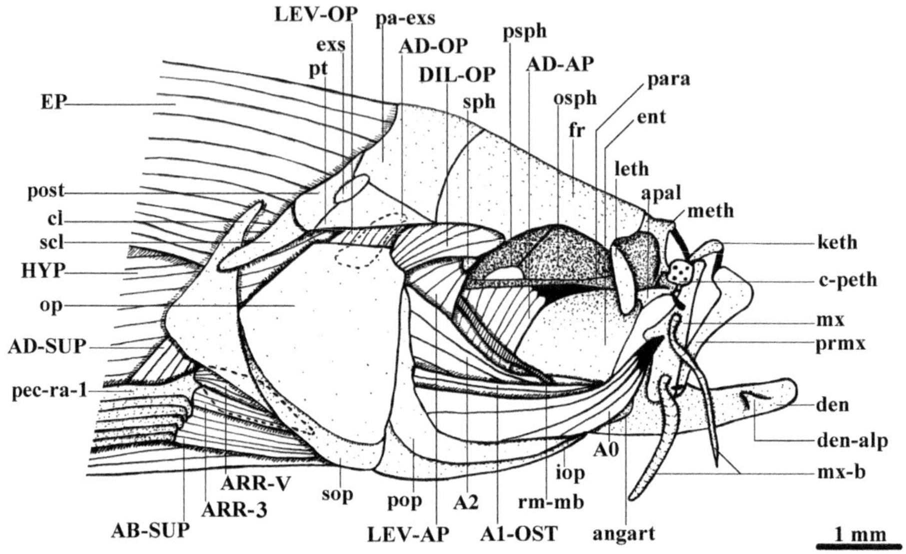

Figure 74 Lateral view of the cephalic musculature of *Danio rerio*. All muscles are exposed, the maxillary barbels and the mesial branch of the ramus mandibularis are also illustrated; the nasals, infraorbitals and postcleithra were removed.

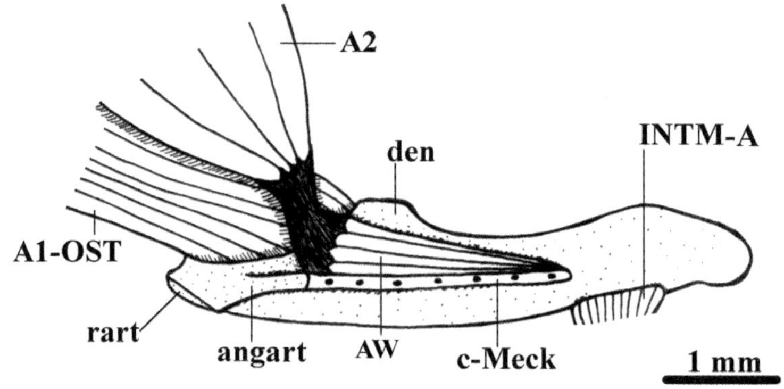

Figure 75 Mesial view of the left mandible and adductor mandibulae of *Danio rerio*, the anterior intermandibularis is also shown; the adductor mandibulae A0 was removed.

have supramaxillae, such as certain characiforms, but this is seemingly a derived feature for the order Characiformes: e.g., Fink and Fink, 1981, 1996).

Whether these and other "problematic fossil Otophysi" (sensu Grande and De Pinna, 2004) are placed in a monophyletic group or not, their placement outside the clade including the four extant otophysan orders

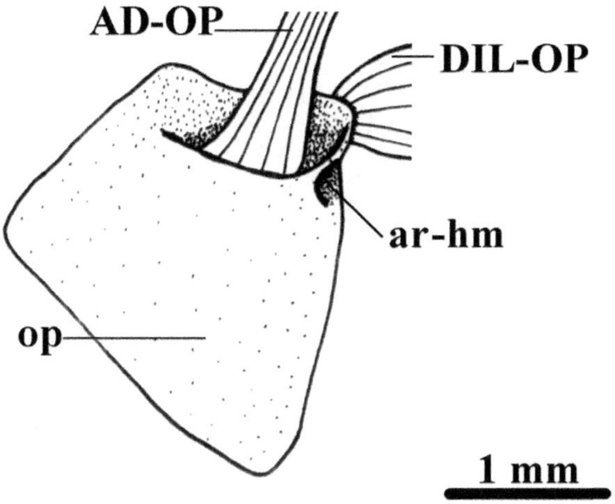

Figure 76 Mesial view of the opercular bone and of the adductor operculi and dilatator operculi of *Danio rerio*.

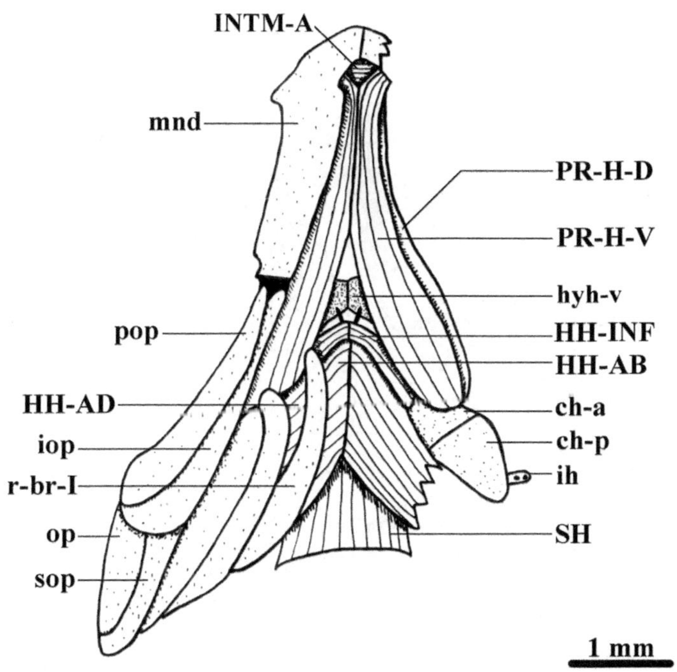

Figure 77 Ventral view of the cephalic musculature of *Danio rerio*. On the right side a portion of the hyohyoidei adductores, as well as of the mandible, was cut, and the opercle, interopercle, subopercle and preopercle are not shown.

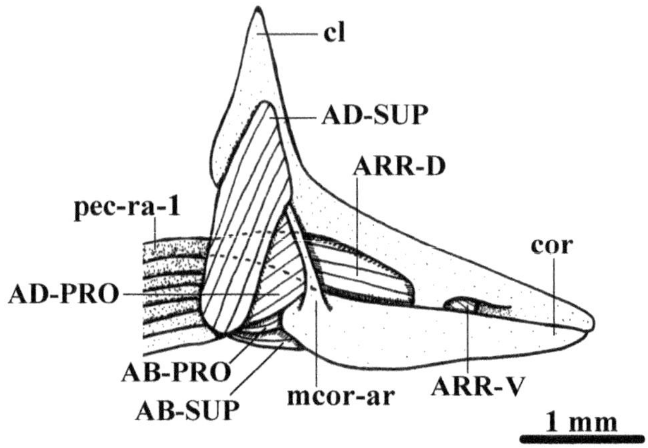

Figure 78 Mesial view of the pectoral girdle musculature of *Danio rerio*.

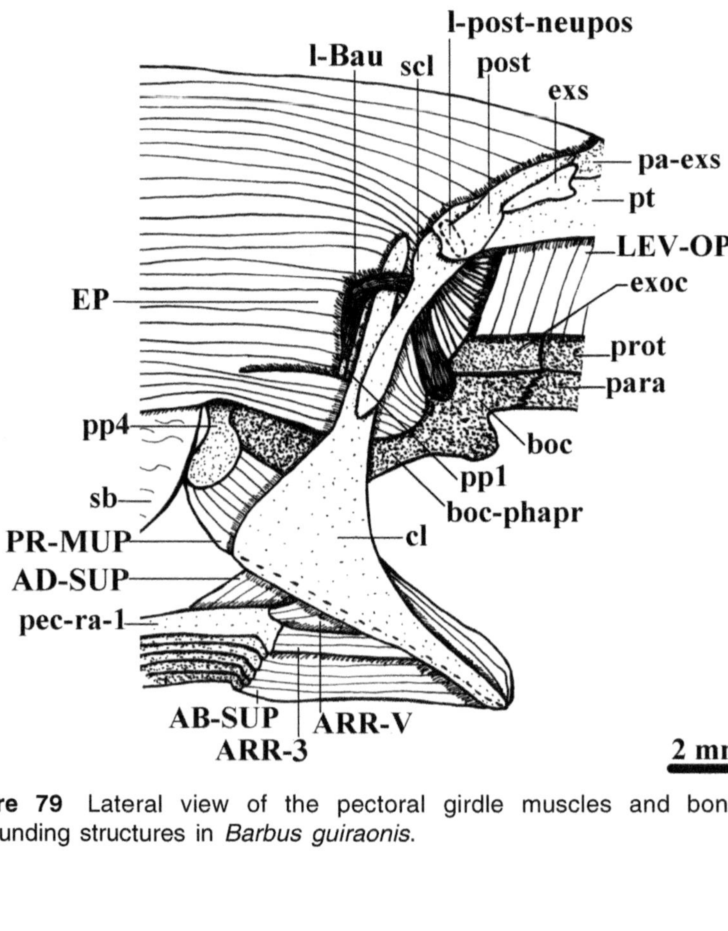

Figure 79 Lateral view of the pectoral girdle muscles and bones and surrounding structures in *Barbus guiraonis*.

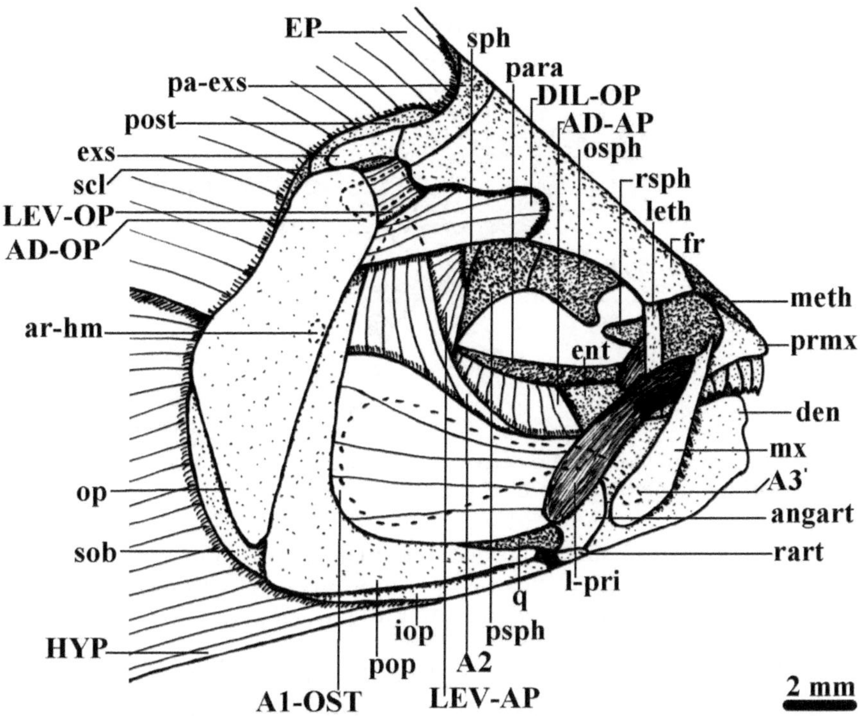

Figure 80 Lateral view of the cephalic musculature of *Brycon guatemalensis*. The pectoral girdle muscles are not illustrated; the postcleithra and the most ventral elements of the pectoral girdle, as well as the nasals and infraorbitals, were removed.

(Figs. 3, 4) has important phylogenetic and evolutionary implications. For example, this indicates that the characteristic Weberian apparatus of the members of these four extant otophysan orders was acquired only once, thus supporting the view of Fink and Fink (1981, 1996), Fink et al. (1984) and Patterson (1984), and other authors and contradicting those of, for example, Gayet (1981, 1985, 1986a). This subject will be discussed in Chapter 4, Section 4.5.

†***Clupavus maroccanus:*** No unambiguous features

†***Santanichthys diasii:*** [92: 0→1]

Clade 46 (†*Chanoides macropoma* + †*Lusitanichthys characiformis*): [115: 0→1]

Although there is a single unambiguous synapomorphy uniting †*Chanoides macropoma* and †*Lusitanichthys characiformis* (115: 0→1, centrum

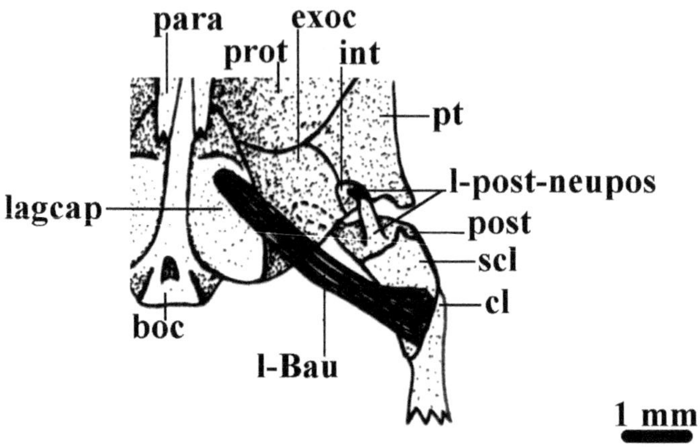

Figure 81 Ventral view of the posterior region of the neurocranium and dorsal elements of the pectoral girdle of *Brycon guatemalensis*.

of third free vertebra markedly shorter than other surrounding centra), it is interesting to notice that this feature is homoplasy-free within the numerous taxa included in the analysis in which this character could be discerned.

†*Chanoides macropoma*: [64: 0→1], [100: 0→1], [277: 1→0]

†*Lusitanichthys characiformis*: [80: 0→1], [110: 1→0], [139: 0→1]

Clade 47 (all extant otophysans included in the cladistic analysis): [104: 1→2], **[107: 1→2]**, **[111: 1→2]**, **[113: 1→2]**

This clade is supported by four synapomorphies: presence of characteristic scaphium (104: 1→2, not acquired outside of clade 47 but reversed in some of its taxa such as *Callichthys* and *Trichomycterus*), presence of characteristic intercalarium (**107: 1→2**), presence of characteristic tripus (**111: 1→2**) and presence of characteristic os suspensorium (**113: 1→2**). Some features with ambiguous distributions may eventually be interpreted as synapomorphies of this clade if a "fast optimization" is chosen (132: 0→1, "ligament between posttemporal and posterior margin of neurocranium" not ossified; 257: 0→1, high mobility between autopalatine and rest of suspensorium) or, alternatively, if a "slow optimization" is chosen (106: 0→1, presence of claustrum as an independent element in adults; 118: 0→1, perilymph system of inner ear peculiarly extended posteriorly, constituting sinus impar; note that the ambiguity of these two latter features is due to the fact that they were coded as '?' in the five ostariophysan fossils included in the cladistic analysis). See comments given above for clade 45.

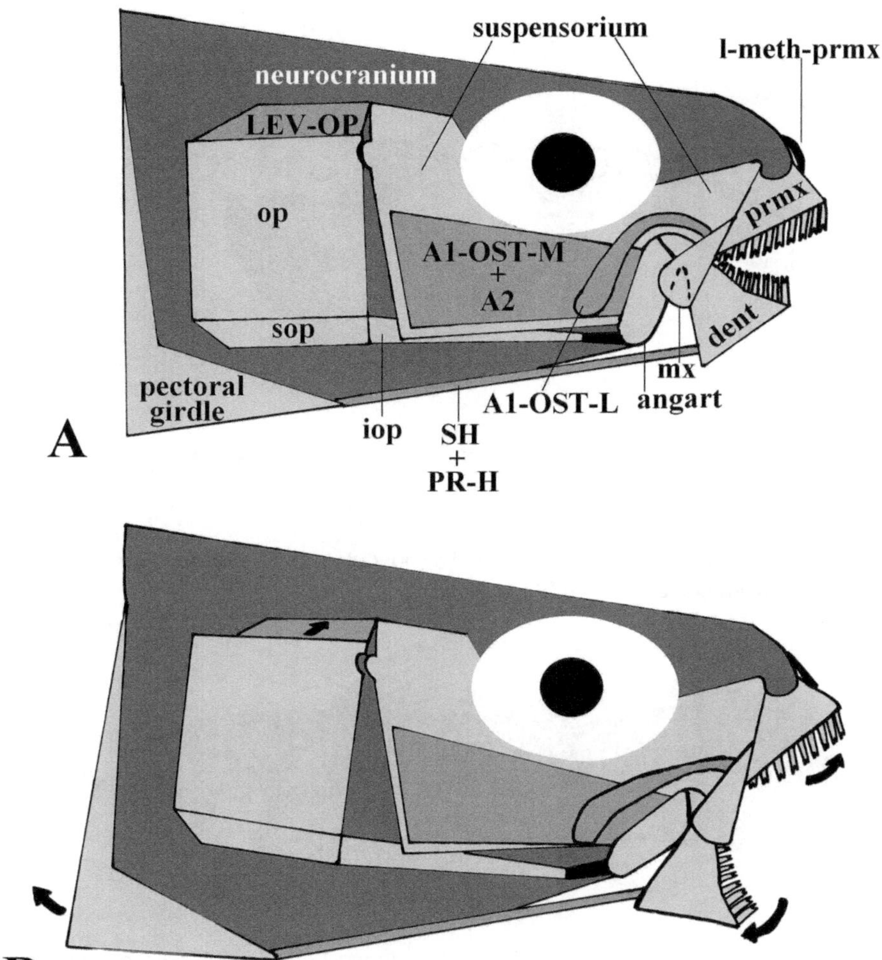

Figure 82 Illustration of the author's hypothesis concerning the mechanisms of mouth closure by the action of the adductor mandibulae and mouth opening by the action of the levator operculi in *Distichodus notospilus* (note that only some cephalic structures are illustrated and that the ventral cephalic muscles and the mechanisms of mouth opening and mouth closure associated with these muscles are not represented; the movements shown are exaggerated in order to facilitate the understanding of the illustration). **A)** The premaxilla is retracted and the mandible is raised because of contraction of the adductor mandibulae. **B)** The premaxilla is protracted and the mandible is lowered because of contraction of the levator operculi.

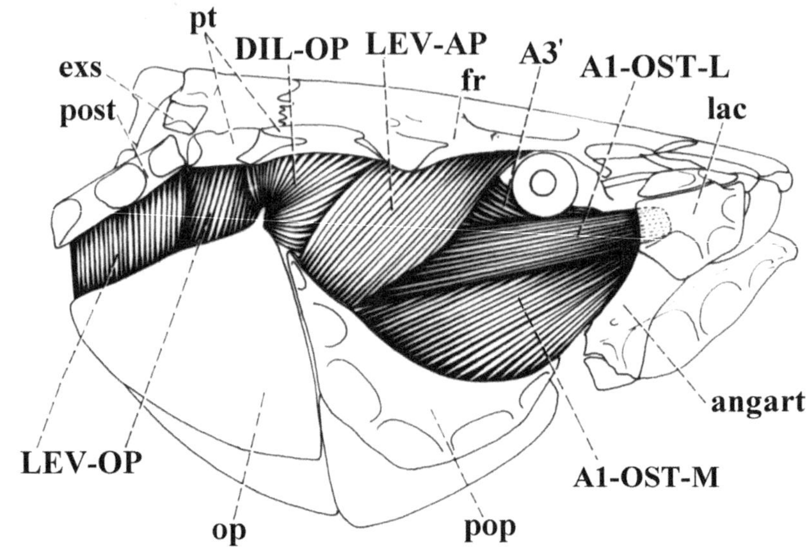

Figure 83 Lateral view of the cephalic musculature of *Sternopygus macrurus* (modified from De la Hoz, 1974).

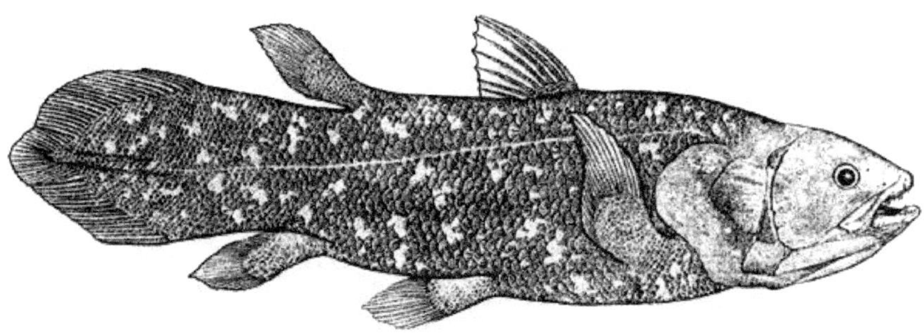

Figure 84 General aspect of a fish of the genus *Latimeria* (Actinistia) (modified from Smith, 1986).

Clade 48 (all cypriniforms included in the cladistic analysis): [27: 1→0], [64: 0→1], [66: 1→0], [85: 0→1], **[98: 0→1]**, [124: 1→0], **[211: 0→1]**, **[253: 0→1]**, **[254: 0→1]**

The grouping of the Cypriniformes included in the cladistic analysis is expected (see Fig. 2) and well supported. Unlike the other features listed above, the arrector dorsalis not subdivided into different sections (27: 1→0) and the dorsomesial limb of posttemporal (or posttemporo-supracleithrum) not markedly thin and mesially extended (124: 1→0) were not listed, at least to my knowledge, as potential cypriniform synapomorphies in previous works of other authors.

Clade 49 (*Catostomus* + *Cobitis*): [187: 0→1]

The assembly of the two taxa examined belonging to the Cobitoidea is expected (see, e.g., Siebert, 1987; Liu et al., 2002; Liu, 2004; Saitoh et al., 2006).

Catostomus: [86: 0→1], [109: 0→1], [129: 1→0], [197: 0→1], [206: 0→1], [290: 0→1]

Cobitis: [65: 0→1], [68: 0→1], [73: 0→1], [97: 1→0], [138: 1→0], **[189: 0→1]**, [193: 0→1], [259: 0→1], **[262: 0→1]**

Clade 50 (*Opsariichthys* + *Danio* + *Barbus*): **[211: 1→2]**, **[249: 0→1]**

The assembly of the three taxa examined belonging to the Cyprinoidea is expected (see, e.g., Siebert, 1987; Liu et al., 2002; Liu, 2004; Saitoh et al., 2006).

Opsariichthys: [138: 1→0], [197: 0→1], [203: 0→1], [259: 0→1]

Clade 51 (*Danio* + *Barbus*): [114: 0→1]

Danio: [302: 0→1], **[336: 0→1]**

Barbus: [313: 0→1]

Clade 52 (all characiform, gymnotiform and siluriform taxa included in the cladistic analysis): [20: 1→0], [67: 0→1], **[105: 0→1]**, [109: 0→1], [129: 1→0], [185: 1→0], [222: 1→0], [238: 1→0], [277: 1→0] , [323: 1→0]

The close relationship between the characiform, gymnotiform and siluriform fishes included in the cladistic analysis is expected (see Fig. 2) and is well supported. It should, however, be noted that among the 10 features supporting this clade, 7 pertain to reversions to the plesiomorphic condition. Although this distribution appears as the most parsimonious one if one strictly follows the principle of parsimony, in such a discussion one should obviously analyze each and every feature in a critical way. And, in this specific case, a hypothesis that the ancestors of this clade suffered a truly "explosive morphological reversion", with more than two thirds of the characters supporting the clade being reversions, does seem, at least at first sight, somewhat unsound. However, this does not necessarily mean that there is something wrong with this clade 52. In fact, the grouping of Characiformes, Gymnotiformes and Siluriformes in a monophyletic unit has been strongly supported by numerous morphological (e.g., Fink and Fink, 1981, 1996; Lauder and Liem, 1983; Arratia, 1992; this study) and molecular (e.g., Dimmick and Larson, 1996; Saitoh et al., 2003; Lavoué et al., 2005; Peng et al., 2006) phylogenetic studies in the last years. What might be happening is that when the principle of parsimony is applied strictly to features such as the absence of premaxillary (238, state 1) and mandibular teeth (323, state 1) in gonorynchiforms, cypriniforms and most ostariophysan fossil taxa included in the analysis, these are interpreted as plesiomorphic for the ostariophysans as a whole, later reversed in this clade 52. However, this could be one of those cases in which evolution might not compulsorily follow strict parsimony. That is, one can suppose, for instance, that such teeth were independently lost in gonorynchiforms, in cypriniforms, and in clade 45, rather than completely lost in clade 52 and then reacquired in some fishes of this latter clade. The principle of parsimony may well be a common and useful rule, and I do think that this is very likely the case (hence my commitment to phylogenetic analyses following an explicit cladistic methodology: Diogo, 2004b; this work), but there is no evidence, so far, that exceptions to this rule are completely impossible in evolution. The example concerning the loss of teeth referred to above might (or might not) be one of those exceptions.

Clade 53 (all characiforms included in the cladistic analysis): [91: 0→1], [92: 0→1], **[108: 0→1]**, [206: 0→1], [231: 0→1], [259: 0→1], [276: 0→1], **[350: 0→1]**

The grouping of the Characiformes examined is expected (see Fig. 2) and well supported. The "majority fools" tree obtained (see Figs. 3, 4) does not corroborate or contradict the grouping of the distichodontid genera *Distichodus* and *Xenocharax* or the close relationship between these distichodontid taxa and the citharinid *Citharinus*, as would be expected from the works of Vari (1979), Orti and Meyer (1997), Buckup (1998), and others.

Xenocharax: [75: 0→1], [197: 0→1], [216: 1→0], [263: 1→0], [276: 1→2]

Distichodus: [1: 0→1], [75: 0→1], [182: 0→1], [187: 0→1], [231: 1→0], **[286: 0→1]**, [313: 0→1]

Citharinus: [85: 0→1], **[103: 0→1]**, [157: 1→0], [193: 0→1], [227: 0→1], [290: 0→1], [313: 0→1]

Brycon: [13: 1→0], [80: 0→1], [124: 1→0], [157: 1→0], [182: 0→1], [197: 0→1], [216: 1→0], [227: 0→1], [276: 1→0], [277: 0→1]

Clade 54 (all gymnotiform and siluriform taxa included in the cladistic analysis): [68: 0→1], [97: 1→0], [114: 0→1], [130: 0→1], [138: 1→0], [217: 0→1], [258: 0→1], **[304: 0→1]**

As seen in Fig. 2, most researchers now accept a sister-group relationship between Gymnotiformes and Siluriformes. However, this sister-group relationship has been mostly supported by morphological evidence (e.g.,

Fink and Fink, 1981, 1996; Lauder and Liem, 1983). Most molecular cladistic analyses published so far support a sister-group relationship between gymnotiforms and characiforms (e.g., Dimmick and Larson, 1996; Saitoh et al., 2003; Peng et al., 2006) or eventually between characiforms and siluriforms (e.g., Lavoué et al., 2005). The present cladistic analysis does provide strong evidence for a clade including siluriforms and gymnotiforms (Figs. 3, 4). This is because it not only corroborated many of the synapomorphies listed by Fink and Fink (1981, 1996) to support such a clade, but also pointed out additional synapomorphies to support this clade, such as the ossification of the ligament connecting the suspensorium to the ethmoid region (258: 0→1) and the interhyal (ossified or not) connected by ligaments to both the hyoid arch and the suspensorium (304: 0→1) (see Section 3.2).

Clade 55 (all gymnotiforms included in the cladistic analysis): [72: 0→1], [106: 1→0], **[184: 0→1]**, [197: 0→1], [247: 0→1], [290: 0→1]

The grouping of the Gymnotiformes included in the cladistic analysis in this clade 55 is expected (see Fig. 2) and well supported. The third and fourth features listed above (184: 0→1, levator arcus palatini markedly lateral to all bundles of adductor mandibulae; 197: 0→1, insertion of a significant part of adductor arcus palatini on lateral surface of suspensorium) concern muscular features that have not been proposed in the literature as potential gymnotiform synapomorphies.

Brachyhypopomus: [1: 0→1], [238: 0→1], [313: 0→1], [323: 0→1]

Clade 56 (*Sternopygus* + *Gymnotus*): [13: 1→0], **[16: 0→1]**, **[212: 0→1]**

De La Hoz (1974), Albert and Campos-da-Paz (1998), Albert (2001), and others have mantained that the sternopygid gymnotiforms are more closely related to hypopomids than to gymnotids. Triques (1993), Gayet et al. (1994), and Alves-Gomez et al. (1995) have instead held that hypopomids are more closely related to gymnotids than to sternopygids. Curiously, the three features listed above, and particularly the two features that appear homoplasy-free within the taxa included in the cladistic analysis, provide support for a closer relationship between the sternopygid *Sternopygus* and the gymnotid *Gymnotus* than between any of these taxa and the hypopomid *Brachyhypopomus*. However, it is evident that only a cladistic analysis including numerous other gymnotiform taxa will help us to discern the relationships between the members of this order.

Sternopygus: [20: 0→1], [217: 1→0], [263: 1→0], **[337: 0→1]**

Gymnotus: [130: 1→0]

Clade 57 (all siluriforms included in the cladistic analysis): [27: 1→0], [85: 0→1], [86: 0→1], **[95: 0→1]**, [121: 0→1], [124: 1→0], [128: 0→1], [136: 0→1], [137: 0→1], [140: 0→1], [157: 1→0], [193: 0→1], [250: 0→1], [272: 1→0], [281: 0→1], [331: 0→1], [241: 0→1], **[351: 0→1]**

The assembly of the Siluriformes included in the cladistic analysis is expected (see Fig. 2) and well supported. Many features listed above have been commonly mentioned in previous works as potential synapomorphies of the siluriforms (e.g., Regan, 1911b; Chardon, 1968; Roberts, 1973; Lundberg, 1975; Howes, 1983a,b, 1985b; Fink and Fink, 1981, 1996; Arratia, 1987, 1992; Schaefer, 1990; Mo, 1991; De Pinna, 1993, 1998; Diogo, 2004a). However, some of the listed features seemingly constitute additional potential synapomorphies to diagnose the order, such as the arrector dorsalis not being subdivided into different sections (27: 1→0), the absence of "ligament between posttemporal and posterior margin of neurocranium" (128: 0→1) and the presence of a peculiar coracoid bridge (140: 0→1). Within this clade 57, the phylogenetic scenario shown in Fig. 3 is essentially similar to that of Diogo (2004a). As Diogo (2004a) already provided a detailed discussion of the different siluriform clades obtained in the present work, as well as a comparison with previous works on catfish phylogeny, I will list the synapomorphies supporting these clades without giving further comments. For a detailed discussion of these clades, see Diogo (2004a).

Diplomystes: [216: 1→0]

Clade 58: [20: 0→1], [186: 1→0], [194: 1→2], [230: 1→0]

Clade 59: [106: 1→0], [177: 0→1], [182: 0→1], [187: 0→1]

Callichthys: [15: 0→1], [127: 0→1], [138: 0→2], [238: 0→1], [341: 1→0], [342: 0→1]

Clade 60: **[278: 0→1]**, [290: 0→1], **[295: 0→1]**, [335: 0→1]

Nematogenys: [140: 1→0], [194:2→1], [197: 0→1], [210: 0→1], **[309: 1→0]**, **[310: 0→1]**, [352: 0→1]

Trichomycterus: [206: 0→1], **[208: 0→1]**

Clade 61: **[7: 0→1]**, [30: 0→1], [352: 0→1]

Cetopsis: **[88: 0→1]**, [106: 1→0], [136: 1→0], **[173: 0→1]**, [177: 0→1], [203: 0→1], [205: 0→1], [288: 1→0], [338: 0→1]

Silurus: **[71: 0→1]**, [341: 1→0]

Clade 62: [29: 1→0], [138: 0→2]

Chrysichthys: [263: 1→0]

Clade 63: **[256: 0→1]**

Bagrus: [139: 0→1]

Pimelodus: [78: 0→1], **[213: 0→1]**, [276: 0→1]

Clade 64 (all sarcopterygians included in the cladistic analysis): [34: 0→1], [121: 0→1], [130: 0→1], [131: 0→1], [141: 0→1], **[146: 0→1]**, [320: 0→1]

As explained in Chapter 1, the Sarcopterygii (lobefins and tetrapods) is a highly speciose and diverse group of osteichthyans estimated to comprise more than 24,000 living species. It is accepted by most authors that this group comprises three main subgroups with extant forms: the Actinistia (including *Latimeria*), the Dipnoi (including *Neoceratodus*, *Lepidosiren* and *Protopterus*) and the Tetrapoda (including numerous amphibian and amniote living taxa) (see Fig. 1). However, there is still controversy regarding the relationships between these three main subgroups: in most recent general textbooks the extant dipnoans are placed as the closest living relatives of tetrapods, but some authors continue to defend the position that tetrapods are more closely related to *Latimeria* than to dipnoans (see above). Examples of features that are often listed in general textbooks as potential synapomorphies of the Sarcopterygii are the fleshy pectoral and pelvic fins having a single basal skeletal element, the presence of muscular lobes at the base of those fins, the peculiar attachment between the fourth and fifth branchial arches, the presence of subdermal anocleithrum and/or the presence of true enamel on tooth surface (e.g., Pough et al., 1996; Lecointre and Le Guyader, 2001; Kardong, 2002). Numerous other features have been proposed as synapomorphies of this clade in more specific works (Rosen et al., 1981; Cloutier and Ahlberg, 1996; Zhu and Schultze, 2001).

In the cladistic analysis of the present work, seven unambiguous features support the monophyly of the clade comprising all sarcopterygians included in that analysis: pectoral muscles extending far into pectoral fin/forelimb (**34: 0→1**, homoplasy-free within the taxa included in the analysis); supracleithrum not present as an independent ossification (121: 0→1, reversed in †*Tiktaalik* and also occurring in some actinopterygian taxa included in the analysis); at least partial ossification of Baudelot's ligament (130: 0→1, reversed in *Lepidosiren*, coded as inapplicable in *Ambystoma* and *Timon* since this ligament is seemingly missing in these taxa; also occurring in some actinopterygian taxa included in the analysis); incorporation of functional anocleithrum in the pectoral girdle (131: 0→1, exclusively present in taxa of this clade 64; reversed in *Lepidosiren* and in the node leading to *Ambystoma* + *Timon*); absence of distinct mesocoracoid arch (**141: 0→1**, not reversed within the taxa of clade 64 in which this character could be discerned, coded as '?' in †*Tiktaalik* and in †*Acanthostega*; also occurring in some actinopterygian taxa included in the analysis); single pectoral fin endoskeletal element ("humerus") articulating with pectoral girdle (146: 0→1, homoplasy-free within the taxa included in the analysis); absence of ossified hypohyals (320: 0→1, also occurring in some actinopterygians included in the analysis; coded as '?' in †*Tiktaalik*, †*Acanthostega*, and *Timon*; see Section 3.2).

Although the seven features listed above support the monophyly of all sarcopterygians included in the cladistic analysis, they do not necessarily represent synapomorphies of the Sarcopterygii as a whole. For instance, an independent, ossified supracleithrum is found in various sarcopterygian fossils that were not included in the analysis, such as *Eusthenopteron*, and the absence of this element does not seem, in fact, to represent a sarcopterygian synapomorphy (e.g., Rosen et al., 1981). However, various features listed above do seem to constitute potential synapomorphies of the Sarcopterygii as a whole, such as the pectoral muscles extending far into pectoral fin/forelimb, the presence of a characteristic anocleithrum and the presence of a single pectoral fin endoskeletal element ("humerus") articulating with the pectoral girdle (e.g., Rosen et al., 1981; Cloutier and Ahlberg, 1996; Zhu and Schultze, 2001).

As explained above, in molecular works by Arnason and colleagues (e.g., Rasmussen and Arnason, 1999a,b; Arnason et al., 2001, 2004), these authors held that tetrapods are the sister-group of a clade including taxa such as lungfishes, cladistians, coelacanths, sharks and teleosts. Thus, according to these authors, the Sarcopterygii sensu this work would not be a monophyletic clade. The results of the present work do not support the phylogenetic scenario defended by these authors. In fact, as also mentioned above, the methodology and results of those molecular works have been severely criticized and questioned by numerous researchers (see, e.g., Meyer and Zardoya, 2003).

***Latimeria*:** [29: 0→1], [101: 0→1], **[102: 0→1]**, **[148: 0→1]**, **[201: 0→1]**, [209: 0→1] , [272: 0→1], [274: 0→1], [317: 0→1]

As mentioned in Chapter 2, *Latimeria* is the only living genus of the Actinistia and was thus chosen to represent the extant members of this clade in the cladistic analysis of the present work. Actinistians were, however, much more diverse in the past, comprising numerous other genera and exhibiting a worldwide distribution (e.g., Cloutier, 1991; Cloutier and Forey, 1991; Forey, 1998; Schultze, 2004). Thus, the nine features listed above are probably a mixture of synapomorphies of clades as diverse as *Latimeria*, the Latimeriidae, the Latimeroidei, the Coelacanthiformes, the Actinistia, or even of a more inclusive group (note: there is much controversy about the names of the different actinistian clades; I follow here the nomenclature used by Schultze, 2004, and others, but some authors, for instance, Clement, 2005, use the name Actinistia to designate an order and thus do not use the name Coelacanthiformes). For example, the peculiar intracranial joints (**102: 0→1**) are by no means found only in *Latimeria*: they are also found in many cladistian as well as non-cladistian osteichthyan fossils (e.g., Rosen et al., 1981; Cloutier and Ahlberg, 1996; Forey, 1998; Zhu and Schultze, 2001). According to Bjerring (1993) and others, a peculiar basicranial muscle (**201: 0→1**) such as that found in *Latimeria* is also found in many cladistian and non-cladistian osteichthyan fossils. However, because of the difficulty of analyzing the presence/absence of muscles in fossils, it is obviously very difficult to appraise the precise taxonomic distribution of this feature within osteichthyans. It should also be noted that the presence of extracleithra (**148: 0→1**) does not occur only in *Latimeria*: this feature is found in numerous other actinistians and may actually constitute a synapomorphy of the Actinistia as a whole (e.g., Cloutier and Ahlberg, 1996). Another feature that was listed above and that was previously proposed as a potential actinistian synapomorphy concerns the presence of more than one well-defined articulatory facet of the palatoquadrate for the mandible (274: 0→1) (e.g., Cloutier and Ahlberg, 1996; note that a similar feature also occurs, independently, in some osteichthyan taxa examined, e.g., *Amia*). For more information about the relationships among the actinistian subgroups as well as about the various potential synapomorphies supporting these subgroups, see, for example, the excellent book edited by Musick et al. (1991) and the remarkable overview of coelacanths provided by Forey (1998).

An illustration of the general shape of a member of the genus *Latimeria* is given in Fig. 84. More detailed drawings of the configuration of the cephalic and pectoral girdle osteological and myological structures of fishes of this genus are given in Figs. 85, 86 and 87.

Clade 65 (all non-actinistian sarcopterygians included in the cladistic analysis): [99: 0→1], **[165: 0→1]**, **[167: 0→1]**, [179: 0→1], **[190: 0→1]**, **[243: 0→1]**, [247: 0→1], **[301: 0→1]**, [302: 0→1], [331: 0→1]

One of the most debated and most controversial questions in vertebrate systematics concerns the identity of the closest living relative of tetrapods (see above). In fact, this question has crucial implications for the

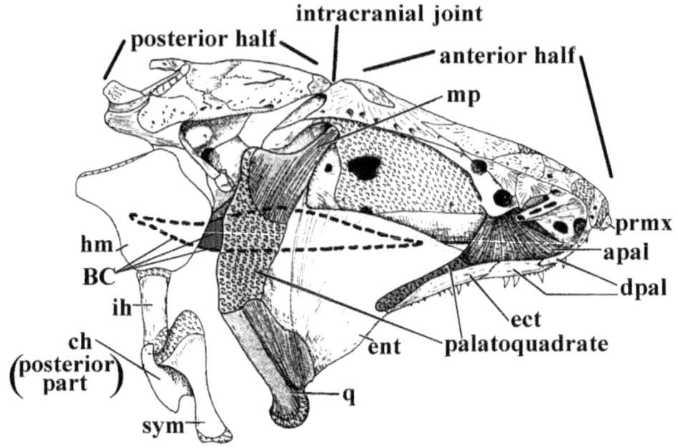

Figure 85 *Latimeria chalumnae*: Lateral view of palatoquadrate and surrounding structures; the anterior part of the ceratohyal is not shown (modified from Arratia and Schultze, 1991; the nomenclature of the structures illustrated basically follows that of these authors).

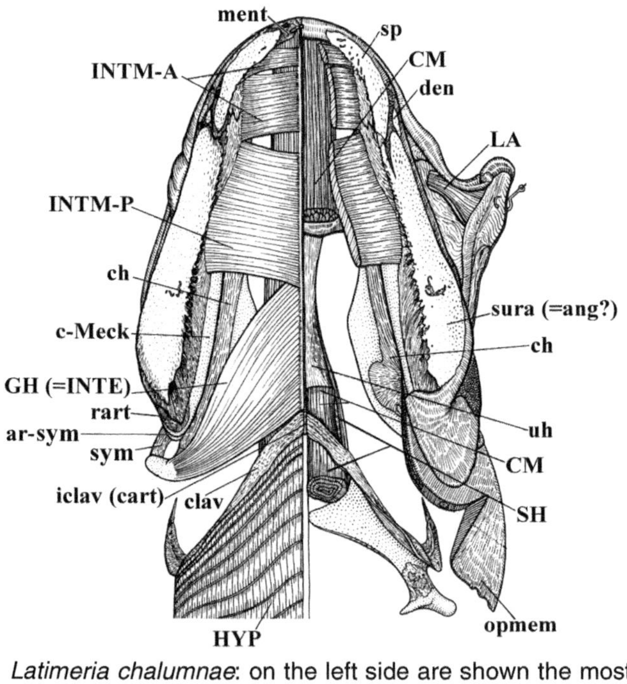

Figure 86 *Latimeria chalumnae*: on the left side are shown the most superficial cephalic muscles, after removal of the gular plate; on the right side most muscles were removed or cut in order to show muscles that are situated more dorsally (modified from Millot and Anthony, 1958; the nomenclature of the structures illustrated basically follows that of these authors).

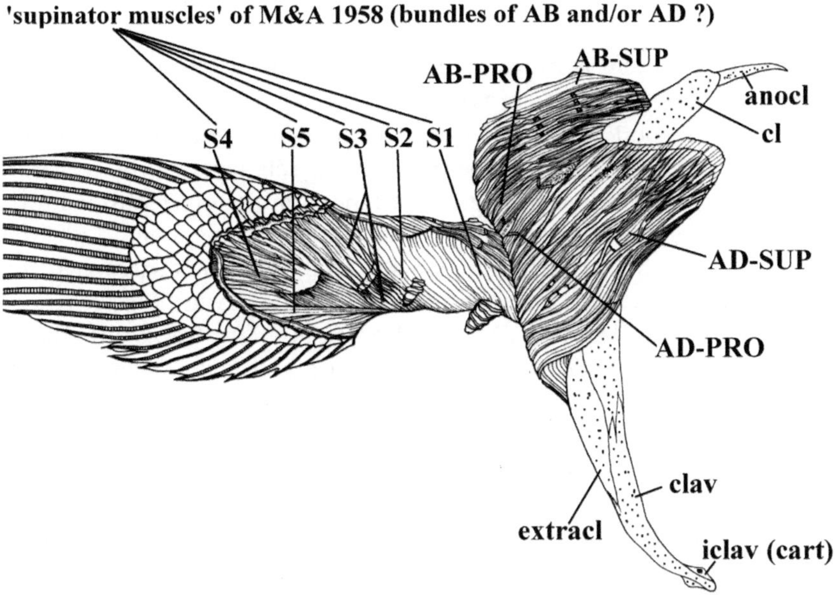

Figure 87 *Latimeria chalumnae*: pectoral girdle and most external (lateral) muscles (modified from Millot and Anthony, 1958; the nomenclature of the structures illustrated basically follows that of these authors).

understanding of the higher-level phylogeny of vertebrates and particularly for discussions on the origin and evolution of early tetrapods. Some authors (e.g., Zhu and Schultze, 2001) maintain that *Latimeria* is the closest extant relative of tetrapods, a view that was widely accepted until the last decades of the last century, especially before the publication of Rosen et al. (1981). However, the view commonly accepted nowadays in general textbooks (e.g., Pough et al., 1996; Lecointre and Le Guyader, 2001; Kardong, 2002; Dawkins, 2004) is that the Tetrapoda are more closely related to the Dipnoi than to the Actinistia, as proposed by Rosen et al. (1981). Examples of features provided in such general textbooks to support the close relationship of dipnoans and tetrapods are those concerning the pubic and ischial processes formed by the pelvic girdle, the partial division of the conus arteriosus of the heart, the presence of two primary joints with unique articulations in the pectoral and pelvic appendages and/or the loss of interhyal bone. It should be noted that different names are sometimes used in such general textbooks to designate the less inclusive clade including tetrapods and dipnoans: for instance, some authors use the name "Rhipidistia" (Lecointre and Le Guyader, 2001), while others use "Choanates" (Pough et al., 1996). In order to avoid the problems of using the name "Choanates", I prefer to use "Rhipidistia" in the present work. Apart from tetrapods and dipnoans, the Rhipidistia includes numerous fossil taxa, although there is much controversy about the precise number, the names and/or the relationships of these taxa.

The phylogenetic results of the cladistic analysis of the present work strongly support the view that the closest living relatives of tetrapods are the dipnoans, and not *Latimeria* (Figs. 3, 4). Ten unambiguous features support this scenario in the analysis: absence of teeth on parasphenoid (99: 0→1, within the taxa included in the cladistic analysis also occurring in some actinopterygian clades but not reversed inside clade 65); presence of adductor mandibulae A2-PVM (**165: 0→1**, homoplasy-free within the taxa included in the cladistic analysis; inside clade 65 this character was coded as '?' in †*Tiktaalik roseae* and †*Acanthostega gunnari*); presence of a "levator hyoideus" in at least some developmental stages (**167: 0→1**, homoplasy-free within the taxa included in the cladistic analysis; inside clade 65 this character was coded as '?' in †*Tiktaalik roseae* and †*Acanthostega gunnari*); absence of recognizable adductor operculi (179: 0→1, almost homoplasy-free within the taxa included in the cladistic analysis, only occurring also in the actinopterygian *Eurypharynx*; inside clade 65 this character was coded as '?' in †*Tiktaalik roseae* and †*Acanthostega gunnari*); absence of recognizable adductor arcus palatini (**190: 0→1**, homoplasy-free within the taxa included in the cladistic analysis; inside clade 65 this character was coded as '?' in †*Tiktaalik roseae* and †*Acanthostega gunnari*); autostylic suspension of mandibular arch (**243: 0→1**, homoplasy-free within the taxa included in the cladistic analysis; inside clade 65 this character was coded as '?' in †*Tiktaalik roseae*); autopalatines missing or being almost completely, or completely, unossified (247: 0→1, within the taxa included in the cladistic analysis it was not reversed inside clade 65 and only occurs, outside of this clade, in taxa of clades 13 and 55; inside clade 65 this character was coded as '?' in †*Tiktaalik roseae*); posterodorsal portion of ceratohyal reaching far dorsally (**301: 0→1**, homoplasy-free within the taxa included in the cladistic analysis; within clade 65 this character was coded as '?' in †*Tiktaalik roseae*); absence of ossified interhyal (302: 0→1, within the taxa included in the cladistic analysis it was not reversed inside clade 65, but it also occurs in some groups outside of it; within clade 65 this character was coded as '?' in †*Tiktaalik roseae*); retroarticulars not present as independent elements (331: 0→1, within the taxa included in the cladistic analysis it was not reversed inside clade 65 but occurs in some groups outside of it).

Of the 10 features listed above supporting this clade 65, special caution should be used concerning the absence of independent retroarticulars and the absence of parasphenoid teeth. The absence of independent retroarticulars is a rather variable feature within osteichthyans, being also present in some other clades examined in the present work (see Section 3.2). Moreover, as stressed by Schultze and Campbell (1986: 30) and others, the lower jaw bones of dipnoans are so peculiar that some authors have inclusively suggested a new nomenclature for these bones. According to Schultze and Campbell (1986: 30) "the prearticular is the only bone that can be homologized with some certainty with that in other osteichthyans." As explained in Section 3.2, there are strong reasons to consider that independent retroarticulars are absent in the two extant dipnoan taxa included in the cladistic analysis. However, a detailed comparative study of the bones of the lower jaw of all known dipnoan fossils and of numerous other fossil and extant sarcopterygians is needed to discern whether or not the absence of independent retroarticulars may constitute a potential feature to support a close relationship between tetrapods and dipnoans. Regarding the absence of parasphenoid teeth, in the present cladistic analysis this feature appears somewhat less homoplasic than the absence of independent articulars. However, it is important to stress that parasphenoid teeth are present in various dipnoan and non-dipnoan rhipidistian fossils (Rhipidistia sensu this work: see above) (e.g., Rosen et al., 1981; Schultze and Campbell, 1986; Schultze, 1986; Campbell and Barwick, 1986). Therefore, the absence of parasphenoid teeth does not seem to provide strong evidence for a close relationship between tetrapods and dipnoans.

However, the other eight features listed above do seem to provide strong support for such a close relationship. In fact, five of these eight features have been commonly referred to by other authors as evidence supporting a close relationship between dipnoans and tetrapods: the presence of a "levator hyoideus" in at least some developmental stages, the autostylic suspension of the mandibular arch, the absence of autopalatines, the posterodorsal portion of ceratohyal reaching far dorsally, and the absence of ossified interhyal (e.g., Rosen et al., 1981; Forey, 1986; Cloutier and Ahlberg, 1996).

The remaining three features concern muscular characters that, at least to my knowledge, have not been listed in previous works as evidence for a close relationship between tetrapods and dipnoans. However, they do seem to provide strong evidence to support such a relationship. This is because, within the taxa included in the cladistic analysis in which these features could be discerned, one of these three features (absence of adductor operculi) is almost homoplasy-free, occurring also only in the rather peculiar actinopterygian *Eurypharynx*, and the other two (absence of adductor arcus palatini and presence of adductor mandibulae A2-PVM) appear as completely homoplasy-free. As stated by Diogo (2004a), in a cladistic analysis such as this one, in which are included numerous terminal taxa for numerous major groups of a clade as diverse and complex as the osteichthyans, the occurrence of features free from homoplasy is rather rare. Therefore, to have not one, but various homoplasy-free features supporting a certain clade in such an analysis—as is the case with this clade 65—does provide strong support for that clade.

As mentioned above, the absence of adductor operculi, the absence of adductor arcus palatini, and the presence of an A2-PVM were coded as '?' in †*Tiktaalik roseae* and in †*Acanthostega gunnari* (as well as in the actinopterygian fossils included in the analysis). This is because the descriptions of these fossils in the literature do not make it possible to discern whether or not these features were found in them. Therefore, one cannot totally exclude the hypothesis that they were eventually missing in these fossils, and thus that they might have eventually been independently acquired in dipnoans and in a node leading for instance to extant tetrapods. However, I think there are strong reasons to consider that this was very likely not the case. †*Tiktaalik roseae* and †*Acanthostega gunnari* seemingly did not have an ossified opercular bone (see Section 3.2) and it is thus likely that they missed an adductor operculi. In fact, among all the osteichthyan taxa I have observed so far, the only taxa seemingly lacking an opercular bone and conserving at the same time a recognizable adductor operculi are the extant acipenseriforms (see Section 3.2 and Chapter 4, Section 4.2). The absence of adductor arcus palatini is also expected in †*Tiktaalik roseae* and †*Acanthostega gunnari*, since, as is the case in the extant dipnoans and tetrapods examined, the mobility of the palatoquadrate in these two taxa was seemingly rather reduced (see Section 3.2 and Chapter 4, Section 4.2). Concerning the presence of A2-PVM, it seems rather unsound that an adductor mandibulae bundle that has precisely the same overall configuration, that occupies precisely the same position, that has precisely the same type of attachments, and that I did not find in any other osteichthyan taxa I have observed, was independently acquired in extant dipnoans and extant tetrapods. However, one cannot completely exclude the hypothesis that this was the case. We can confidently determine whether these three muscular features were or not found in †*Tiktaalik roseae* and †*Acanthostega gunnari*, as well as in other fossils phylogenetically closer to tetrapods than to dipnoans, only if and when detailed studies allow us to directly check their presence or absence in those fossils.

Besides the 10 features listed above, there are various other interesting features with ambiguous distributions that, if a "fast optimization" is chosen, also support a close relationship between dipnoans and tetrapods, such as the absence of adductor mandibulae Aù (187: 0→1, within the taxa of clade 65 in which this feature could be discerned, only reversed in *Timon*, but occurring in various taxa outside this clade; coded as '?' in †*Tiktaalik roseae* and †*Acanthostega gunnari*), the absence of recognizable levator arcus palatini (202: 0→1, within the taxa of clade 65 in which this feature could be discerned, only reversed in *Timon*, and not acquired in any taxon outside this clade; coded as '?' in †*Tiktaalik roseae* and †*Acanthostega gunnari*), the pterygoids/dermopalatines/pterygopalatines peculiarly attached, or fused, to their counterparts in the midline (244: 0→1, within the taxa of clade 65 in which this feature could be discerned, only reversed in †*Tiktaalik roseae* and *Ambystoma ordinarium*, and not acquired in any taxon outside this clade), and the presence of depressor mandibulae (166: 0→1, within the taxa of clade 65 in which this feature could be discerned, only reversed in *Neoceratodus*, and not acquired in any taxon outside this clade; coded as '?' in †*Tiktaalik roseae* and †*Acanthostega gunnari*). Actually, the two latter features have already been discussed in previous studies of other authors and listed in those studies to precisely support a close relationship between tetrapods and dipnoans; thus, they may eventually constitute useful evidence to support this close relationship (see, e.g., Rosen et al., 1981; Forey, 1986).

With respect to the ambiguous features concerning the absence of adductor mandibulae Aù and of recognizable levator arcus palatini, their distributions within clade 65 appear ambiguous because, as explained in Section 3.2, I have tentatively coded *Timon* as CS-0 for both these characters. It should be noted that some authors have stated that some rhipidistian fossils (sensu this work) may well have had an Aw (e.g., Lauder, 1980a). Moreover, it should be stressed that the presence/absence of Aù is a rather variable character within the osteichthyan taxa included in the present cladistic analysis. Therefore, the absence of this bundle does not seem to constitute strong evidence to support the hypothesis that tetrapods are more closely related to dipnoans than to actinistians. However, the other feature, that concerning the absence of levator arcus palatini, may well be a valid synapomorphy to support such a hypothesis. As explained in Section 3.2 and Chapter 4, Section 4.2, *Timon* was coded as CS-0 for this character not because it has a true muscle levator arcus palatini, but because the peculiar protractor and levator pterygoidei found in this lizard, as well as in many other amniotes, seem to be the result of a modification of this muscle, or at least of the muscular portion from which it originates embryologically. In other words, one cannot really state that the levator arcus palatini was completely lost in amniotes such as lizards (and that is why *Timon* was coded as CS-0); instead, this muscle, or at least the muscular portion from which it originates, seems to have been peculiarly modified in the protractor and levator pterygoidei found in those taxa (see Section 3.2 and Chapter 4,

Section 4.2). It is important to note that the presence/absence of a true levator arcus palatini is a rather consistent, almost homoplasy-free character within osteichthyans: within all the taxa included in the cladistic analysis in which this character could be discerned, a true levator arcus palatini is only absent in dipnoans, tetrapods and acipenseriforms such as *Acipenser* and *Psephurus*, in which the muscle was peculiarly modified in a protractor hyomandibulae (see Section 3.2 and Chapter 4, Section 4.2). The constant presence of a true levator arcus palatini in the vast majority of actinopterygians and in *Latimeria*, and its absence in extant dipnoans and in extant tetrapods, thus seems to provide further evidence to support the hypothesis that the closest living relatives of tetrapods are the extant dipnoans and not *Latimeria*.

In summary, from the discussion above it can be said that 11 of the 14 features discussed may constitute potential synapomorphies to support a close relationship between dipnoans and tetrapods, namely: (1) the presence of adductor mandibulae A2-PVM; (2) the presence of a "levator hyoideus" in at least some developmental stages; (3) the absence of recognizable adductor operculi; (4) the absence of a recognizable adductor arcus palatini; (5) the autostylic suspension of the mandibular arch; (6) the autopalatines missing or almost completely, or completely, unossified; (7) the posterodorsal portion of ceratohyal reaching far dorsally; (8) the absence of ossified interhyal; (9) the absence of true levator arcus palatini; (10) the pterygoids/dermopalatines/pterygopalatines peculiarly attached, or fused, to their counterparts in the midline; and (11) the presence of depressor mandibulae. Overall, I thus consider that the cladistic analysis of the present work does provide strong support for such a close relationship. The fact that 6 of these 11 features concern muscular characters, and particularly the fact that three of these six myological features appear completely homoplasy-free in a work such as this, seems to reinforce the idea that myological characters may provide useful data for phylogenetic studies (see Diogo, 2004a,b; see also Chapter 4, Section 4.6).

Clade 66 (*Lepidosiren* + *Neoceratodus*): [1: 0→1], [125: 0→1], [126: 0→1], [186: 0→1], **[242: 0→1]**, [287: 0→1], [325: 0→1]

As explained in Chapter 2, the Dipnoi includes only three extant genera, which are usually placed in two different groups: the Neoceratodontidae, including *Neoceratodus*, and the Lepidosirenidae, including *Lepidosiren* and *Protopterus*. *Neoceratodus* and *Lepidosiren* were thus chosen to represent each of these two groups, respectively. However, as the actinistians (see above), the dipnoans were much more diverse in the past, comprising numerous other genera and exhibiting a worldwide distribution (e.g., Schultze, 2004).

Therefore, the seven features listed above are not necessarily dipnoan synapomorphies. This issue was amply debated in works such as Miles (1977), Schultze and Campbell (1986), Schultze (1986) and Campbell and Barwick (1986), in which are listed and discussed various potential synapomorphies of the Dipnoi. Some aspects of the anatomy of extant dipnoans are illustrated in Figs. 88, 89, 90, 91, 92, 93, 94 and 95. An illustration of the general shape of a fish of the genus *Protopterus* is given in Fig. 88. More detailed drawings of the configuration of the cephalic and pectoral girdle osteological and myological structures of fishes of the genera *Lepidosiren* and *Neoceratodus* are given in Figs. 89, 90, 91, 92, 93, 94 and 95; the last four figures are original illustrations by the author.

Figure 88 General aspect of a fish of the genus *Protopterus* (Dipnoi) (modified from Owen, 1841).

Lepidosiren: [20: 0→1], [130: 1→0], [131: 1→0], **[156: 0→1]**, [164: 0→1], **[267: 0→1]**, **[327: 0→1]**, [329: 0→1]

Neoceratodus: [274: 0→1]

Clade 67 (†Tiktaalik roseae + tetrapod taxa included in the cladistic analysis): [149: 0→1], **[150: 0→1]**, [281: 0→1], [284: 0→1], [324: 0→1], [330: 0→1]

As expected (Daeschler et al., 2006; Shubin et al., 2006), †*Tiktaalik roseae* appears more closely related to tetrapods than to the other sarcopterygian taxa included in the cladistic analysis (Figs. 3, 4). The six features supporting this clade 67 in the analysis are as follows: the marked robusticity and mobility of distal elements of pectoral fins/forelimbs (**149: 0→1**, homoplasy-free within the taxa included in the cladistic analysis); the glenoid orientation with lateral component (**150: 0→1**, homoplasy-free within the taxa included in the cladistic analysis); the absence of subopercle as an independent element (281: 0→1, not reverted

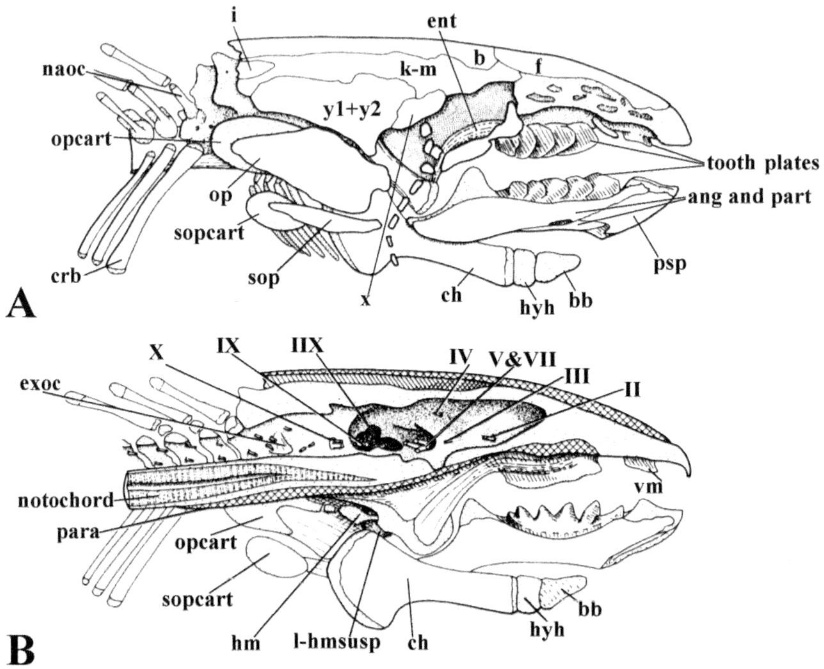

Figure 89 *Neoceratodus forsteri*: Skull in lateral view **(A)** and in a sagittal section **(B)** (modified from Miles, 1977; the nomenclature of the structures illustrated basically follows that of this author).

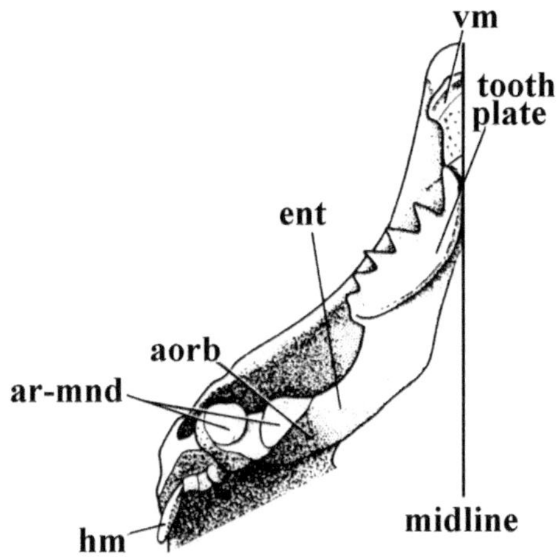

Figure 90 *Neoceratodus forsteri*: Ventral view of one side of the skull (modified from Miles, 1977; the nomenclature of the structures illustrated basically follows that of this author).

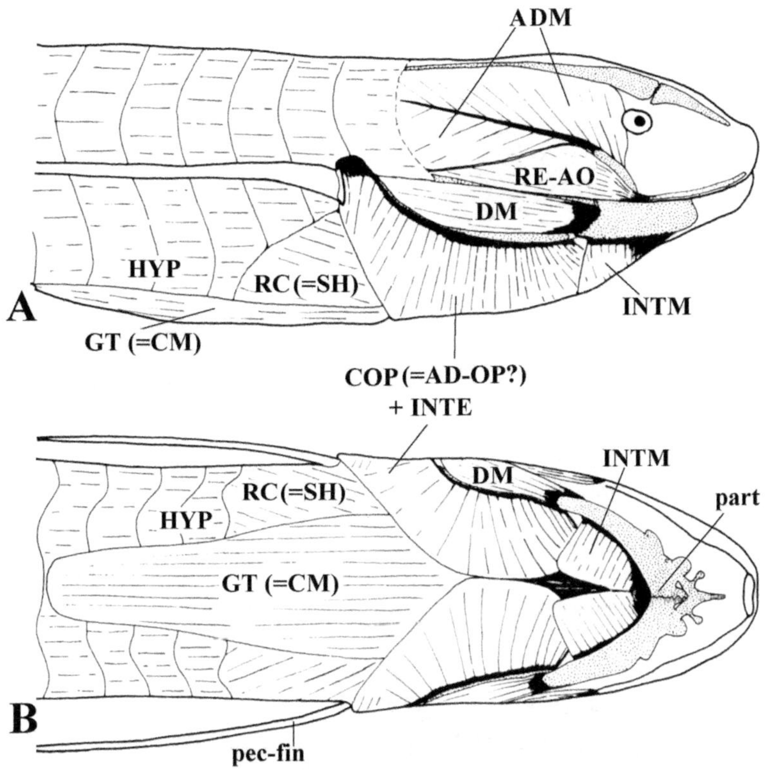

Figure 91 *Lepidosiren paradoxa*: Lateral **(A)** and ventral **(B)** views of the superficial cranial musculature (modified from Bemis and Lauder, 1986; the nomenclature of the structures illustrated basically follows that of these authors).

within the taxa of clade 67 included in the analysis in which this feature could be discerned, but occurring in a few taxa outside this clade); the absence of the opercular bone an independent element (284: 0→1, within the taxa included in the analysis in which this feature could be discerned, this feature was not reversed inside clade 67 but occurs in a few taxa outside this clade); the presence of surangular as an independent ossification (324: 0→1, within the taxa included in the analysis this feature was reversed in *Ambystoma ordinarium* and also occurs in a few taxa outside clade 67); articular mainly fused with angular and/or retroarticular (324: 0→1, within the taxa included in the analysis this feature was not reversed inside clade 67 but also occurs in various taxa outside this clade). It should, however, be noted that features such as the presence of surangulars as independent ossifications are seemingly found in certain dipnoan fossils and, thus, do not seem to provide strong evidence to support a closer relationship between

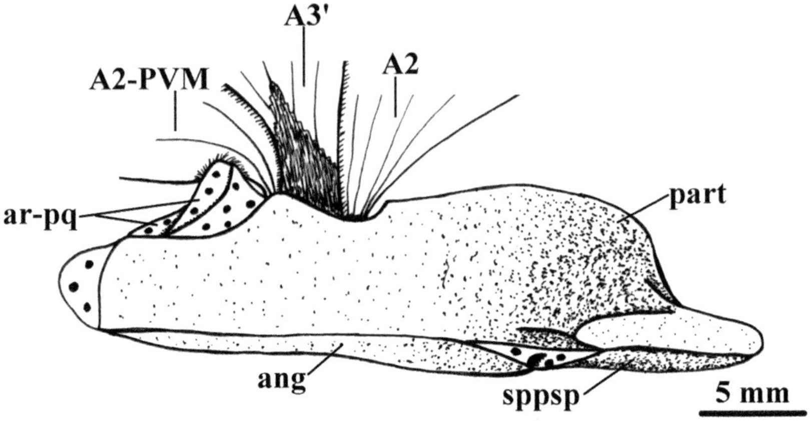

Figure 92 *Neoceratodus forsteri*: Mesial view of adductor mandibulae and mandible; the mandibular tooth-plates are not illustrated.

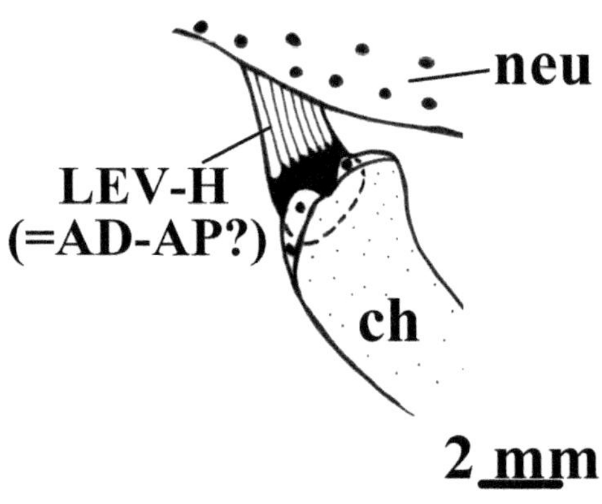

Figure 93 *Lepidosiren paradoxa*: Lateral view of muscle levator hyoidei; the ceratohyal was cut.

tetrapods and †*Tiktaalik roseae* than between tetrapods and rhipidistian taxa such as the Dipnoi (see, e.g., Schultze and Campbell, 1986; Schultze, 1986). However, the first four features listed above, particularly those two appearing homoplasy-free in the cladistic analysis, seem to provide strong evidence to support the hypothesis that tetrapods are more closely related to †*Tiktaalik roseae* than to taxa such as dipnoans (see Daeschler et al., 2006; Shubin et al., 2006).

†*Tiktaalik roseae*: [121: 1→0]

The discovery of †*Tiktaalik roseae* was reported in the press with much enthusiasm (and in many cases also with much sensationalism). Many

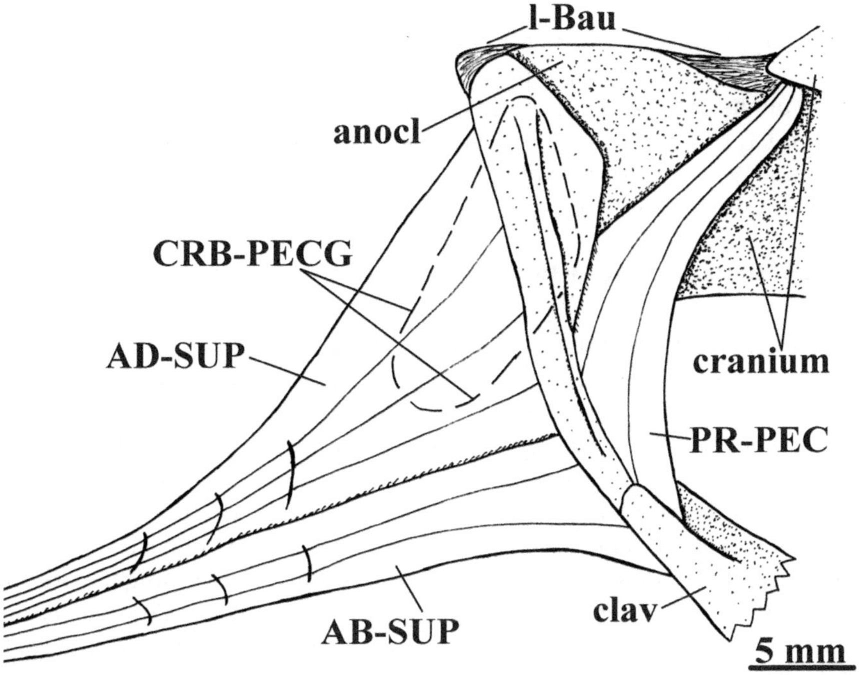

Figure 94 *Neoceratodus forsteri*: Lateral view of pectoral girdle musculature; the clavicle was cut.

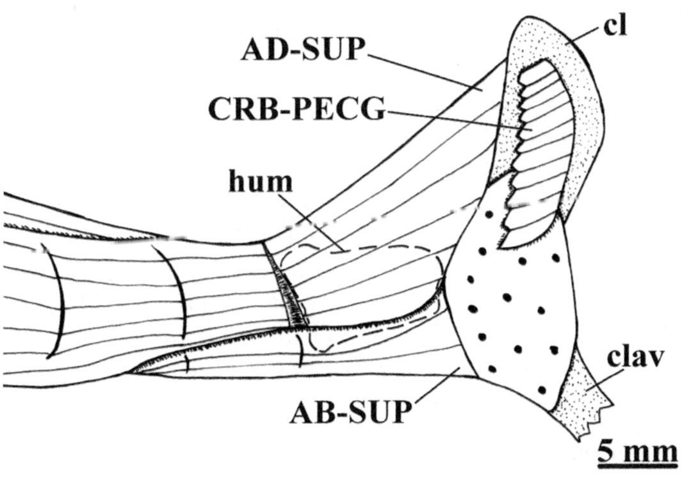

Figure 95 *Neoceratodus forsteri*: Mesial view of pectoral girdle musculature; the anocleithrum and Baudelot's ligament were removed; the clavicle and the muscle between the cranial rib and the pectoral girdle were cut.

newspapers announced that the "missing link" between tetrapods and sarcopterygian fishes such as *Eusthenopteron* had finally been discovered (see, e.g., Fig. 96). Obviously, the discovery of †*Tiktaalik* does not allow us to clarify all the questions concerning the puzzling, and fascinating, origin of tetrapods. However, the information provided by this fossil, and particularly by its remarkable pectoral structures, does increase our understanding of this subject (see Section 3.2).

Clade 68 (tetrapod taxa included in the cladistic analysis): [134: 0→1], **[151: 0→1]**

As expected (Daeschler et al., 2006; Shubin et al., 2006) †*Acanthostega gunnari* appears more closely related to extant tetrapods than to the other sarcopterygian taxa included in the cladistic analysis (Figs. 3, 4). The exact limits of the clade Tetrapoda and its definition have been, and continue to be, the subject of much controversy. In most general textbooks, the definition of this clade is usually related to the presence of digits (**151: 0→1**). However, this is far from being agreed upon by all researchers. Although it is obviously very difficult to choose among the numerous studies discussing the origin, comparative anatomy, phylogeny, evolution and systematics of early tetrapods, I would advise the reader to start with recent works such as Clack (2002, 2006) and Ruta et al. (2003) and then, for more detailed information on these issues, pass to more specific studies such as those cited in these works. For the purposes of this book, the term "tetrapods" will correspond to that used in Clack's (2002: 68) excellent volume, that is, "to animals with four legs bearing digits" (**151: 0→1**). The other feature listed above supporting the grouping of †*Acanthostega gunnari* and the extant tetrapods included in the cladistic analysis concerns the absence of cleithrum as an independent element (134: 0→1). However, it should be noted that †*Acanthostega gunnari*, as well as other early tetrapod fossils, does have a cleithrum: this fossil species was coded as CS-1 in character 134 because its cleithrum is apparently fused with the scapulocoracoid and, thus, despite being present, is not present as an independent ossification (see Fig. 97). Thus, it is important to emphasize that some tetrapods do have a cleithrum, and that the complete absence of this pectoral girdle element does not constitute a synapomorphy of the Tetrapoda as a whole (see, e.g., Coates, 1996; Clack, 2002; Ruta et al., 2003).

†*Acanthostega gunnari*: No unambiguous features

As explained in Chapter 2, †*Acanthostega gunnari* is one of the best-conserved and best-described early tetrapod fossils (see Fig. 97). Many details about the anatomy of the members of this species, about their

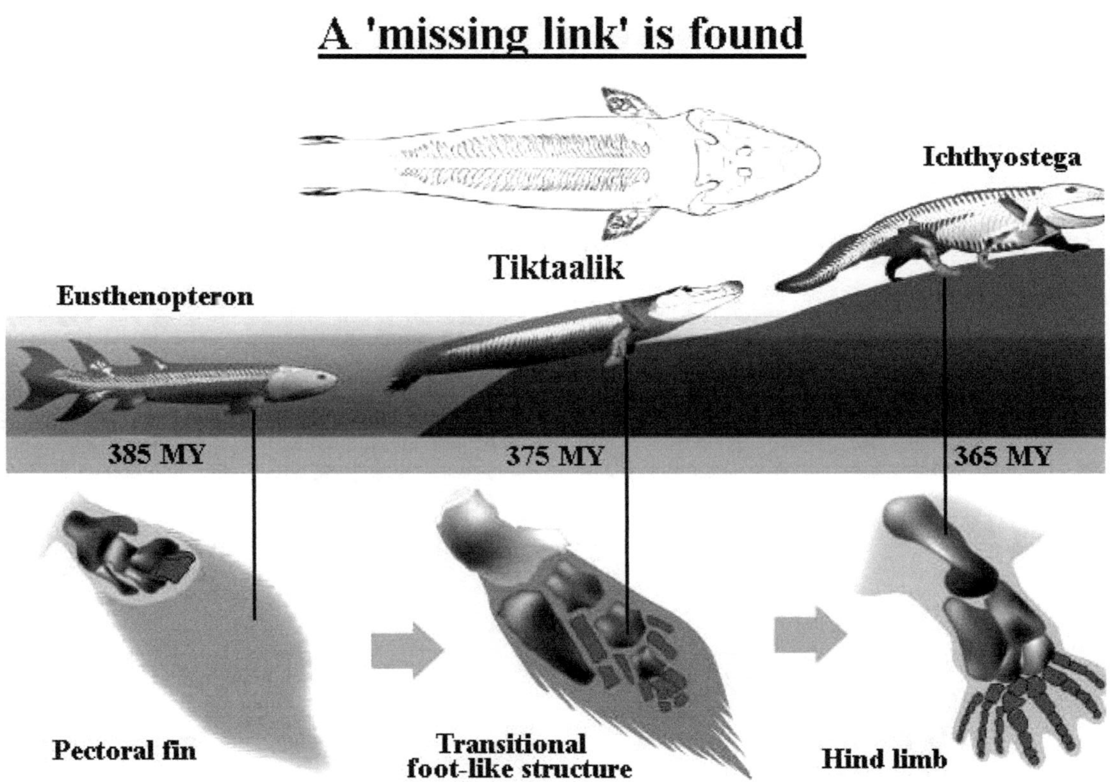

Figure 96 Enthusiastic (and many times sensational) reports in the press regarding the discovery of †*Tiktaalik roseae* (modified from the *New York Times*, April 6, 2006; the illustration of this fossil provided in the top of the figure is modified from Daeschler et al., 2006).

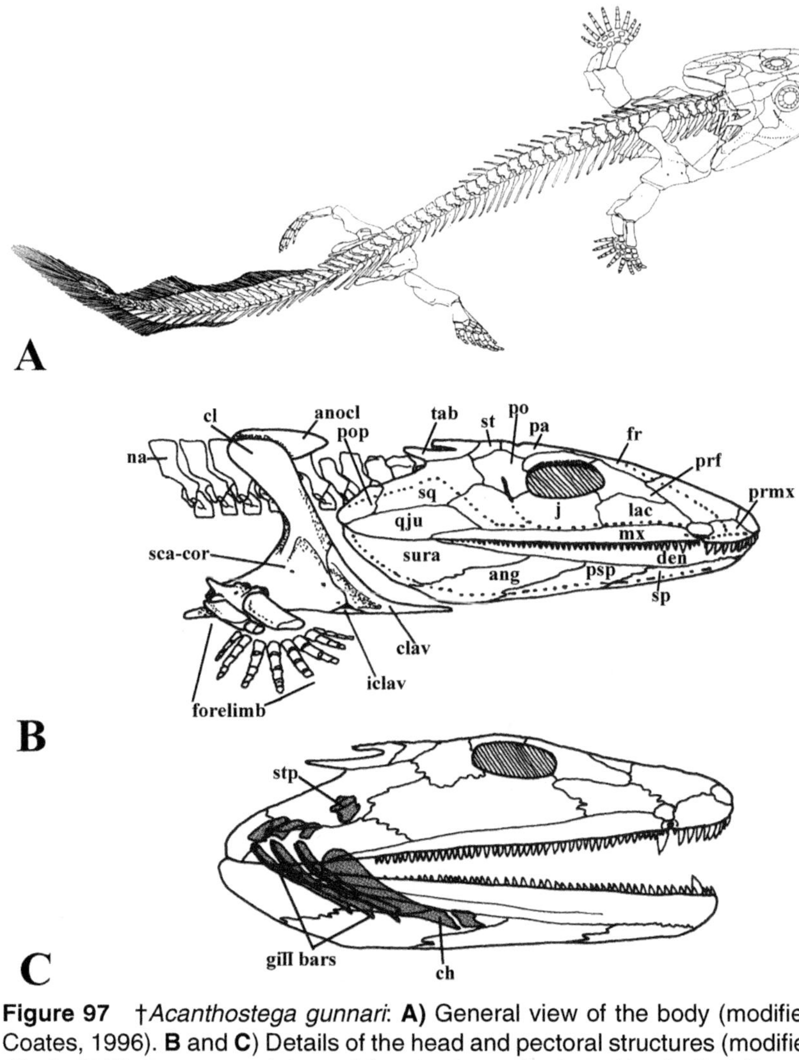

Figure 97 †*Acanthostega gunnari*: **A)** General view of the body (modified from Coates, 1996). **B** and **C**) Details of the head and pectoral structures (modified from Clack, 2002; the nomenclature of the structures illustrated basically follows that of this author).

discovery, and about the implications of the information provided by them for our understanding of the comparative anatomy, phylogeny and evolution of early tetrapods are given in the excellent work of Clack (2002). Some of these issues will be discussed in Chapter 4. An illustration of the general shape of †*Acanthostega gunnari* is given in Fig. 97A; more detailed illustrations of its cephalic region and pectoral girdle are given in Fig. 97B and 97C.

Clade 69 (extant tetrapod taxa included in the cladistic analysis): [127: 0→1], [131: 1→0], [245: 0→1], [263: 0→1], [287: 0→1]

As expected, the extant amphibian *Ambystoma ordinarium* and the extant amniote *Timon lepidus* are grouped together (Figs. 3, 4). Of course, there are seemingly various tetrapod nodes between the node leading to the less inclusive clade containing extant tetrapods and fossil tetrapods such as †*Acanthostega* and the node leading to the less inclusive clade containing extant Amniota and extant Amphibia. Examples of early tetrapod fossils that are often considered more closely related to extant tetrapods than is the taxon †*Acanthostega* are †*Greererpeton* and †*Crassigyrinus* (Clack, 2002; Ruta et al., 2003). One of the features listed above, the absence of anocleithrum (131: 0→1), is for instance listed in general textbooks such as Pough et al. (1996) as a feature providing evidence for a closer relationship between extant tetrapods and †*Crassigyrinus* than between the former and †*Acanthostega*, although this is far from being agreed upon by all researchers (see, e.g., Ruta et al., 2003).

There are many muscular features that are found in extant amphibians and extant amniotes and in no other living vertebrates (e.g., the numerous peculiar muscles of the shoulder and the forelimb: see Section 3.2, Chapter 4, Section 4.4) and that are referred to by some authors as potential synapomorphies of the less inclusive clade including these two extant groups. However, the information available on rhipidistian fossils does not allow us to discern appropriately whether at least some of these myological features were eventually present in early tetrapod fossils such as †*Acanthostega* or even in non-tetrapod rhipidistian fossils such as †*Tiktaalik*. The muscular features that are exclusively found in the extant tetrapods included in the cladistic analysis of the present work were thus coded as '?' in †*Acanthostega* and †*Tiktaalik* (as well as in the actinopterygian fossils included in the analysis). Consequently, those muscular features appeared with an ambiguous distribution in the analysis. It would be interesting to undertake a detailed study including a large number of rhipidistian fossils, both tetrapod and non-tetrapod, in order to assess whether it is eventually possible to check the presence or absence of at least some of those muscular features in some of these fossils. Even if the number of features that can be examined is rather small, it would provide precious information about the evolution of these features, their taxonomic distribution and phylogenetic implications, and the functional morphology of the fossils in which they could be analyzed.

As explained above, the main goal of the present work is not to provide a detailed discussion of all the potential synapomorphies of derived osteichthyans clades such as the Amniota or the Amphibia, since such a discussion has been provided in numerous recent works. Some of these potential synapomorphies will nevertheless be discussed in Chapter 4, which, as mentioned above, will mainly concern certain aspects of the comparative anatomy, higher-level phylogeny and/or macroevolution of the osteichthyans as a whole. As was done for the other major osteichthyan clades listed above, and with the aim of helping the comparisons and discussions provided in the following Sections, some aspects of the anatomy of extant tetrapods are illustrated in Figs. 98 to 114. An illustration of the general shape of some amphibian salamanders is given in Fig. 98; more detailed drawings of the configuration of the cephalic and pectoral girdle osteological and myological structures of certain salamanders are given in Figs. 99, 100, 101, 102, 103, 104, 105 and 106. Figure 107 provides an illustration of the general shape of an amniote, namely of a lizard; more detailed drawings of the configuration of the osteological and myological structures of the cephalic and pectoral regions of certain lepidosaur amniotes are given in Figs. 108, 109, 110, 111, 112, 113 and 114. Figures 99 and 100 are original illustrations by the author.

Figure 98 General aspect of certain amphibian salamanders (Amphibia) (modified from Slevin, 1928).

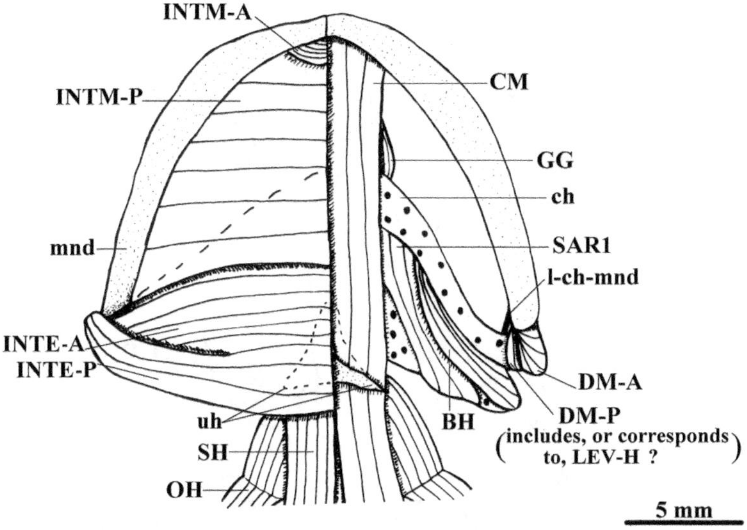

Figure 99 *Ambystoma ordinarium*: Ventral view of the cephalic musculature; on the right side most of the ventral muscles were removed.

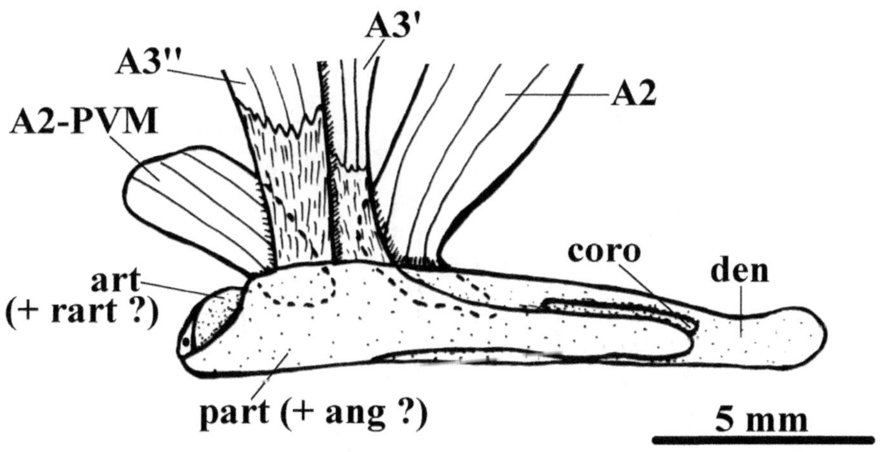

Figure 100 *Ambystoma ordinarium*: Mesial view of the mandible and adductor mandibulae; the mandibular teeth are not illustrated.

Ambystoma ordinarium: **[35: 0→1]**, **[50: 0→1]**, **[51: 0→1]**, [64: 0→1], [85: 0→1], **[152: 0→1]**, [260: 0→1], **[298: 0→1]**, [317: 0→1], [324: 1→0], [329: 0→1]

Timon: **[61: 0→1]**, **[62: 0→1]**, **[63: 0→1]**, [78: 0→1], [101: 0→1], [164: 0→1], **[169: 0→1]**, **[198: 0→1]**, **[199: 0→1]**, [224: 0→1], [306: 0→1]

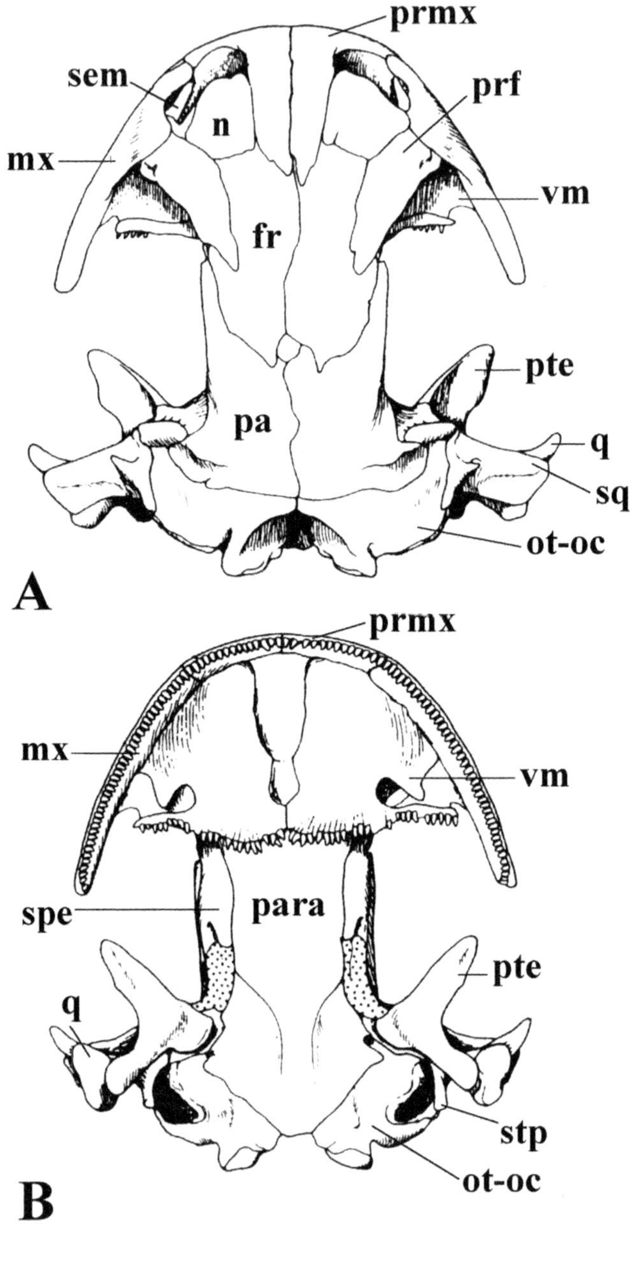

Figure 101 *Ambystoma maculatum*: Dorsal **(A)** and ventral **(B)** views of the neurocranium and palatoquadrate (modified from Carroll and Holmes, 1980; the nomenclature of the structures illustrated basically follows that of these authors).

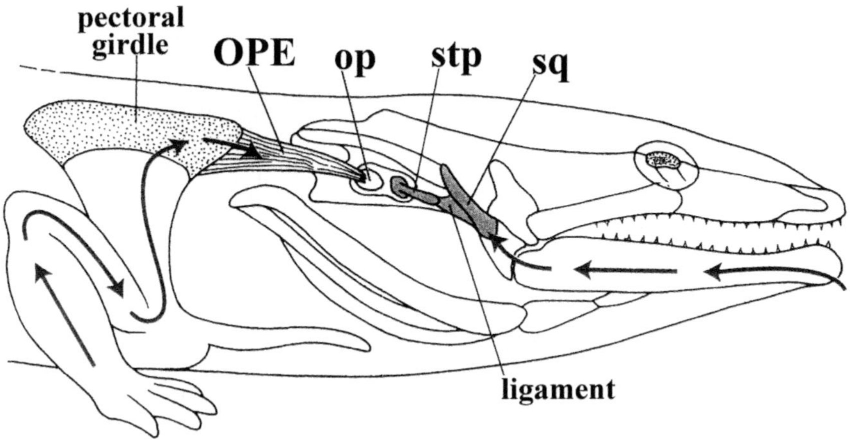

Figure 102 Illustration of the "opercular mechanism" found in certain salamanders (modified from Lecointre and Le Guyader, 2001; the nomenclature of the structures illustrated basically follows that of these authors). Vibrations are transmitted from the forelimbs to the pectoral girdle, and from this latter to the "opercle" by means of the muscle opercularis; the "opercle" then transmits these vibrations to the inner ear. As the stapes is connected to the squamosal by a ligament, some authors argue that vibrations can also be transmitted from the mandible to the oval fenestra, via the squamosal, the ligament, and the stapes.

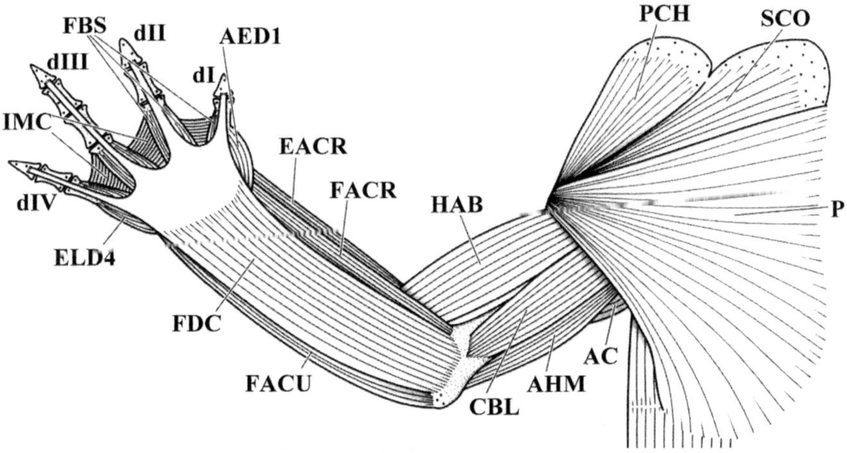

Figure 103 *Taricha torosa*: Ventral view of the superficial musculature of the forelimb; anterior is towards the top of the figure (modified from Walthall and Ashley-Ross, 2006; the nomenclature of the structures illustrated basically follows that of these authors).

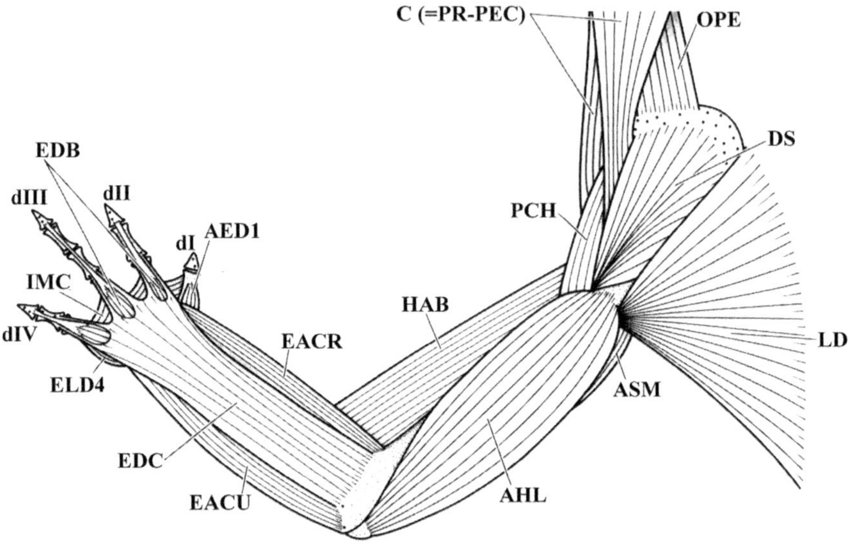

Figure 104 *Taricha torosa*: Dorsal view of the superficial musculature of the forelimb; anterior is towards the top of the figure (modified from Walthall and Ashley-Ross, 2006; the nomenclature of the structures illustrated basically follows that of these authors).

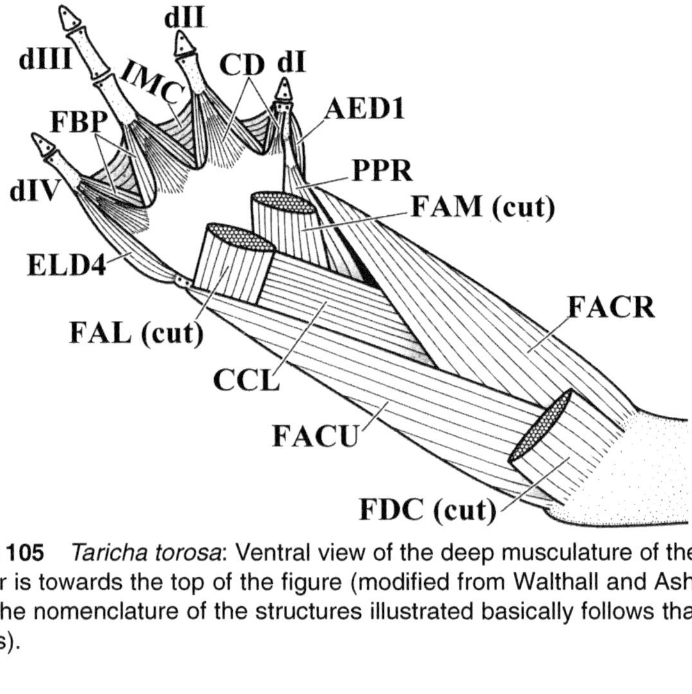

Figure 105 *Taricha torosa*: Ventral view of the deep musculature of the forearm; anterior is towards the top of the figure (modified from Walthall and Ashley-Ross, 2006; the nomenclature of the structures illustrated basically follows that of these authors).

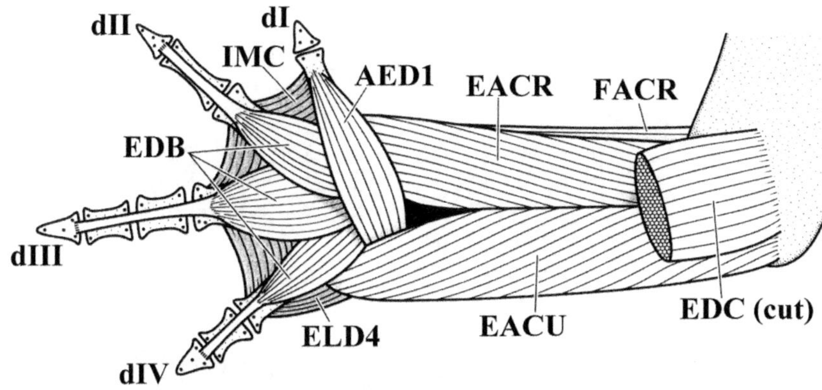

Figure 106 *Taricha torosa*: Dorsal view of the deep musculature of the forearm; anterior is towards the top of the figure (modified from Walthall and Ashley-Ross, 2006; the nomenclature of the structures illustrated basically follows that of these authors).

Figure 107 General aspect of a lepidosaur lizard (Amniota) (modified from Conant, 1975).

3.2 LIST OF CHARACTERS INCLUDED IN THE CLADISTIC ANALYSIS

The 356 morphological characters included in the cladistic analysis, concerning essentially the configuration of the bones, muscles, cartilage, and ligaments of the cephalic region, anterior vertebrae and pectoral girdle (see Chapter 2), are listed below. It should be noted that various aspects concerning the methodology followed here such as the use of "multistate" characters and "inapplicable" entries, which basically follows that employed by Diogo (2004a) (see Chapter 2), have received further philosophical support in works such as Fitzhugh (2006). The textual information given in this Section is complemented by a large number of anatomical figures. Because of size restrictions, it is obviously not possible to include anatomical drawings illustrating all the numerous morphological

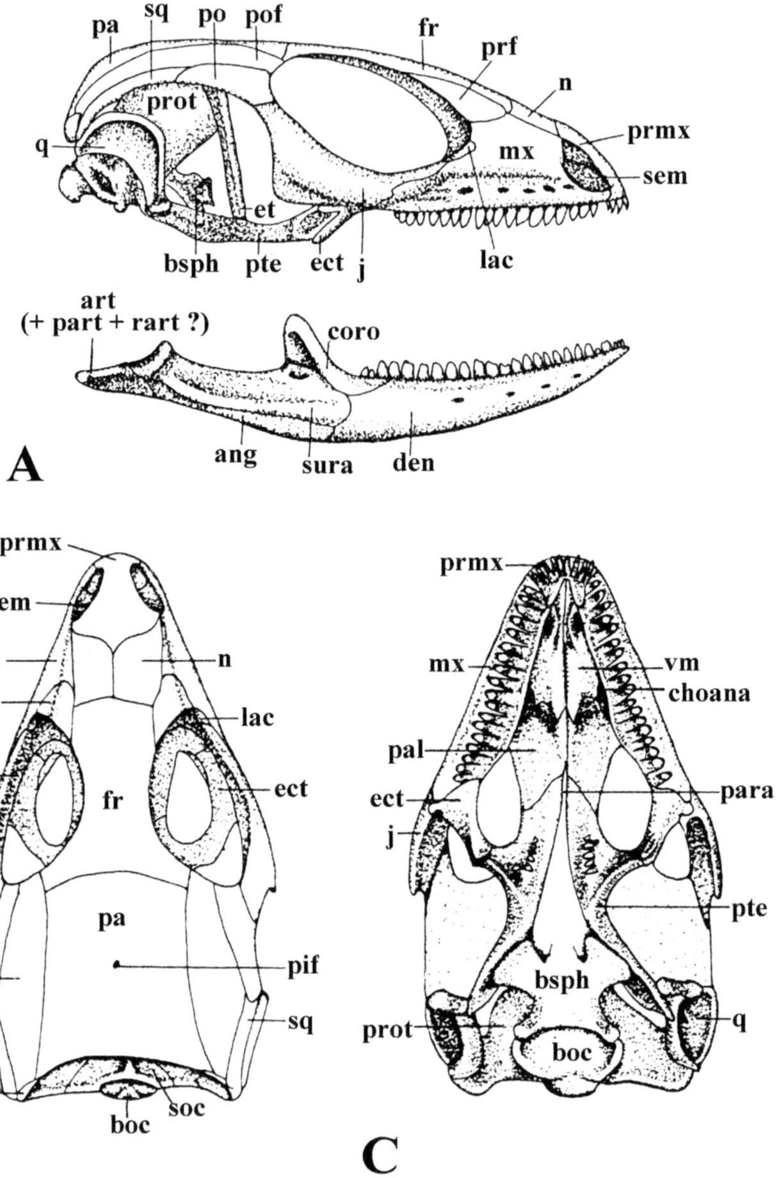

Figure 108 *Lacerta virilis*: Lateral view of the skull **(A)** and dorsal **(B)** and ventral **(C)** views of the neurocranium and palatoquadrate (modified from Bels et al., 1993; the nomenclature of the structures illustrated basically follows that of these authors).

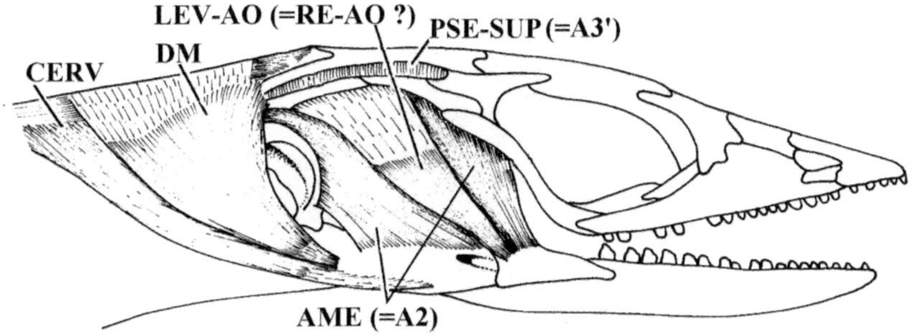

Figure 109 *Euspondylus acutirostris*: Lateral view of the cephalic musculature (modified from Montero et al., 2002; the nomenclature of the structures illustrated basically follows that of these authors).

features mentioned in this list of characters. Apart from the numerous drawings included in this work, an effort was thus made to supply, for most of the characters, references to previous works in which the configuration corresponding to their derived states has been illustrated. In this way, I also wish to pay tribute to previous works done on osteichthyan comparative anatomy and phylogeny, without which it would not have been possible to undertake a study such as this one. It can thus be said that anatomical illustrations of both osteological and myological structures are given for the vast majority of the main osteichthyan groups analyzed and discussed in the book. It is hoped that the information given in these anatomical illustrations, combined with the information given in the text, will be a useful contribution to the knowledge of osteichthyan comparative anatomy.

Ventral Cephalic Musculature

1. *Presence of a single section of intermandibularis (character inspired from, e.g., Edgeworth, 1935; Millot and Anthony, 1958; Jarvik, 1963; Greenwood, 1971; Winterbottom, 1974; Lauder, 1980a; Bemis, 1986; Bemis and Lauder, 1986; Bartsch, 1994; Kardong, 2002)*. The plesiomorphic condition for the osteichthyan groups included in the analysis is seemingly that in which the muscle intermandibularis is divided into distinct anterior and posterior divisions; such a condition is found in numerous non-osteichthyan gnathostomes, in numerous actinopterygians and in numerous sarcopterygians (e.g., Fig. 86) [State 0]. In taxa of CS-1 the intermandibularis has a single section (e.g., Fig. 91) [State 1]. In most teleosts the intermandibularis is divided into anterior and posterior divisions, but the posterior one is integrated in the protractor hyoidei (see below).

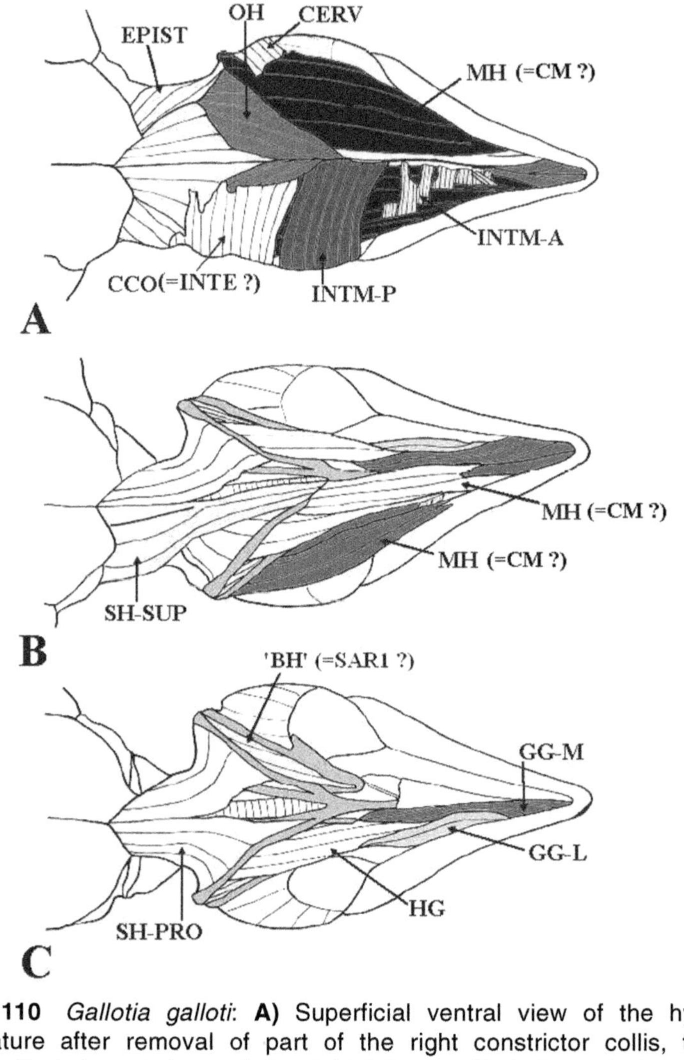

Figure 110 *Gallotia galloti*: **A)** Superficial ventral view of the hyolingual musculature after removal of part of the right constrictor collis, the right intermandibularis anterior and the right intermandibularis posterior; **B)** Deep ventral view after removal of the constrictor colli, the entire intermandibularis group, the left and right omohyoideus, episternocleidomastoideus and mandibulohyoideus 1 and the right mandibulohyoideus 2 and 3 and sternohyoideus superficialis; **C)** Deepest view after further removal of the left sternohyoideus superficialis, the remaining portions of the mandibulohyoideus, and the right genioglossus pars medialis and lateralis and hyoglossus; the left genioglossus pars medialis has been pulled aside to show the left genioglossus pars lateralis (modified from Herrel et al., 2005; the nomenclature of the structures illustrated basically follows that of these authors).

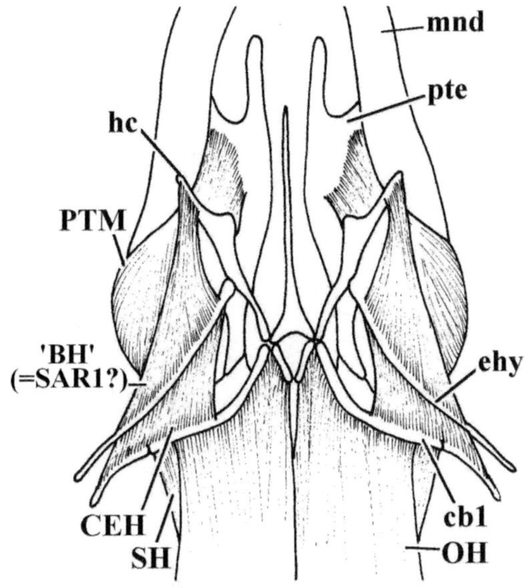

Figure 111 *Euspondylus acutirostris*: Ventral view of the deep ventral cephalic muscles (modified from Montero et al., 2002; the nomenclature of the structures illustrated basically follows that of these authors).

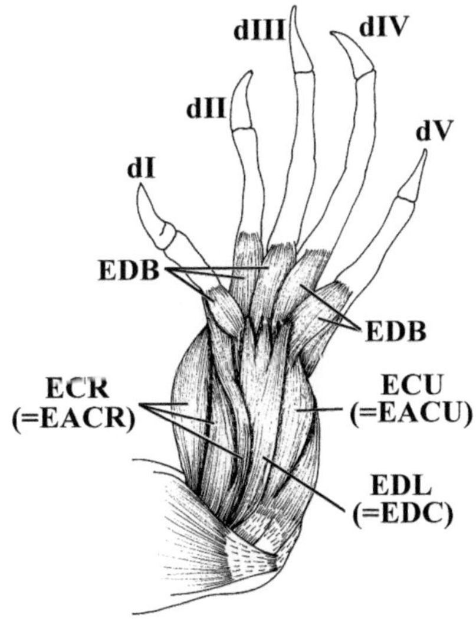

Figure 112 *Phymaturus* sp.: Dorsal view of the superficial musculature of the forearm; anterior is towards the top of the figure (modified from Abdala and Moro, 2006; the nomenclature of the structures illustrated basically follows that of these authors).

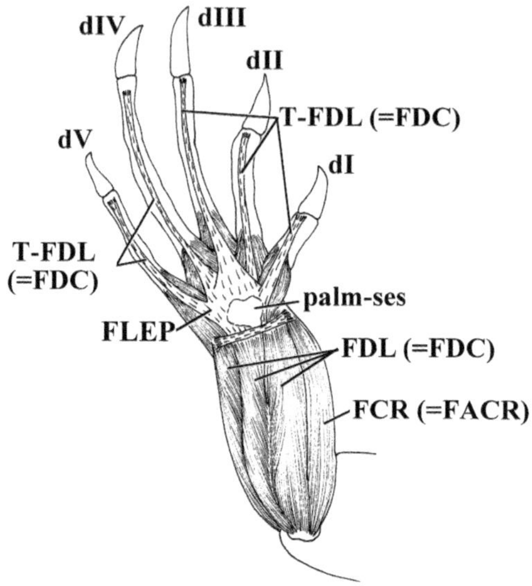

Figure 113 *Phymaturus* sp.: Ventral view of the superficial musculature of the forearm; anterior is towards the top of the figure (modified from Abdala and Moro, 2006; the nomenclature of the structures illustrated basically follows that of these authors).

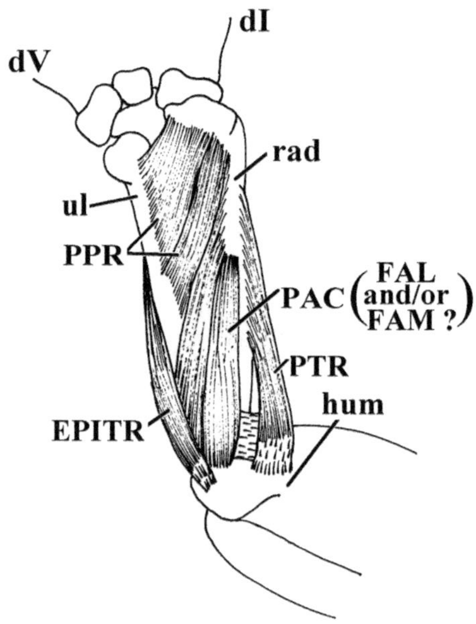

Figure 114 *Phymaturus* sp.: Ventral view of the deep musculature of the forearm; anterior is towards the top of the figure (modified from Abdala and Moro, 2006; the nomenclature of the structures illustrated basically follows that of these authors).

2. *Some fibers of intermandibularis attaching on peculiar "lateral ethmoid processes" (inspired from, e.g., Carroll and Wainwright, 2003).* Contrary to taxa of CS-0 [State 0], such a configuration is found in taxa of CS-1 (e.g., Carroll and Wainwright, 2003: fig. 3B) [State 1].
3. *Fibers of intermandibularis and interhyoideus forming protractor hyoidei (ordered multistate character) (inspired from, e.g., Edgeworth, 1935; Jarvik, 1963; Greenwood, 1971; Winterbottom, 1974; Miyake et al., 1992).* Plesiomorphically no fibers of the intermandibularis form, together with the interhyoideus, a protractor hyoidei (e.g., Fig. 6) [State 0]. In taxa of CS-1 part of the fibers of the intermandibularis form, together with the interhyoideus, the muscle protractor hyoidei (e.g., Fig. 34), but this latter muscle is deeply associated with the remaining fibers of the intermandibularis [State 1]. In taxa of CS-2 the protractor hyoidei is a well-differentiated, distinct muscle, its fibers not being deeply mixed with those of other muscles (e.g., Fig. 40) [State 2]. It should be noted that Winterbottom (1974) used the name "intermandibularis anterior" only for those teleostean taxa in which the intermandibularis posterior is not incorporated on the protractor hyoidei. In the present work, I use the name "intermandibularis anterior" even in those cases where the intermandibularis posterior is incorporated on the protractor hyoidei. Vrba (1968) stated that in members of the genus *Elops* the intermandibularis posterior and the interhyoideus do not form a protractor hyoidei. However, in the *Elops* specimens examined in the present work, as well as in those examined by Winterbottom (1974) and others, these two myological structures do form a protractor hyoidei, although there is a significant association between the anterior portion of this latter muscle and the anterior intermandibularis. Greenwood (1971) stated that in the osteoglossomorph notopterids, including the members of the genus *Xenomystus*, the intermandibularis posterior and interhyoideus do not combine to form a protractor hyoidei (e.g., Greenwood, 1971: fig. 8). In the *Xenomystus* specimens analyzed in the present work it is not completely clear that this is the case. This is because the muscle named "intermandibularis posterior" in Greenwood's (1971) page 21 does appear to have a myocommata dividing its anterior and posterior portions, which is very similar to the myocommata dividing the anterior and posterior portions (i.e., intermandibularis posterior and interhyoideus portions) of the protractor hyoidei in many other teleosts. Also, in the *Xenomystus* specimens examined the muscle named "interhyoideus" in Greenwood's (1971) page 21 is very similar to the "hyohyoideus inferior" (sensu Winterbottom, 1974, and the present work) of many other teleosts (see below). In fact, unlike the interhyoideus of teleosts such as *Albula* and *Mormyrus*, the muscle

named "interhyoideus" in Greenwood's (1971) page 21 does not attach anteriorly on the lower jaw: instead, it fuses anteroventromesially with its counterpart and inserts on to the hypohyals through a short tendon, as is the case with the hyohyoideus inferior of many other teleosts. It is, however, important to stress that Greenwood (1971) stated that the muscle named "intermandibularis posterior" in his description of *Xenomystus* was innervated by a branch of the mandibular V nerve, and not of both nerves V and VII, as is usually the case with the protractor hyoidei of other teleosts. The data available in the literature and the observations of the present work, therefore, do not allow us to discern the configuration of the ventral musculature of *Xenomystus*. Is the muscle named "intermandibularis posterior" in Greenwood's (1971) page 21 in fact a protractor hyoidei? Or is it an intermandibularis posterior, the "interhyoideus" of that author being thus an interhyoideus and the hyohyoideus inferior being thus completely lost or completely fused with the hyohyoideus abductor? Or is the "intermandibularis posterior" of Greenwood's page 21 an intermandibularis posterior, the "interhyoideus" of Greenwood's (1971) page 21 being actually a hyohyoideus inferior and the interhyoideus being thus missing? I plan to analyze these questions in a future work. For the time being, *Xenomystus* is coded here as '?'.

4. *Interhyoideus missing (inspired from, e.g., Tchernavin, 1947a,b).* In taxa of CS-0 the interhyoideus may be present as an independent structure or may be integrated in the protractor hyoidei (see above) [State 0]; in *Eurypharynx* the interhyoideus is missing [State 1]. Although some authors do not refer to the interhyoideus in their descriptions of the ventral muscles of lacertid lizards (e.g., Herrel et al., 2005), this muscle is found in the *Timon* specimens examined in the present work; it seemingly forms, together with the hyohyoideus of the present work, the "constrictor colli" of certain authors (see Fig. 110).
5. *Significant part of interhyoideus associated with mandible.* Contrary to taxa of CS-0 (e.g., Fig. 6) [State 0], in taxa of CS-1 a significant part of this muscle (or of the protractor hyoidei, in those cases in which the interhyoideus is incorporated into the protractor hyoidei: see above) attaches on the mandible (e.g., Fig. 35) [State 1].
6. *Intermandibularis exclusively attaching to angulars and/or articulars (inspired from, e.g., Howes, 1985a).* The intermandibularis of *Phractolaemus* [State 1] is quite different from that of taxa of CS-0 [State 0], as it exclusively connects the angulo-articulars of each side of the fish (see Howes, 1985a: figs. 20 and 21).
7. *Differentiation of protractor hyoidei into pars dorsalis, pars ventralis and pars lateralis (inspired from, e.g., Diogo and Vandewalle, 2003; Diogo,*

2004a). In taxa of CS-0 this muscle is constituted by a single mass of fibers or by ventral and dorsal bundles (e.g., Fig. 77) [State 0]; in taxa of CS-1 it is characteristically divided into a pars lateralis, a pars dorsalis and a pars ventralis (e.g., Diogo, 2004a: fig. 3-41) [State 1].

8. *Insertion of protractor hyoidei high on the coronoid process (inspired from, e.g., Howes, 1985a)*. Contrary to taxa of CS-0 [State 0], such a configuration is found in *Gonorynchus* (e.g., Fig. 62) [State 1].
9. *Presence of specialized glossal muscles (e.g., hyoglossus and/or genioglossus) associated with tongue in at least some ontogenetic stages (inspired from, e.g., Edgeworth, 1935; Jarvik, 1963; Larsen and Guthrie, 1975; Carroll and Holmes, 1980; Bemis et al., 1983; Lauder and Shaffer, 1985, 1988; Shaffer and Lauder, 1985a,b; Smith, 1988; Reilly and Lauder, 1989, 1990, 1991; Lauder and Reilly, 1990; Bauer, 1992; Herrel et al., 2001, 2005)*. Contrary to taxa of CS-0 [State 0], taxa of CS-1 exhibit this feature (e.g., Fig. 110) [State 1].
10. *Presence of well-differentiated, broad muscle omohyoideus (inspired from, e.g., Edgeworth, 1935; Gasc, 1968; Larsen and Guthrie, 1975; Herrel et al., 2001, 2005; Montero et al., 2002; Abdala and Moro, 2003)*. Contrary to taxa of CS-0 [State 0], taxa of CS-1 exhibit this feature (e.g., Figs. 99, 110) [State 1].
11. *Presence of well-differentiated, broad muscle branchiohyoideus in at least some ontogenetic stages (inspired from, e.g., Edgeworth, 1935; Jarvik, 1963; Jollie, 1982; Lauder and Shaffer, 1985, 1988; Herrel et al., 2001, 2005; Montero et al., 2002; Abdala and Moro, 2003; Ericsson and Olsson, 2004; Ericsson et al., 2004)*. Contrary to taxa of CS-0 [State 0], specimens of *Ambystoma ordinarium* exhibit such a feature (e.g., Figs. 99, 111) [State 1]. According to Lauder and Shaffer (1988) and others, this muscle becomes lost after metamorphosis in various salamanders, such as those of the species *Ambystoma tigrinum*. The metamorphosed specimens of *Ambystoma ordinarium* examined in the present work do, however, have a distinct muscle branchiohyoideus (Fig. 99). As explained by Edgeworth (1935), the structure often named "branchiohyoideus" in amniotes is seemingly homologous to the branchial muscle subarcualis rectus 1 of amphibians (see Fig. 99), and not to the hyoid muscle branchiohyoideus of the present work (which he designated as "branchiohyoideus externus"). Thus, the "branchiohyoideus" of *Timon lepidus* seemingly does not correspond to the branchiohyoideus of *Ambystoma ordinarium*. However, by way of precaution, I prefer to code *Timon* as '?'. I will return to this subject in Chapter 4, Section 4.2.
12. *At least some muscular fibers can be assigned to the hyohyoideus inferior and the presence of this muscle confirmed*. The plesiomorphic condition for the osteichthyan taxa included in the analysis is seemingly that in

which there is no major differentiation between the hyohyoideus inferior and the hyohyoideus superior (e.g., Fig. 15) [State 0]. In taxa of CS-1, although fibers of the hyohyoideus inferior may be mixed with those of other myological structures such as the hyohyoideus superior, at least some muscular fibers can be confidently assigned to the hyohyoideus inferior and the presence of this structure thus confirmed (e.g., Fig. 77) [State 1]. Edgeworth's (1935: fig. 210) illustration of *Amia* seems to suggest that the members of this taxon display a condition such as that of CS-0. However, in the *Amia* specimens examined in the present work there were some fibers originating on the anterior ceratohyal and running forward to attach on the hypohyals; these fibers thus do not contact the branchiostegal rays at all. This could indicate that the members of this genus have at least some fibers that can be assigned confidently to the hyohyoideus inferior. However, until more data is available, I prefer to prudently code *Amia* as '?'.

13. *Each side of hyohyoideus inferior mainly mixing mesially with its counterpart on midline and/or attaching anteriorly to the anterior region of the side of the hyoid arch from which it originates.* Among those taxa included in the analysis with a recognizable hyohyoideus inferior the two sides of this structure mainly overlap each other and run from the anteromesial surface of the hyoid arch to the hyoid arch (and eventually the branchiostegal rays) of the opposite side of the animal (e.g., Fig. 29) [State 0]. In taxa of CS-1 each side of the hyohyoideus inferior mainly attaches anteriorly on a median aponeurosis and/or on the anterior region of the hyoid arch side from which it originates (e.g., Fig. 77) [State 1].
14. *Hyohyoideus superior, if present as a distinct muscle, having peculiar configuration associated with peculiar branchial specializations (inspired from, e.g., Greenwood, 1971, 1973).* Unlike taxa of CS-0 [State 0], *Mormyrus* exhibits such a feature (e.g., Fig. 35) [State 1].
15. *Hyohyoideus superior, if present as a distinct muscle, with significant part of its fibers attaching on pectoral girdle.* Unlike taxa of CS-0 [State 0], taxa of CS-1 exhibit such a feature [State 1].
16. *Presence of muscle "hyohyoideus ventralis" (inspired from, e.g., De la Hoz and Chardon, 1984).* Contrary to taxa of CS-0 [State 0], taxa of CS-1 exhibit such a feature [State 1] (e.g., De la Hoz and Chardon, 1984: fig. 16].
17. *Muscle coracomandibularis not recognizable as independent element (inspired from, e.g., Pollard, 1892; Danforth, 1913; Luther, 1913; Edgeworth, 1911, 1923; 1935; Fox, 1963, 1965; Wiley, 1979a,b; Lauder, 1980ac; Bemis, 1986; Bemis and Lauder, 1986; Miyake et al., 1992; Bartsch,*

1994; Wilga et al., 2000). In basal bony fishes there is a recognizable muscle coracomandibularis usually running from the pectoral girdle to the mandible (e.g., Fig. 86) [State 0]. Such a muscle is not recognizable as an independent element in taxa of CS-1 (e.g., Fig. 35) [State 1].

18. *Peculiar configuration of coracomandibularis, in which this muscle is modified in a "branchiomandibularis" (inspired from, e.g., Pollard, 1892; Danforth, 1913; Luther, 1913; Edgeworth, 1935; Wiley, 1979a,b; Lauder, 1980a; Miyake et al., 1992; Wilga et al., 2000)*. Unlike taxa of CS-0 [State 0], taxa of CS-1 exhibit such a feature (e.g., Fig. 6) [State 1].
19. *Sternohyoideus missing (inspired from, e.g., Tchernavin, 1947a,b)*. Unlike taxa of CS-0 [State 0], *Eurypharynx* exhibits such a feature [State 1].
20. *Significant part of anteroventromesial portion of hypaxialis continuous with posteroventromesial portion of sternohyoideus (thus almost completely, or completely, covering the anteroventromesial surface of the pectoral girdle in ventral view) (inspired from, e.g., Edgeworth, 1935; Kesteven, 1942-1945; Lauder, 1980a)*. In basal osteichthyans the anteroventromesial portion of the hypaxialis is usually not continuous with the posteroventromesial portion of the sternohyoideus (e.g., Fig. 6), these two myological structures being usually separated by the anteroventromesial surface of the pectoral girdle (e.g., Lauder, 1980a) [State 0]. In taxa of CS-1 a significant part of the anteroventromesial portion of the hypaxialis is continuous with the posteroventromesial portion of the sternohyoideus (e.g., Fig. 50) [State 1].
21. *Sternohyoideus consolidated into a single median muscle (inspired from, e.g., Lauder, 1980a)*. Unlike taxa of CS-0 [State 0], taxa of CS-1 exhibit such a feature (e.g., Fig. 77) [State 1].

Musculature of Pectoral Girdle and of Pectoral Fins/Forelimbs

22. *Presence of recognizable muscle "arrector 3"*. Unlike taxa of CS-0 (e.g., Fig. 8) [State 0], taxa of CS-1 have a recognizable muscle "arrector 3", which usually situates ventrally to the arrector ventralis and attaches on the ventral condyle of the first pectoral ray (e.g., Fig. 74) [State 1]. It should be noted that this muscle was, curiously, not described by Winterbottom (1974). One possible explanation is the fact that it is usually rather small and found only in certain teleostean taxa examined, being absent in taxa such as elopomorphs and osteoglossomorphs, as well as in all non-teleosts analyzed. As explained in Chapter 2, the names used in the present work to

designate some pectoral girdle muscles differ from those used in previous works of the author such as Diogo et al. (2001a) and Diogo (2004a). Thus, in order to facilitate comparisons with those previous works, it is worth noting that the "arrector ventralis", "arrector dorsalis", "abductor superficialis 1", "abductor superficialis 2", "adductor superficialis 1", "adductor superficialis 2" and "abductor profundus" of those previous works correspond respectively to the "arrector 3", "arrector ventralis", "abductor superficialis", "abductor profundus", "adductor superficialis", "adductor profundus" and "arrector dorsalis" of the present work.

23. *Presence of well-developed, distinct muscle coracoradialis (inspired from, e.g., Winterbottom, 1974).* Unlike taxa of CS-0 (e.g., Fig. 8) [State 0], taxa of CS-1 exhibit such a feature (e.g., Fig. 44) [State 1].
24. *Presence of recognizable arrector ventralis.* Unlike taxa of CS-0 (e.g., Fig. 8) [State 0], taxa of CS-1 exhibit such a feature (e.g., Fig. 74) [State 1].
25. *Presence of recognizable arrector dorsalis.* Unlike taxa of CS-0 (e.g., Fig. 8) [State 0], taxa of CS-1 exhibit such a feature (e.g., Fig. 78) [State 1].
26. *Abductor superficialis and/or abductor profundus, if present, being hypertrophied, a significant part of them also originating on mesial surface of pectoral girdle (inspired from, e.g., Greenwood and Thomson, 1960; Le Danois, 1967; Winterbottom, 1974).* Unlike taxa of CS-0 [State 0], taxa of CS-1 exhibit such a feature [State 1] (e.g., Greenwood and Thompson, 1960: figs. 6, 8) [State 1].
27. *Arrector dorsalis subdivided into different sections (ordered multistate character).* Unlike taxa of CS-0 (e.g., Fig. 8) [State 0], taxa of CS-1 exhibit such a feature (e.g., Fig. 44) [State 1].
28. *Arrector dorsalis attaching on both the first and second pectoral rays.* Among those taxa exhibiting an arrector dorsalis the plesiomorphic condition seems to be that in which this structure attaches only on the first pectoral ray (e.g., Figs. 17, 23) [State 0]. In taxa of CS-1 this muscle attaches on both the first and second pectoral rays (e.g., Fig. 30) [State 1].
29. *Presence of recognizable protractor pectoralis (inspired from, e.g., Edgeworth, 1935; Monath, 1965; Greenwood and Lauder, 1981; Walthall and Ashley-Ross, 2006).* The presence of a recognizable protractor pectoralis (see Fig. 94) is seemingly plesiomorphic for the osteichthyan taxa included in the cladistic analysis (e.g., Greenwood and Lauder, 1981) [State 0]. In taxa of CS-1 there is no recognizable protractor pectoralis [State 1]. In some members of the genera *Galaxias*, *Polypterus* and *Acipenser* the protractor pectoralis is missing, while in others this muscle is seemingly present (e.g., Edgeworth, 1935; Greenwood and Lauder, 1981); these three genera were thus coded as '?'.

30. *Significant part of mesial portion of arrector ventralis passing through coracoid-cleithrum foramen*. Unlike taxa of CS-0 [State 0], taxa of CS-1 exhibit this feature (e.g., Diogo, 2004a: fig. 3-56) [State 1].
31. *Arrector ventralis divided into two well-developed, well-differentiated bundles widely separated by large horizontal lamina of coracoid (or scapulo-coracoid) (inspired from, e.g., Diogo, 2004a)*. Although in some taxa examined a significant part of the mesial portion of the arrector ventralis may pass through the "coracoid-cleithrum foramen" (see above), this muscle is usually not widely separated into two well-developed, well-differentiated sections by a large horizontal lamina of the coracoid (or scapulo-coracoid) (see Diogo et al., 2001a) [State 0]. In taxa of CS-1 the composite formed by the scapula plus the coracoid presents a posterodorsal, large laminar projection subdividing the arrector ventralis into two well-developed, well-distinguished divisions (e.g., Diogo, 2004a: fig. 3-40) [State 1].
32. *Arrector ventralis peculiarly divided into well-developed posterodorsal and anteroventral bundles, both originating on ventrolateral surface of pectoral girdle*. Unlike taxa of CS-0 [State 0], the *Xenomystus* specimens examined exhibit such a feature [State 1].
33. *Presence of peculiar "muscle connecting well-developed cranial rib to posterodorsal portion of pectoral girdle" (inspired from, e.g., McMahon, 1969; Bemis, 1986)*. Unlike taxa of CS-0 [State 0], taxa of CS-1 exhibit this feature (e.g., Figs. 94, 95) [State 1].
34. *Pectoral muscles extending far into pectoral fin/forelimb (inspired from, e.g., Romer, 1924; Millot and Anthony, 1958; Kardong and Zalisko, 1998; Kardong, 2002)*. As noticed by these and other authors, contrary to the seemingly plesiomorphic condition for osteichthyans (e.g., Fig. 17) [State 0], in taxa of CS-1 the pectoral muscles extend far into the pectoral fin/forelimb, thus giving to these latter a so-called "lobed" or "fleshy" appearance [State 1] (e.g., Figs. 87, 94, 95, 103, 104, 105, 106, 112, 113, 114). Although the pectoral muscles of †*Tiktaalik roseae* and †*Acanthostega gunnari* are not preserved, in this specific case I prefer to make an exception to the procedure usually followed in this Section, in which fossils are generally coded as '?' for muscular characters. This is because the processes, crests and/or other details of the configuration of the pectoral fin/forelimb elements of these two fossil taxa (described and illustrated in the literature) seem to indicate that, unlike the other fossils included in the present analysis, their pectoral muscles did extend far into their fins/forelimbs. Therefore, I tentatively code †*Tiktaalik roseae* and †*Acanthostega gunnari* as CS-1; this primary homology hypothesis will, of course, be checked against the phylogenetic results obtained in this cladistic analysis, which includes numerous other characters.

35. *Presence of peculiar muscle "opercularis" (inspired from, e.g., Edgeworth, 1935; Monath, 1965; Larsen and Guthrie, 1975; Walthall and Ashley-Ross, 2006).* Unlike in taxa of CS-0 [State 0], in taxa of CS-1 there is a peculiar muscle "opercularis" (sensu, e.g., Walthall and Ashley-Ross, 2006: fig. 5), which is probably related to a highly peculiar mechanism of sound transmission (see Fig. 102) [State 1].
36. *Presence of distinct pectoralis (inspired from, e.g., Romer, 1944; Kardong and Zalisko, 1998; Kardong, 2002; Walthall and Ashley-Ross, 2006).* Unlike taxa of CS-0 (e.g., Fig. 8) [State 0], taxa of CS-1 exhibit such a feature (e.g., Fig. 103) [State 1]. It should be noted that, regarding the muscles of the forelimb that are seemingly found in both amphibians such as *Ambystoma ordinarium* and amniotes such as *Timon lepidus*, I will follow the nomenclature used by Walthall and Ashley-Ross (2006) for salamanders. In this specific case, the name "pectoralis" is also commonly used for researchers working with amniotes. However, as will be seen below, there are cases in which the names used by Walthall and Ashley-Ross (2006) for salamanders are rarely, or never, used by researchers working on amniotes. In such cases I will thus provide at least an example of a name that is often used by those researchers working on amniotes, in order to facilitate the comparisons between amphibians such as salamanders and amniotes such as lizards (see Chapter 4, Sections 4.2, 4.4).
37. *Presence of distinct supracoracoideus (inspired from, e.g., Romer, 1944; Kardong and Zalisko, 1998; Kardong, 2002; Walthall and Ashley-Ross, 2006).* Unlike taxa of CS-0 [State 0], taxa of CS-1 exhibit such a feature (e.g., Fig. 103) [State 1].
38. *Presence of distinct procoracohumeralis (inspired from, e.g., Romer, 1944; Kardong and Zalisko, 1998; Kardong, 2002; Walthall and Ashley-Ross, 2006).* Unlike in taxa of CS-0 [State 0], in those taxa of CS-1 there is a distinct procoracohumeralis (see Fig. 103), which seemingly corresponds to the "deltoides claviculars" of, for example, Romer (1944) [State 1].
39. *Presence of distinct coracobrachialis longus (inspired from, e.g., Romer, 1944; Kardong and Zalisko, 1998; Kardong, 2002; Walthall and Ashley-Ross, 2006).* Unlike in taxa of CS-0 [State 0], in those taxa of CS-1 there is a distinct coracobrachialis longus (e.g., Fig. 103) [State 1].
40. *Presence of anconaeus group (inspired from, e.g., Romer, 1944; Kardong and Zalisko, 1998; Kardong, 2002; Walthall and Ashley-Ross, 2006).* Unlike in taxa of CS-0 [State 0], in those taxa of CS-1 there is a peculiar anconaeus group of muscles (e.g., Figs. 103, 104, 105), which seemingly corresponds to the "triceps" of, for example, Romer (1944) [State 1].

41. *Presence of distinct latissimus dorsi (inspired from, e.g., Romer, 1944; Kardong and Zalisko, 1998; Kardong, 2002; Walthall and Ashley-Ross, 2006)*. Unlike in taxa of CS-0 [State 0], in those taxa of CS-1 there is a distinct latissimus dorsi (e.g., Fig. 104) [State 1].
42. *Presence of distinct dorsalis scapulae (inspired from, e.g., Romer, 1944; Kardong and Zalisko, 1998; Kardong, 2002; Walthall and Ashley-Ross, 2006)*. Unlike in taxa of CS-0 [State 0], in those taxa of CS-1 there is a distinct dorsalis scapulae (e.g., Fig. 104), which seemingly corresponds to the "deltoides scapularis" of, for example, Romer (1944) [State 1].
43. *Presence of distinct humeroantebrachialis (inspired from, e.g., Romer, 1944; Kardong and Zalisko, 1998; Kardong, 2002; Walthall and Ashley-Ross, 2006)*. Unlike in taxa of CS-0 [State 0], in taxa of CS-1 there is a distinct humeroantebrachialis (e.g., Fig. 104), which seemingly corresponds to part, or all, of the "biceps" of, for example, Romer (1944) [State 1].
44. *Presence of distinct flexor digitorum communis (inspired from, e.g., Moro and Abdala, 2004; Abdala and Moro, 2006; Walthall and Ashley-Ross, 2006)*. Unlike in taxa of CS-0 [State 0], in taxa of CS-1 there is a distinct flexor digitorum communis (e.g., Fig. 103), which seemingly corresponds to the "flexor digitorum longus" of, for example, Abdala and Moro (2006) (e.g., Fig. 113) [State 1].
45. *Presence of distinct flexor antebrachii et carpi radialis (inspired from, e.g., Moro and Abdala, 2004; Abdala and Moro, 2006; Walthall and Ashley-Ross, 2006)*. Unlike in taxa of CS-0 [State 0], in those taxa of CS-1 there is a distinct flexor antebrachii et carpi radialis (e.g., Fig. 103), which seemingly corresponds to the "flexor carpi radialis" of, for example, Abdala and Moro (2006) (e.g., Fig. 113) [State 1].
46. *Presence of distinct flexor antebrachii et carpi ulnaris (inspired from, e.g., Moro and Abdala, 2004; Abdala and Moro, 2006; Walthall and Ashley-Ross, 2006)*. Unlike in taxa of CS-0 [State 0], in taxa of CS-1 there is a distinct flexor antebrachii et carpi ulnaris (e.g., Fig. 103), which seemingly corresponds to the "flexor carpi ulnaris" of, for example, Abdala and Moro (2006) [State 1].
47. *Presence of distinct extensor digitorum communis (inspired from, e.g., Moro and Abdala, 2004; Abdala and Moro, 2006; Walthall and Ashley-Ross, 2006)*. Unlike in taxa of CS-0 [State 0], in taxa of CS-1 there is a distinct extensor digitorum communis (e.g., Fig. 104), which seemingly corresponds to the "extensor digitorum longus" of, for example, Abdala and Moro (2006) (e.g., Fig. 112) [State 1].
48. *Presence of distinct extensor antebrachii et carpi radialis (inspired from, e.g., Moro and Abdala, 2004; Abdala and Moro, 2006; Walthall and Ashley-Ross, 2006)*. Unlike in taxa of CS-0 [State 0], in those taxa of CS-1 there

is a distinct extensor antebrachii et carpi radialis (e.g., Fig. 104), which seemingly corresponds to the "extensor carpi radialis" of, for example, Abdala and Moro (2006) (e.g., Fig. 112) [State 1].

49. *Presence of distinct extensor antebrachii et carpi ulnaris (inspired from, e.g., Moro and Abdala, 2004; Abdala and Moro, 2006; Walthall and Ashley-Ross, 2006).* Unlike in taxa of CS-0 [State 0], in those taxa of CS-1 there is a distinct extensor antebrachii et carpi ulnaris (e.g., Fig. 104), which seemingly corresponds to the "extensor carpi ulnaris" of, for example, Abdala and Moro (2006) (e.g., Fig. 112) [State 1].
50. *Presence of distinct contrahentium caput longum (inspired from, e.g., Walthall and Ashley-Ross, 2006).* Unlike in taxa of CS-0 [State 0], in those taxa of CS-1 there is a distinct contrahentium caput longum (e.g., Fig. 105) [State 1]. This muscle is seemingly missing in *Timon lepidus* and, thus, this species is coded as CS-0. However, it is worth noting that a muscle apparently corresponding to the contrahentium caput longum might eventually be present in some other lizards (Virginia Abdala, pers. comm.).
51. *Presence of distinct flexor accessorius lateralis and/or flexor accessorius medialis (inspired from, e.g., Moro and Abdala, 2004; Abdala and Moro, 2006; Walthall and Ashley-Ross, 2006).* Unlike taxa of CS-0 [State 0], those taxa of CS-1 exhibit distinct flexor accessorius lateralis and/or flexor accessorius medialis (e.g., Fig. 105), which may correspond to part, or all, of the "pronator accessorius" of, for example, Abdala and Moro (2006) (e.g., Fig. 114) [State 1].
52. *Presence of distinct pronator profundus (inspired from, e.g., Moro and Abdala, 2004; Abdala and Moro, 2006; Walthall and Ashley-Ross, 2006).* Unlike in taxa of CS-0 [State 0], in taxa of CS-1 there is a distinct pronator profundus (e.g., Figs. 105, 114) [State 1].
53. *Presence of distinct flexores breves superficiales (inspired from, e.g., Moro and Abdala, 2004; Abdala and Moro, 2006; Walthall and Ashley-Ross, 2006).* Unlike in taxa of CS-0 [State 0], in taxa of CS-1 there are distinct flexores breves superficiales (e.g., Fig. 103), which seemingly correspond to the "flexores digiti brevis superficialis" of, for example, Abdala and Moro (2006) [State 1].
54. *Presence of distinct intermetacarpales (inspired from, e.g., Moro and Abdala, 2004; Abdala and Moro, 2006; Walthall and Ashley-Ross, 2006).* Unlike in taxa of CS-0 [State 0], in taxa of CS-1 there are distinct intermetacarpales (e.g., Fig. 103), which may eventually be divided into intermetacarpales I and intermetacarpales II (e.g., Abdala and Moro, 2006) [State 1].
55. *Presence of distinct extensores digitorum breves (inspired from, e.g., Moro and Abdala, 2004; Abdala and Moro, 2006; Walthall and Ashley-Ross,*

2006). Unlike in taxa of CS-0 [State 0], in taxa of CS-1 there are distinct extensores digitorum breves (e.g., Fig. 104), which seemingly correspond to the "extensores digiti brevis" of, for example, Abdala and Moro (2006) (e.g., Fig. 112) [State 1].

56. *Presence of distinct abductor et extensor digiti 1 (inspired from, e.g., Moro and Abdala, 2004; Abdala and Moro, 2006; Walthall and Ashley-Ross, 2006)*. Unlike in taxa of CS-0 [State 0], in taxa of CS-1 there is a distinct abductor et extensor digiti 1 (e.g., Fig. 103), which seemingly correspond to the "abductor longus pollici" of, for example, Abdala and Moro (2006) [State 1].
57. *Presence of distinct extensor lateralis digiti IV / abductor digitorum V (inspired from, e.g., Moro and Abdala, 2004; Abdala and Moro, 2006; Walthall and Ashley-Ross, 2006)*. Unlike taxa of CS-0 [State 0], *Ambystoma ordinarium* exhibits a distinct extensor lateralis digiti IV (e.g., Fig. 103). This muscle seemingly corresponds to the abductor digitorum V of lizards such as *Timon lepidus* (sensu, e.g., Abdala and Moro, 2006). Thus, these two taxa are coded as CS-1 [State 1]. It should be noted that since *Timon lepidus* has five fingers on the manus, and not four as does *Ambystoma ordinarium*, I consider that in this specific case it may be appropriate to retain the name "abductor digitorum V", commonly used by researchers working with lizards, and not the name "extensor lateralis digiti IV", for the muscle of *Timon lepidus*.
58. *Presence of distinct contrahentes digitorum (inspired from, e.g., Moro and Abdala, 2004; Abdala and Moro, 2006; Walthall and Ashley-Ross, 2006)*. Unlike in taxa of CS-0 [State 0], in taxa of CS-1 there are distinct contrahentes digitorum (e.g., Fig. 105) [State 1].
59. *Presence of distinct flexores breves profundi (inspired from, e.g., Moro and Abdala, 2004; Abdala and Moro, 2006; Walthall and Ashley-Ross, 2006)*. Unlike in taxa of CS-0 [State 0], in taxa of CS-1 there are distinct flexores breves profundi (e.g., Fig. 105), which seemingly correspond to the "flexores digiti brevis profundus" of, for example, Abdala and Moro (2006) [State 1].
60. *Presence of distinct epitrochleoanconeus (inspired from, e.g., Moro and Abdala, 2004; Abdala and Moro, 2006)*. Unlike in taxa of CS-0 [State 0], in *Timon lepidus* there is a well-differentiated epitrochleoanconeus (e.g., Fig. 114) [State 1]. It is worth noting that this muscle was not described by Walthall and Ashley-Ross (2006) in the newt *Taricha*. However, a muscle apparently corresponding to the epitrochleoanconeus of lizards is seemingly found in at least some members of *Ambystoma* (Virginia Abdala, pers. comm.). Since my observations on *Ambystoma ordinarium* did not allow me to discern whether or not such a muscle is present in members of this *Ambystoma* species, I prefer to prudently code this species as '?'.

61. *Presence of distinct abductor brevis pollicis (inspired from, e.g., Moro and Abdala, 2004; Abdala and Moro, 2006).* Unlike in taxa of CS-0 [State 0], in *Timon lepidus* there is a distinct abductor brevis pollicis (e.g., Moro and Abdala, 2004) [State 1].
62. *Presence of distinct dorsometacarpalis (inspired from, e.g., Moro and Abdala, 2004; Abdala and Moro, 2006).* Unlike in taxa of CS-0 [State 0], in *Timon lepidus* there is a distinct dorsometacarpalis (e.g., Moro and Abdala, 2004: fig. 3) [State 1]. This muscle is seemingly missing in the specimens examined of *Ambystoma ordinarium.*
63. *Presence of distinct flexor digitorum V transversus I and flexor digitorum transversus II (inspired from, e.g., Moro and Abdala, 2004; Abdala and Moro, 2006).* Unlike in taxa of CS-0 [State 0], in *Timon lepidus* there are distinct muscles flexor digitorum V transversus I and flexor digitorum V transversus II (e.g., Moro and Abdala, 2004: fig. 6) [State 1]. These muscles are seemingly missing in the specimens examined of *Ambystoma ordinarium.*

Neurocranium, Anterior Vertebrae and Related Structures

64. *Presence of "kinethmoid" bone (inspired from, e.g., Fink and Fink, 1981, 1996).* Unlike taxa of CS-0 [State 0], taxa of CS-1 exhibit such a feature (e.g., Fig. 74) [State 1].
65. *Mesethmoid, if present, fused with prevomer/vomer (inspired from, e.g., Trewavas, 1932; Ramaswami, 1953; Smith, 1989a,b; Belouze, 2002; Taverne, 2004).* Unlike taxa of CS-0 [State 0], taxa of CS-1 exhibit such a feature (e.g., Ramaswami, 1953: fig. 1) [State 1].
66. *Anterior margin of prevomer/vomer situates well posteriorly to anterior margin of mesethmoid (or rostrodermethmoid and/or supraethmoid) (inspired from, e.g., Fink and Fink, 1981, 1996; Begle, 1992; Johnson and Patterson, 1996).* Among those taxa having a mesethmoid (or rostrodermethmoid and/or supraethmoid), the plesiomorphic condition seems to be that in which the anterior surface of the prevomer/vomer situates ventrally or even anteroventrally to the anterior surface of this element (or of the rostrodermethmoid and/or supraethmoid) (e.g., Fig. 32) [State 0]. In the taxa of CS-1 the anterior margin of the prevomer/vomer situates well posteriorly to the anterior margin of the mesethmoid (e.g., Diogo, 2004a: fig. 3-66) [State 1].
67. *Mesethmoid (or rostrodermethmoid and/or supraethmoid), if present, exhibiting peculiar anterolateral arms supporting and/or articulating with premaxillae (inspired from, e.g., Fink and Fink, 1981; 1996).* Unlike taxa of

CS-0 [State 0], taxa of CS-1 exhibit such a feature (e.g., Diogo, 2004a: fig. 3-67) [State 1]. In the specimens examined of the clupeiform genera *Engraulis* and *Thryssa*, the mesethmoid has anterolateral processes similar to those of, for example, some otophysans coded as CS-1, but these processes do not support and / or articulate with the premaxillae (unlike those of *Ethmalosa*, coded as CS-1). As pointed out by Gayet (1985: 109), it is difficult to appropriately discern this character in †*Lusitanichthys characiformis*. However, the illustrations provided by Gayet (1981, 1985) suggest that in this fossil species the anterolateral margins of the mesethmoid are mainly related to the proximal surfaces of the maxillae, and not to the premaxillae as is the case in taxa of CS-1 ("anterolaterally, the mesethmoid presents a profound depression that receives the articular process of the maxilla": Gayet, 1981, page 175). Thus, †*Lusitanichthys characiformis* does not seem to present a configuration such as that of taxa of CS-1, being therefore coded as CS-0.

68. *Posterodorsal portion of mesethmoid (or rostrodermethmoid and/or supraethmoid) appearing compressed anteroposteriorly when seen in a dorsal view of the neurocranium (inspired from, e.g., Fink and Fink, 1981, 1996).* Unlike taxa of CS-0 [State 0], taxa of CS-1 exhibit such a feature (e.g., Diogo, 2004a: fig. 3-12) [State 1].
69. *Mesethmoid, if present, peculiarly shaped, being markedly compressed anteroposteriorly and expanded transversally (inspired from Thys van den Audenaerde, 1961).* Unlike taxa of CS-0 [State 0], *Phractolaemus* exhibits such a feature (e.g., Thys van den Audenaerde, 1961: fig. 13) [State 1].
70. *Lateral ethmoids, if present, exhibiting remarkably large, peculiar lateral extensions (inspired from, e.g., Gayet, 1993; Grande, 1994; Grande and Poyato-Ariza, 1999).* Unlike taxa of CS-0 [State 0], taxa of CS-1 exhibit such a feature (e.g., Thys van den Audenaerde, 1961: fig. 15).
71. *Presence of peculiar contact between lateral ethmoid and autosphenotic (inspired from, e.g., Bornbusch, 1991).* Unlike taxa of CS-0 [State 0], *Silurus* exhibits such a feature (e.g., Bornbusch, 1995: fig. 5B) [State 1].
72. *Lateral ethmoid, if present, exhibiting anteroventrolateral, anteroventrally pointed process (inspired from, e.g., De la Hoz and Chardon, 1984).* Unlike taxa of CS-0 [State 0], taxa of CS-1 exhibit such a feature (e.g., De la Hoz and Chardon, 1984: figs. 1, 4) [State 1].
73. *Lateral ethmoid, if present, being highly modified and loosely attached to the rest of the neurocranium (inspired from, e.g., Lekander, 1949; Ramaswami, 1952a,b,c,d, 1953, 1955a,b, 1957; Albert 2001).* Unlike taxa of CS-0 [State 0], taxa of CS-1 exhibit such a feature (e.g., Ramaswami, 1953: fig. 1) [State 1].
74. *Presence of ossified prevomer/vomer.* Unlike in taxa of CS-0 (e.g., Fig. 10) [State 0], taxa of CS-1 lack an ossified prevomer / vomer [State 1]; it is

worth noting that I agree with Jollie (1980), Findeis (1997) and other authors in that extant acipenseriforms seemingly have no ossified structures corresponding to the prevomer/vomers of taxa of CS-0; *Acipenser* and *Psephurus* are therefore tentatively coded as CS-1.

75. *Prevomer/vomer, if present, not exhibiting two well-developed, anteriorly directed anterolateral processes (inspired from Thys van den Audenaerde, 1961; Lenglet, 1974).* Unlike taxa of CS-0 [State 0], taxa of CS-1 exhibit such a feature (e.g., Thys van den Audenaerde, 1961: fig. 13; Lenglet, 1974: fig. 6) [State 1].
76. *Prevomer/vomer unpaired (inspired from, e.g., Regan, 1923; Mayhew, 1924; Bonebrake and Brandon, 1971; Patterson, 1973, 1975, 1982; Benton, 1985; De Pinna, 1996; Grande and Bemis, 1998; Lee, 1998; Arratia, 1999).* Unlike taxa of CS-0 [State 0], taxa of CS-1 exhibit such a feature (e.g., Fig. 60) [State 1]. Patterson (1975), De Pinna (1996) and others have documented that certain specimens of the genera *Hiodon* and *Osmerus* may exhibit a condition such as that of CS-0, while others may exhibit a condition such as that of CS-1; *Hiodon* and *Osmerus* are thus coded as '?'.
77. *Presence of large, strong "pseudocartilaginous ligament" between prevomer/vomer, autopalatines and/or mandibles (inspired from, e.g., Thys van den Audenaerde, 1961; Howes, 1985a).* Unlike taxa of CS-0 [State 0], *Phractolaemus* exhibits such a feature (e.g., Fig. 64; Thys van den Audenaerde, 1961: fig. 19) [State 1].
78. *Absence of prevomerine/vomerine teeth (inspired from, e.g., Ridewood, 1904a, 1905a,b; Thys van den Audenaerde, 1961; Pasleau, 1974; Howes, 1985a; Matsui and Rosenblatt, 1987; Poyato-Ariza, 1996; Johnson and Patterson, 1996; Grande and Poyato-Ariza, 1999; Diogo, 2004a).* Unlike taxa of CS-0 [State 0], taxa of CS-1 exhibit such a feature (e.g., Fig. 60) [State 1]. In Sanford's (2000) table II, it is stated that *Alepocephalus* exhibits prevomerine/vomerine teeth. This is apparently due to a printing error or to another type of error, since all the specimens of this genus examined in the present work, as well as those examined by Gegenbaur (1878), Gosline (1969), and other authors, do not have prevomerine/vomerine teeth.
79. *Many small stellate bones making up lateral supports for characteristic, peculiarly elongated "paddle" (inspired from, e.g., Grande and Bemis, 1991, 1996; Bemis et al., 1997).* Unlike taxa of CS-0 [State 0], *Psephurus* exhibits such a feature (e.g., Fig. 12) [State 1].
80. *Presence of rhinosphenoid (inspired from, e.g., Weitzman 1962, 1964; Vari, 1979; Fink and Fink, 1981, 1996; Gayet, 1981, 1985; Buckup, 1998; Cavin, 1999).* Unlike taxa of CS-0 [State 0], taxa of CS-1 exhibit such a feature (e.g., Fig. 80) [State 1]. The presence/absence of this feature in

†*Lusitanichthys characiformis* has been the subject of controversy (e.g., Gayet, 1985: 107-108). In my opinion, the position of the "rhinosphenoid" of the illustrations of Gayet (1981, 1985) is in fact somewhat similar to that of the rhinosphenoid of taxa of CS-1. Moreover, in a paper published after Gayet's (1981, 1985) works, Cavin (1999) described a new species of †*Lusitanichthys*, †*L. africanus* that, according to Cavin, also appears to have a rhinosphenoid (see Cavin, 1999: fig. 2). Thus, I consider that one should not exclude, *a priori*, the hypothesis that the structure named "rhinosphenoid" by Gayet (1981, 1985) and Cavin (1999) might eventually be homologous to the rhinosphenoid of certain characiforms such as those of the genus *Brycon*. These structures will thus be tentatively coded here as primary homologues. It should be noted that Cavin (1999: 692) stated that he "observed a trace of a subrectangular bone behind the lateral ethmoid on several specimens of †*Clupavus maroccanus* that could be interpreted as a rhinosphenoid". As Taverne's (1977a, 1995) descriptions of †*Clupavus maroccanus* make no reference to an eventual presence of a rhinosphenoid in the members of this latter species, I prudently code this species as '?'.

81. *Absence of ossified orbitosphenoid (inspired from, e.g., Marshall, 1962; Fink and Fink, 1981, 1996; Forey et al., 1996; Gayet, 1993; e.g., Johnson and Patterson, 1996; Poyato-Ariza, 1996; Arratia, 1997; Grande and Poyato-Ariza, 1999; Hilton and Bemis, 1999; Sanford, 2000; Hilton 2003).* Unlike taxa of CS-0 (e.g., Fig. 32) [State 0], taxa of CS-1 exhibit such a feature [State 1]. In Sanford's (2000) table II, it is stated that *Alepocephalus* and *Argentina* lack orbitosphenoids. However, all the specimens of these genera examined in the present work, as well as those analyzed by Gegenbaur (1878), Chapman (1942), Gosline (1969), and other authors, have ossified orbitosphenoids. Some authors documented that orbitosphenoids are present in *Esox* (e.g., Jollie, 1975; Arratia, 1997, 1999). However, the "orbitosphenoids" of Jollie's (1975) fig. 10 clearly do not seem to correspond to the orbitosphenoids of the present work, but rather to pterosphenoids. In fact, in the specimens of *Esox* examined in this work, as well as those analyzed by, for example, Johnson and Patterson (1996) and Sanford (2000), the orbitosphenoids are missing. The genus *Thymallus* was also coded by Arratia (1997) as having orbitosphenoids, but the *Thymallus* specimens analyzed in the present work, as well as in works such as Norden (1961) and Sanford (2000), lack ossified orbitosphenoids. As stressed by Hilton and Bemis (1999), the ossification of the cranial bones of *Acipenser* are highly variable: for instance, the lateral ethmoids are ossified in some adult members of this genus but not in others. However, in all adult *Acipenser* specimens examined in the

present work, as well as in those described by Hilton and Bemis (1999) and by, for example, Parker (1882) and Bemis et al. (1997), the orbitosphenoids seem to be ossified (e.g., Hilton and Bemis, 1999: fig. 11). It is worth noting that among researchers working with tetrapods (e.g., Carroll and Holmes, 1980) the name "sphenethmoids" is often used to refer to the orbitosphenoids of the present work. †*Acanthostega gunnari* thus seemingly has an orbitosphenoid sensu the present work (which is fused with both the parasphenoid and the basisphenoid: e.g., Clack, 1998, 2002); this fossil species is therefore tentatively coded as CS-0.

82. *Basisphenoid absent as independent ossification (inspired from, e.g., Ridewood, 1904a,b,c, 1905a,b; Marshall, 1962; Taverne, 1972, 1977b, 1978, 1979, Fink and Fink, 1981, 1996; Jollie, 1986; Grande and Bemis, 1991; Forey et al., 1996; Johnson and Patterson, 1996; Arratia, 1997, 1999; Sanford, 2000; Hilton, 2003).* According to Jollie (1986: 373) and other authors the presence of ossified basisphenoids (e.g., Fig. 46) is a "basic osteichthyan feature" [State 0]. In the specimens examined of those genera of CS-1 the basisphenoids are absent as independent ossifications [State 1]. Sanford (2000) coded *Esox* as not having basisphenoids, but these structures were found in the *Esox* specimens examined in the present work and in studies such as Jollie (1975) and Johnson and Patterson (1996). In the *Timon* specimens examined there was no visible suture between the basisphenoid and the parasphenoid, that is, these two bones seemingly form a compound (as is also the case in the lizard shown in Fig. 108); *Timon* is thus coded as CS-1.
83. *Frontals widely separated from each other along the dorsal midline (inspired from, e.g., Swinnerton, 1903; D'Aubenton, 1961; Grande, 1994; Grande and Poyato-Ariza, 1999).* Unlike in taxa of CS-0 [State 0], taxa of CS-1 exhibit such a feature (e.g., Grande and Poyato-Ariza, 1999: fig. 4B) [State 1]. The homologies and identity of the bones of the skull roof of dipnoans have been, and continue to be, the subject of much controversy (e.g., Huxley, 1876; Miles, 1977; Schultze and Campbell, 1986; Schultze, 1986, 2004; Campbell and Barwick, 1986; Marshall, 1986; Bartsch, 1994; Kemp, 1999; Cavin et al., in press); by way of precaution, some authors opt in fact to use letters to designate these bones (see Fig. 89). Therefore, in those characters concerning the presence/absence and/or the configuration of bones such as the frontals and/or the parietals (sensu this work) I prefer to prudently code *Neoceratodus* and *Lepidosiren* as '?'. Since some members of the genus *Acipenser* exhibit a condition such as that of CS-1 while others seemingly exhibit a condition such as that of CS-0, this genus is also coded as '?'.

84. *Presence of both independent, clearly distinguishable autopterotic and dermopterotic bones (inspired from, e.g., Gosline, 1969; Patterson, 1973; Johnson and Patterson, 1996)*. Unlike taxa of CS-0 [State 0], taxa of CS-1 exhibit such a feature (e.g., Fig. 38) [State 1]. As explained by, for example, Patterson (1973) and Grande and Bemis (1998), it is difficult to discern whether or not *Lepisosteus* has ossified dermopterotics plus ossified autopterotics; this taxon is thus prudently coded as '?'.
85. *Presence of vertical, complete laminar bony connection mesially to the eye, usually between the frontals/parietals, dorsally, and the parasphenoid, ventrally*. Unlike other taxa included in the analysis, in which there is a relatively wide non-ossified space between the dorsal surface of the parasphenoid and the ventral margins of the frontals and/or parietals mesially to the eye [State 0], the specimens examined of taxa of CS-1 exhibit such a vertical, complete laminar bony connection (usually, but not always, made by the orbitosphenoid) [State 1]. Since some specimens of the genera *Pantodon*, *Amia*, *Acipenser* and *Psephurus* exhibit a configuration similar to that of CS-0 (e.g., Taverne, 1978: fig. 31; Jollie, 1984a: fig. 13B), while others exhibit a configuration similar to that of CS-1 (e.g., Kershaw, 1970: fig. 3; Patterson, 1973: fig. 9B), these four genera were coded as '?'.
86. *Autosphenotics, if present, occupying a significant portion of the dorsal surface of the cranial roof*. In taxa of CS-0 the autosphenotics are usually widely covered by other bones of the skull and/or by muscles such as the levator arcus palatini and/or dilatator operculi; therefore, only a small part of their dorsal surface actually constitutes the cranial roof (e.g., Fig. 53) [State 0]. In taxa of CS-1 the autosphenotics do constitute a significant part of the dorsal surface of the cranial roof (e.g., Diogo, 2004a: fig. 3-67) [State 1]. Since some *Gonorynchus* specimens seem to display a condition such as CS-1 (e.g., Grande and Poyato-Ariza, 1999: fig. 4A), while others seemingly display a condition such as CS-0 (e.g., Pasleau, 1974: fig. 25), this genus was coded as '?'.
87. *Frontals fused along midline (inspired from, e.g., Smith, 1989b; Gayet, 1993; Forey et al., 1996; Grande and Poyato-Ariza, 1999)*. Unlike in taxa of CS-0 [State 0], in taxa of CS-1 there is a complete fusion of the frontals along the midline (e.g., Smith, 1989b: fig. 505E) [State 1]. Forey et al. (1996) coded the elopomorph saccopharyngiforms as having the frontals fused along the midline; however, as explained by Tchernavin (1947a), this is the case in, for example, *Saccopharynx* but not in *Eurypharynx*, which has a configuration such as CS-0.
88. *Lateral ethmoid and pterosphenoid, if present, contacting in a peculiar way (inspired from, e.g., Diogo, 2004a)*. Unlike taxa of CS-0 [State 0], specimens of *Cetopsis* exhibit such a feature (see, e.g., Diogo, 2004a: fig. 3-46) [State 1].

89. *Presence of peculiar prootic and/or pterotic bullae lodging diverticulum of swimbladder (inspired from, e.g., Greenwood, 1968; Grande, 1985a,b; Di Dario, 2004; Grande and De Pinna, 2004).* Unlike taxa of CS-0 [State 0], taxa of CS-1 exhibit such a feature (e.g., Grande, 1985a: fig. 31) [State 1].
90. *Saccular and lagenar otoliths situated more posteriorly and principally nearer to midline (inspired from, e.g., Rosen and Greenwood, 1970; Fink and Fink, 1981, 1996; Grande and De Pinna, 2004).* Contrary to taxa of CS-0 [State 0], taxa of CS-1 exhibit such a feature (e.g., Fig. 69) [State 1]. Filleul and Maisey (2004) reported a lagenar capsule situated posteriorly on the exoccipital and on the basioccipital of †*Santanichthys diasii*; this thus seems to indicate that this fossil had a configuration such as CS-1.
91. *Presence of peculiar, large "auditory foramen" in adults (inspired from, e.g., Weitzman, 1962, 1964; Fink and Fink, 1981, 1996).* Unlike taxa of CS-0 [State 0], taxa of CS-1 exhibit such a feature (e.g., Weitzman, 1962: fig. 4).
92. *Markedly large, globular "lagenar capsule" (inspired from, e.g., Weitzman, 1962, 1964; Fink and Fink, 1981, 1996).* Unlike taxa of CS-0 [State 0], taxa of CS-1 exhibit such a feature (e.g., Weitzman, 1962: fig. 4) [State 1]. Gayet (1985: 115) considered that it was not possible to discern this character in the specimens of †*Lusitanichthys characiformis* that she analyzed. However, in fig. 23 of Gayet (1985) the posterior portion of the basioccipital is visible, and a markedly large, globular "lagenar capsule" such as that described for CS-1 does not seem to be present.
93. *Presence of a peculiar "recessus lateralis" (inspired from, e.g., Greenwood, 1968; Grande, 1985a,b; Arratia, 1997, 1999; Chang and Maisey, 2003; Di Dario, 2004).* Unlike taxa of CS-0 [State 0], taxa of CS-1 exhibit this feature (e.g., Fig. 46; Di Dario, 2004: figs. 3, 4) [State 1]. Taverne (1977a) described a "recessus lateralis" in †*Clupavus maroccanus*, but Taverne (1977) stated that such a feature was seemingly absent in this fossil species; this species is thus prudently coded as '?'.
94. *Exoccipitals, if present, exhibiting peculiar posterolateral expansion (inspired from Poyato-Ariza, 1996; Grande and Poyato-Ariza, 1999).* Unlike in taxa of CS-0 [State 0], in *Chanos* the exoccipitals are expanded posterolaterally and extend well over the first neural arch [State 1].
95. *Presence of compound bone formed by parietals (or parieto-extrascapulars) and chondral supraoccipital (inspired from, e.g., Lundberg, 1975; Fink and Fink, 1981, 1996).* Unlike taxa of CS-0 [State 0], taxa of CS-1 exhibit such a feature (e.g., Diogo, 2004a: fig. 3-67) [State 1].
96. *Parietals (or parieto-extrascapulars), if present, having their main bodies widely separated from each other in dorsal view (inspired from, e.g., Fink and*

Fink, 1981, 1996; Grande 1985a,b; Poyato-Ariza, 1996; Taverne, 1995; Johnson and Patterson, 1996; Grande and Poyato-Ariza, 1999). Unlike taxa of CS-0 [State 0], taxa of CS-1 exhibit such a feature (e.g., Fig. 38) [State 1]. Since some specimens of *Esox, Thymallus* and *Coregonus* display a condition such as CS-1, while others seem to display a condition such as CS-0 (e.g., Norden, 1961; Shaposhnikova, 1967; Sanford, 2000; Grande et al., 2004; this study), these three genera are coded as '?'. Also, since some members of *Astronesthes* have ossified parietals but others do not (e.g., Weitzman, 1967a,b), this genus was also coded as '?'.

97. *Presence of independent, ossified intercalar (inspired from, e.g., Taverne, 1972, 1977b, 1978; Fink and Fink, 1981, 1996; Jollie, 1986; Forey et al., 1996; Gardiner et al., 1996; Hilton, 2003)*. Unlike taxa of CS-0 [State 0], taxa of CS-1 exhibit this feature (e.g., Fig. 60) [State 1]. Chapman (1934), Wilson and Veilleux (1982), and others have described members of *Umbra* in which the intercalar is apparently missing; in the *Umbra* specimens examined in the present work the intercalar was seemingly present, this taxon is thus prudently coded as '?'.
98. *Basioccipital, if present, exhibiting enlarged ventral "pharyngeal process" (inspired from, e.g., Takahasi, 1925; Weisel, 1960; Winterbottom, 1974; Vandewalle, 1975; Howes, 1978; Taverne and De Vos, 1997)*. Unlike taxa of CS-0 [State 0], taxa of CS-1 exhibit such a feature (e.g., Fig. 79) [State 1].
99. *Absence of teeth on parasphenoid (inspired from, e.g., Bridge, 1878; Pollard, 1892; Ridewood, 1904b; Forey, 1973a; Taverne, 1974, 1999; Grande, 1985a,b; Bartsch, 1993, 1994; Clack, 1998; Arratia, 1997, 1999; Chang and Maisey, 2003)*. Unlike taxa of CS-0 [State 0], taxa of CS-1 exhibit such a feature (e.g., Fig. 60) [State 1]. Although Daeschler et al. (2006) do not refer directly to this character, their fig. 2 seems to indicate that the parasphenoid of †*Tiktaalik roseae* does not have teeth on the parasphenoid; I am thus tentatively coding this fossil taxon as CS-1.
100. *Presence of "dermal basipterygoid process" (inspired from, e.g., Gosline, 1969; Taverne, 1972, 1977b, 1978, 1999; Patterson, 1973, 1984; Grande, 1985a,b; Gardiner and Schaeffer, 1989; Gardiner et al., 1996, 2005; Johnson and Patterson, 1996; Arratia, 1997, 1999; Chang and Maisey 2003; Cloutier and Arratia, 2004; Moritz and Britz, 2005)*. Unlike taxa of CS-0 [State 0], taxa of CS-1 [State 1] exhibit a "dermal basipterygoid process". Taverne (1977a) stated that such a "dermal basipterygoid process" was lacking in †*Clupavus maroccanus*, but the same author (1995) stated that it was seemingly present in this fossil species. Therefore, I prefer to prudently code this species as '?'. Although "there is considerable difficulty in distinguishing the dermal parasphenoid from the endocranial basisphenoid" in †*Acanthostega gunnari*, Clack (1998: 73) suggests that the basipterygoid process of

this fossil taxon is mainly associated with the basisphenoid; therefore, I am tentatively coding it here as CS-0. This seems also to be the case in *Timon*, which is thus also tentatively coded as CS-0.

101. *Ossification of chondral supraoccipital (inspired from, e.g., Tchernavin, 1947a; Patterson, 1973; Forey et al., 1996; Grande and Bemis, 1998; Arratia, 1999; Cloutier and Arratia, 2004; Johanson et al., 2005).* According to Cloutier and Arratia (2004) and others, the plesiomorphic condition of the osteichthyan taxa included in the cladistic analysis of the present work is seemingly that in which there is no chondral supraoccipital [State 0]. I am tentatively following this polarity here. Taxa of CS-1 have an ossified, chondral supraoccipital (e.g., Fig. 25) [State 1]. Apart from most teleosts examined, I have coded as CS-1 the lizard *Timon* and the coelacanth *Latimeria*: the ossified supraoccipital of lacertids such as *Timon* seems to be chondral (e.g., Lee, 1998); Millot and Anthony (1958) described the ossified supraoccipital of *Latimeria* as a chondral element.
102. *Presence of peculiar "intracranial joints" (inspired from, e.g., Thomson, 1967; Alexander, 1973; Anthony, 1980; Lauder, 1980c; Rosen et al., 1981; Forey, 1986; Schultze, 1986; Schultze and Campbell, 1986; Robineau, 1987; Bemis and Northcutt, 1991; Adamicka and Ahnelt, 1992; Bjerring, 1993; Cloutier and Ahlberg, 1996; Clack, 2002; Bernstein, 2003; Clement, 2005; Shubin et al., 2006).* Unlike taxa of CS-0 [State 0], *Latimeria* exhibits such a feature (e.g., Fig. 85) [State 1]. †*Tiktaalik roseae* is coded as '?' since Daeschler et al. (2006) coded this taxon as '?' in a character concerning the eventual presence of an "endoskeletal intracranial joint".
103. *Presence of "highly ossified triangular pars sustentaculum complex" (inspired from, e.g., Vari, 1979).* Unlike taxa of CS-0 [State 0], *Citharinus* exhibits such a feature (e.g., Vari, 1979: fig. 32) [State 1].
104. *Presence of scaphium as an independent element (ordered multistate character) (inspired from, e.g., Sagemehl, 1885; Rosen and Greenwood, 1970; Fink and Fink, 1981, 1996; Gayet, 1981, 1985; Patterson, 1984).* Unlike taxa of CS-0 [State 0], taxa of CS-1 and CS-2 exhibit this feature. It is, however, worth noting that, as pointed out by Patterson (1984), Fink and Fink (1996) and others, in taxa of CS-1 (e.g., Fig. 72A) [State 1] the scaphium does not seem to be as transformed from the plesiomorphic condition found in taxa of CS-0 as it is in taxa of CS-2 (e.g., Fig. 71B) [State 2]. See Chapter 4, Section 4.5.
105. *Anterior margin of neural arch of third free vertebra closely approaching posterior border of neurocranium (inspired from, e.g., Fink and Fink, 1981, 1996).* In taxa of CS-1 (e.g., Fig. 71C) [State 1] the anterior margin of the neural arch of the third free vertebra approaches the posterior border of the neurocranium more closely than in taxa of CS-0 [State 0].

106. *Presence of claustrum as an independent element (inspired from, e.g., Sagemehl, 1885; Rosen and Greenwood, 1970; Fink and Fink, 1981, 1996; Albert and Campos-da-Paz, 1998; Albert, 2001; De Pinna and Grande, 2003; Grande and De Pinna, 2004).* Unlike taxa of CS-0 [State 0], taxa of CS-1 exhibit this feature (e.g., Fig. 71D; see Chapter 4, Section 4.5) [State 1]. According to De Pinna and Grande (2003) and Grande and De Pinna (2004) the claustrum of taxa of CS-1 may be homologous with the "accessory neural arch" present in some other teleosts. Nevertheless, even if this is the case, the modification of the "accessory neural arch" in the characteristic "claustrum" of taxa of CS-1 is a rather peculiar, derived feature.
107. *Presence of intercalarium (ordered multistate character) (inspired from, e.g., Sagemehl, 1885; Rosen and Greenwood, 1970; Fink and Fink, 1981, 1996; Gayet, 1981, 1985; Patterson, 1984).* Unlike taxa of CS-0 [State 0], taxa of CS-1 and CS-2 exhibit this feature. However, as pointed out by Patterson (1984), Fink and Fink (1996) and others, in taxa of CS-1 (e.g., Fig. 72A) [State 1] the intercalarium does not seem to be as transformed from the plesiomorphic condition found in taxa of CS-0 as it is in the taxa of CS-2 (e.g., Fig. 71C) [State 2]. See Chapter 4, Section 4.5.
108. *Presence of prominent "anterodorsal process" of neural arch of third free vertebra (inspired from, e.g., Fink and Fink, 1981, 1996).* Unlike in taxa of CS-0 [State 0], in taxa of CS-1 there is a prominent "anterodorsal process" of the neural arch of third free vertebra, which usually projects lateral to the ascending process of the intercalarium (e.g., Fig. 71C) [State 1].
109. *No distinct parapophyses of first free vertebrae (or they are eventually present but completely fused with ribs) (inspired from, e.g., Rosen and Greenwood, 1970; Fink and Fink, 1981, 1996; Grande and De Pinna, 2004).* Unlike taxa of CS-0 [State 0], taxa of CS-1 exhibit this feature (e.g., Fig. 71C) [State 1]. Since some *Chanos* specimens display a condition such as CS-0 while others display a condition such as CS-1, this genus is coded as '?'.
110. *At least some parapophyses of two first free vertebrae fused to centra (inspired from, e.g., Rosen and Greenwood, 1970; Fink and Fink, 1981, 1996; Gayet, 1981, 1985; Grande and Poyato-Ariza, 1999; Grande and De Pinna, 2004).* Unlike taxa of CS-0 [State 0], the specimens examined of taxa of CS-1 exhibit this feature (e.g., Fig. 71B) [State 1]. Since some *Pantodon* specimens seemingly display a condition such as CS-1 (e.g., Taverne, 1978) while others seem to exhibit a condition such as CS-0 (e.g., some of the specimens examined in the present work), this genus is coded as '?'.
111. *Presence of tripus (ordered multistate character) (inspired from, e.g., Sagemehl, 1885; Rosen and Greenwood, 1970; Fink and Fink, 1981, 1996;*

Patterson, 1984). Unlike taxa of CS-0 [State 0], taxa of CS-1 and CS-2 exhibit this feature. However, it is worth noting that in taxa of CS-1 (e.g., Fig. 72B) [State 1] the tripus does not seem to be as transformed from the plesiomorphic condition found in taxa of CS-0 as it is in taxa of CS-2 (e.g., Fig. 71D) [State 2]. See Chapter 4, Section 4.5.

112. *Modification/widening of rib and/or parapophysis of third free vertebral centrum (inspired from, e.g., Sagemehl, 1885; Rosen and Greenwood, 1970; Fink and Fink, 1981, 1996; Gayet, 1993; Grande, 1994; Grande and Poyato-Ariza, 1999; Grande and De Pinna, 2004).* In taxa of CS-0 the ribs and/or parapophyses of the third free vertebra are roughly similar to those of the following vertebrae or are eventually missing [State 0]. In taxa of CS-1 the parapophyses and/or ribs of the third free vertebra are noticeably different from, and/or broader than, those of the following vertebrae (e.g., Fig. 73A) [State 1]. It should be noted that taxa with either a "rudimentary" or a "true" paired tripus (see above) are coded as CS-1, since the tripus seems to be at least partly constituted by enlarged parapophyses and/or ribs of the third free vertebra (see Fig. 73B) (e.g., Rosen and Greenwood, 1970; Fink and Fink, 1981, 1996; Patterson, 1984; Grande and De Pinna, 2004). Fink and Fink (1996) suggested that *Kneria* and *Parakneria* might be CS-1, although they recognized that the ribs/parapophyses of the third free vertebra of these two taxa are only slightly broader than those of the following vertebrae. In the specimens examined of *Kneria* and *Parakneria*, as well as in those illustrated by, for example, Lenglet (1974: figs. 17, 18, 19), the ribs/parapophyses of the third free vertebra are not considerably larger than those of the following vertebrae; these two genera are thus tentatively coded as CS-0. Grande and Poyato-Ariza (1999) seem to suggest that *Grasseichthys* and *Cromeria* should be coded as CS-1, but the illustrations of Grande (1994: figs. 6, 9) seemingly indicate that in these two latter genera the ribs and/or parapophyses of the third free vertebra are not considerably broader than those of the following vertebrae. My observations of *Grasseichthys* and *Cromeria* did not make it possible to discern whether these specimens display a condition such as CS-0 or such as CS-1; therefore, these two genera are coded as '?'.

113. *Presence of os suspensorium in adults (ordered multistate character) (inspired from, e.g., Sagemehl, 1885; Rosen and Greenwood, 1970; Fink and Fink, 1981, 1996; Patterson, 1984).* Unlike taxa of CS-0 [State 0], taxa of CS-1and CS-2 exhibit this feature. However, it is worth noting that the configuration of taxa of CS-1 (e.g., Fig. 72B) does not seem to be as derived from the plesiomorphic condition found in taxa of CS-0 as it is in taxa of CS-2 (e.g., Fig. 71D) [State 2]. See Chapter 4, Section 4.5.

114. *Highly modified, ovoid peculiar anterolateral face of "transverse process" of fourth free vertebra (inspired from, e.g., Fink and Fink, 1981, 1996).* Unlike taxa of CS-0 [State 0], taxa of CS-1 display this feature (e.g., Fig. 71D) [State 1]. Although Fink and Fink (1981, 1996) did not mention the presence of this feature in cypriniforms, some fishes of this order, such as the *Danio* and *Barbus* specimens examined in the present work, actually exhibit a configuration that is rather similar to that found in other taxa of CS-1 (e.g., Fig. 79). The presence of this feature in *Danio* and *Barbus* is thus tentatively coded here as CS-1 and will be discussed below in view of the results of this cladistic analysis.

115. *Centrum of third free vertebra markedly shorter than that of surrounding centra (inspired from, e.g., Patterson, 1984; Gayet, 1981, 1985).* Unlike taxa of CS-0 [State 0], taxa of CS-1 display this feature (e.g., Figs. 71A, 72A,B) [State 1].

116. *"Swimbladder" with a silvery peritoneal tunic covering at least part of its anterior portion (inspired from, e.g., Rosen and Greenwood, 1970; Fink and Fink, 1981, 1996; Chardon et al., 2003; Grande and De Pinna, 2004).* Unlike taxa of CS-0 [State 0], taxa of CS-1 display this feature (e.g., Rosen and Greenwood, 1970: fig. 4) [State 1]. Although in *Gonorynchus* the swimbladder is almost completely missing, there are remains of a silvery peritoneal tunic associated with structures of the anterior free vertebrae. Therefore, unlike those taxa in which there are no remains at all of the swimbladder, coded as "Inapplicable" here, *Gonorynchus* is coded as CS-1. It should be noted that although CS-1 is often described in the literature for taxa such as clupeiforms and ostariophysans (e.g., Grande and De Pinna, 2004), this character state might eventually be found in other teleosts. This is the case, for example, of certain elopiforms of the genus *Megalops*. However, owing to the different descriptions of Beauford (1909), Greenwood (1970) and others concerning the precise configuration of the swimbladder of the members of this genus, and because my own observations of this taxon did not allow me to discern this character, I prefer to prudently code *Megalops* as '?'.

117. *Swimbladder markedly divided into an anterior and a posterior chamber (inspired from, e.g., O'Connell 1955; Rosen and Greenwood, 1970; Chardon, 1968; Fink and Fink, 1981, 1996; Chardon et al., 2003; Grande and De Pinna, 2004).* Unlike taxa of CS-0 [State 0], taxa of CS-1 display this feature (e.g., Rosen and Greenwood, 1970: fig. 4) [State 1]. Although some authors (e.g., Gayet, 1986a, 1993) argue that the swimbladder of Siluriformes is not markedly divided into anterior and posterior chambers, this is in fact the case in a great number of catfishes (as those coded as CS-1). This includes the plesiomorphic diplomystids, in which the swimbladder is markedly divided into

these two chambers by a transverse septum (e.g., Fig. 69; Chardon, 1968; Chardon et al., 2003). Although CS-1 is often described in the literature for taxa such as clupeiforms and ostariophysans, this character state might eventually also be present in other teleosts, such as some members of the genus *Megalops;* for the reasons explained just above, I prefer to prudently code this genus as '?'.

118. *Presence of a sinus impar (inspired from, e.g., Chardon, 1968; Rosen and Greenwood, 1970; Fink and Fink, 1981, 1996; Chardon et al., 2003; Grande and De Pinna, 2004).* Unlike taxa of CS-0 [State 0], taxa of CS-1 display this feature (e.g., Fig. 69) [State 1].
119. *Presence of peculiarly large gap ("precervical gap") filled mainly with connective tissue between first free vertebra and neurocranium (inspired from, e.g., Rosen, 1985; Johnson and Patterson, 1996).* Unlike taxa of CS-0 [State 0], taxa of CS-1 display this feature (e.g., Rosen, 1985: figs. 14, 15) [State 1]. This feature has been the subject of much controversy. For instance, Rosen (1985) stated that specimens of non-neoteleostean taxa such as *Osmerus* might also have a "precervical" gap, a statement contradicted by, among others, Johnson and Patterson (1996: 278), who wrote that "in osmeroids the articulation between the occipital condyle and V1 (the first free vertebra) is normally close". What can be said here is that within the non-aulopiform and non-stomiiform taxa examined in the present work there were some specimens that appeared to have a "precervical gap", but that this gap was not as large and as distinct as that found in the specimens of these two neoteleostean orders, which were thus coded as CS-1. That is why, to avoid confusion, I prefer to explicitly define CS-1 here as "presence of peculiarly large gap filled mainly with connective tissue between first free vertebra and neurocranium", since, at least within the specimens analyzed in the present work in which this character could be discerned, this only really applies to those taxa coded as CS-1.

Pectoral Girdle and Fins

120. *Absence of posttemporal (inspired from, e.g., Tchernavin, 1947a; Smith, 1989a,b; Meunier and Geistdoerfer, 1991; Forey et al., 1996; Shubin et al., 2006).* Unlike taxa of CS-0 (e.g., Fig. 10) [State 0], taxa of CS-1 display this feature (e.g., Fig. 108) [State 1].
121. *Supracleithrum not present as an independent element (inspired from, e.g., Romer, 1924; Tchernavin, 1947a; Diogo et al., 2001a; Diogo, 2004a; Shubin et al., 2006).* Unlike taxa of CS-0 (e.g., Fig. 10) [State 0], taxa of CS-1 display this feature (e.g., Fig. 108) [State 1].

122. *Supracleithrum firmly attached to posttemporal (inspired from, e.g., Grande and Bemis, 1991, 1996; Bemis et al., 1997).* Unlike taxa of CS-0 (e.g., Fig. 10) [State 0], taxa of CS-1 display this feature (e.g., Findeis, 1997: fig. 8) [State 1].
123. *Main body of posttemporal (or posttemporo-supracleithrum) lying considerably far from neurocranium, with almost no contact between these two structures.* Unlike taxa of CS-0 [State 0], taxa of CS-1 the main body of the posttemporal (or posttemporo-supracleithrum) lies considerably far from the neurocranium, with almost no contact between these two osteological structures: their association is a rather feeble one made essentially through the "ligament between the posttemporal and the posterior margin of the neurocranium" described below and/or eventually through thin/small extrascapulars that only make a rather loose connection between these two osteological structures (e.g., Fig. 38) [State 1].
124. *Dorsomesial limb of posttemporal (or posttemporo-supracleithrum) markedly thin and mesially extended.* Unlike taxa of CS-0 [State 0], taxa of CS-1 display this feature (e.g., Fig. 73) [State 1]. Because *Astronesthes* specimens (e.g., those examined in the present study) exhibit a configuration such as CS-1 while others seemingly exhibit a configuration such as CS-0 (e.g., Weitzman, 1967b: fig. 3), this genus is coded as '?'.
125. *No ossification of scapula (inspired from, e.g., Jollie, 1984a,b).* Unlike taxa of CS-0 [State 0], taxa of CS-1 display this feature (e.g., Fig. 23) [State 1]. Since some specimens of the genera *Acipenser* and *Psephurus* exhibit a configuration such as CS-1 while others seemingly exhibit a configuration such as CS-0, these two genera are coded as '?'.
126. *No ossification of coracoid (inspired from, e.g., Jollie, 1984a,b).* Unlike taxa of CS-0 [State 0], taxa of CS-1 display this feature (e.g., Fig. 23) [State 1]. Because some specimens of the genera *Acipenser* and *Psephurus* exhibit a configuration such as CS-1 and others seemingly exhibit a configuration such as CS-0, these two genera are coded as '?'. There is much controversy and many contradictions in the literature about whether the single endoskeletal element of the pectoral girdle of salamanders such as *Ambystoma ordinarium* is a scapulocoracoid or a scapula (e.g., Lauder and Shaffer, 1985: fig. 4; Kardong and Zalisko, 1998: fig. 5.17C; Kardong, 2002: fig. 9.19N). If it is a scapula, there would be no ossified coracoid. By way of precaution, I thus prefer to code *Ambystoma ordinarium* as '?'.
127. *Absence of Baudelot's ligament (inspired from, e.g., Patterson and Johnson, 1995).* Baudelot's ligament is present in the great majority of the sarcopterygian and actinopterygian taxa examined (e.g., Figs. 10, 94)

[State 0]. The plesiomorphic condition for the osteichthyan taxa included in the cladistic analysis is seemingly that in which this ligament is present; its absence is coded as CS-1 [State 1]. Patterson and Johnson (1995) and others stated that gonorynchiforms do not have a Baudelot's ligament. However, most gonorynchiforms examined exhibit a paired, well-developed ossification that is commonly considered a "cephalic rib" but is strikingly similar to the ossified Baudelot's ligament of certain teleosts such as catfish, connecting the cleithrum and/or supracleithrum to the posteromesial surface of the neurocranium (usually the exoccipital and/or basioccipital) (e.g., Fig. 66A). A potential homology between such gonorynchiform peculiar "cephalic ribs" and the "Baudelot's ligament" of taxa as catfishes has in fact already been proposed by Ridewood (1905b) and other authors. There is no reason, *a priori*, to postulate that these structures cannot be primary homologous. For the moment, and until more data is available, I prefer to code all gonorynchiform taxa examined as '?', with exception to *Grasseichthys*, which lacks "cephalic ribs". Patterson and Johnson (1995) stated that in the elopomorph *Notacanthus* the Baudelot's ligament is also missing. However, the *Notacanthus* specimens examined do have this ligament, which is peculiarly shaped, but is nevertheless similar to that found in other elopomorph taxa as *Anguilla* and *Conger*: it is markedly thin transversally and markedly broad anteroposteriorly, attaching to various anterior vertebrae. The descriptions of Millot and Anthony (1958: 76) seem to indicate that a Baudelot's ligament is present in *Latimeria*; this taxon is thus tentatively coded as CS-0. Although a Baudelot's ligament cannot be directly discerned in †*Tiktaalik roseae* and †*Acanthostega gunnari*, I tentatively code these taxa as CS-0 because my observations and comparisons indicate that at least part of the anocleithrum (sensu this work) may well be the result of an ossification of Baudelot's ligament (see below). Thus, the presence of an anocleithrum in these two fossil taxa seems to indicate that they may have an at least partly ossified ligament of Baudelot (see below). I did not find a distinct Baudelot's ligament in the specimens examined of *Timon* and *Ambystoma*.

128. *Absence of "ligament between posttemporal and posterior margin of neurocranium" (usually intercalar)*. Such a ligament is present in the great majority of the taxa examined, including taxa such as *Polypterus* (e.g., Fig. 10) [State 0]. Its absence is tentatively coded as CS-1 [State 1]. It corresponds to the "posttemporal-intercalar" ligament of, for example, Taverne (1974), but since in some cases (e.g., when the intercalar is missing: see above) it may attach to skull bones other than

the intercalar, I prefer to use the less restrictive name "ligament between posttemporal and posterior margin of neurocranium".

129. *Baudelot's ligament attaching proximally on anterior free vertebrae, and eventually also on neurocranium (inspired from, e.g., Fink and Fink, 1981, 1996; Jollie, 1984a,b; Patterson and Johnson, 1995).* The plesiomorphic condition for the osteichthyan groups included in the cladistic analysis seems to be that found in taxa such as *Polypterus, Acipenser, Psephurus, Amia, Lepidosiren* and *Neoceratodus*, as well as in various other taxa examined, in which the proximal portion of this ligament does not contact the anterior free vertebrae, attaching instead exclusively on the neurocranium (e.g., Fig. 94) [State 0]. In taxa of CS-1 the proximal portion of Baudelot's ligament attaches partly, or completely, on the anterior free vertebrae [State 1]. In some cypriniforms (e.g., *Danio*), apart from the Baudelot's ligament running from the supracleithrum/cleithrum to the anterior free vertebrae, there is also a ligament running from the supracleithrum/cleithrum to the posteromesial margin of the skull (e.g., Fig. 79). This has actually led Cubbage and Mabee (1996: 151), for example, to state that in *Danio* "the Baudelot's ligament attaches to the skull" and not to the "parapophyses of the anterior most centrum". Nevertheless, *Danio* is also coded as CS-1, as are *Opsariichthys* and *Cobitis* (but not *Catostomus*), since CS-0 concerns taxa in which there is no contact at all between any part of the Baudelot's ligament and the anterior free vertebrae, which is not the case in *Danio, Opsariichthys* and *Cobitis*.

130. *At least partial ossification of Baudelot's ligament (inspired from, e.g., Taverne, 1972, 1977b, 1978; Jollie, 1984a,b; Fink and Fink, 1981, 1996; Patterson and Johnson, 1995; Harold and Weitzman, 1996).* In most taxa examined presenting a Baudelot's ligament, this ligament is not ossified (e.g., Fig. 10) [State 0]. In taxa of CS-1 this ligament is at least partly ossified (e.g., Diogo, 2004a: fig. 3-42) [State 1]. *Neoceratodus* is coded here as CS-1, because my observations and my interpretation of the information provided in the literature seem to indicate that the anocleithra of the members of this taxon, as well as of at least some other sarcopterygian taxa, are probably the result of an ossification of Baudelot's ligaments (see below). In fact, in the *Neoceratodus* specimens examined each anocleithrum is embedded in this ligament, being situated between the dorsal surface of the cleithrum, where the ligament originates, and the posteroventral region of the cranium, where the ligament inserts (e.g., Fig. 94). It is actually important to stress that the posterior portion of the sarcopterygian anocleithrum usually lies in the position where the Baudelot's ligament of most osteichthyans usually originates, between the

supracleithrum and the cleithrum. The anterior margin of the anocleithrum usually abuts, or lies near to, the posterior region of the cranium; the anterior attachment of the Baudelot's ligament of many osteichthyans is also on the posterior region of the cranium. The analysis of the descriptions provided in the literature by, for example, Bemis (1986) concerning the Baudelot's ligaments (his "girdle suspensory ligaments") and/or the anocleithra of sarcopterygians such as *Protopterus* seem to support the view that the sarcopterygian anocleithra may well be the result of an ossification of these ligaments (see, e.g., Bemis, 1986: fig. 4A). Unlike in *Neoceratodus* and *Protopterus*, in adult *Lepidosiren* the anocleithra are missing, and I could not find any type of ossification of the Baudelot's ligaments; therefore, *Lepidosiren* is coded as CS-0. Since some *Hiodon* specimens seemingly exhibit a configuration such as CS-0, while others exhibit a configuration such as CS-1 (e.g., Taverne, 1977b), this genus is coded as '?'.

131. *Incorporation of functional unit in the pectoral girdle, the anocleithrum (inspired from, e.g., Jarvik, 1980; Rosen et al., 1981; Schultze, 1986; Forey, 1986; Clack and Coates, 1995; Cloutier and Ahlberg, 1996; Clack, 2002, 2006)*. Unlike taxa of CS-0 [State 0], taxa of CS-1 exhibit this feature (e.g., Fig. 94) [State 1]. Zhu and Schultze (2001), Cloutier and Arratia (2004) and others hypothesized that the anocleithrum of taxa of CS-1 is homologous with the postcleithrum found in some taxa of CS-0. If future studies eventually corroborate that the anocleithrum results from an at least partial ossification of the Baudelot's ligament (see above), such hypothesis of homology should be re-examined. Be that as it may, as stressed by, for example, Rosen et al. (1981) and Forey (1986), only in taxa of CS-1 are the anocleithra characteristically "incorporated as functional units in the pectoral girdle", this condition thus being seemingly derived within bony fishes.

132. *Non-ossification of "ligament between posttemporal and posterior margin of neurocranium"*. In most taxa examined presenting this ligament, the ligament is at least partly ossified (e.g., Fig. 10) [State 0]. In the few taxa of CS-1 the ligament is not ossified [State 1]. Taverne (1972, 1977b, 1978) stated that in the osteoglossomorphs *Hiodon*, *Xenomystus* and *Mormyrus* there is no "process of the posttemporal for the intercalar" (that is, there is no ossification of this ligament). However, in the specimens of *Hiodon*, *Xenomystus* and *Mormyrus* examined (e.g., Fig. 36), the ligament is actually ossified. The presence of this ossification can be overlooked in an analysis that does not include the observation of soft structures such as ligaments. This is because the specimens of these three genera display a peculiar configuration in which the anteroventrolateral surface of the posttemporal is bifurcated

anteriorly into a shorter, lateral arm that is essentially a tubular structure carrying a sensorial canal, and a longer, mesial arm that extends well anterior to the lateral one and that is attached by a thick ligament to the intercalar, or eventually to the autopterotic in *Mormyrus*, in which the interhyal is missing (e.g., Figs. 32, 33, 36). This latter arm thus seems to be homologous to the ossified "ligament between the posttemporal and the posterior margin of neurocranium" of the present work (e.g., Fig. 36). In the *Pantodon* specimens examined the "ligament between the posttemporal and the posterior margin of the neurocranium" is very thick but is not ossified; this genus is thus coded as CS-1.

133. *Posttemporal peculiarly bifurcated anteroventrolaterally into a shorter, lateral arm carrying a sensorial canal and a longer, mesial arm that corresponds to the ossified "ligament between posttemporal and posterior margin of neurocranium" of the present work.* As explained above, although many other taxa analyzed have an ossified "ligament between the posttemporal and the posterior margin of the neurocranium", such a peculiar configuration of the posttemporal is only found in the specimens examined of the osteoglossomorph genera *Hiodon*, *Xenomystus* and *Mormyrus*. The presence of such a peculiar configuration is thus coded here as CS-1 [State 1]; its absence is coded as CS-0 [State 0].
134. *Cleithrum not present as an independent element (inspired from, e.g., Tchernavin, 1947a; Lecuru, 1968a,b; Coates, 1996; Clack, 2002; Kardong, 2002; Vickaryous and Hall, 2006).* Unlike in taxa of CS-0 [State 0], in taxa of CS-1 the cleithrum is not present as an independent, ossified element (e.g., Fig. 108) [State 1]. †*Acanthostega gunnari* is included in CS-1 because in this fossil there is seemingly no complete suture separating the cleithrum and the scapulocoracoid (e.g., Fig. 97B).
135. *Presence of deep, long, curved fossa on lateral surface of cleithrum.* Unlike taxa of CS-0 [State 0], *Phractolaemus* exhibits this feature (e.g., Fig. 65) [State 1].
136. *Cleithrum markedly bifurcated dorsally into well-developed anterodorsal and posterodorsal arms for articulation with supracleithrum (or posttemporo-supracleithrum) (inspired from, e.g., Alexander, 1965; Taverne, 1972, 1977b, 1978; Gosline, 1977; Brosseau, 1978a,b; Diogo et al., 2001a; Diogo, 2004a).* Unlike taxa of CS-0 [State 0], taxa of CS-1 display this feature (e.g., Diogo, 2004a: fig. 3-71) [State 1].
137. *Presence of compound bone scapulocoracoid (or eventually cleithro-scapulocoracoid) (inspired from, e.g., Romer, 1944; Lecuru, 1968a,b; Coates, 1996; Diogo et al., 2001a; Clack, 2002; Kardong, 2002; Vickaryous and Hall, 2006).* Unlike taxa of CS-0 [State 0], taxa of CS-1 display this

feature (e.g., Fig. 971) [State 1]. Some taxa of CS-0 eventually exhibit a scapulocoracoid compound; they are, however, not listed as CS-1 since, unlike in taxa of CS-1, their scapulocoracoid is not ossified. There is much controversy on whether salamanders such as *Ambystoma ordinarium* have an ossified scapulocoracoid element or not (see above); this taxon is thus coded as '?'. †*Tiktaalik roseae* is also coded as '?', since in some cases Shubin et al. (2006: 764) refer to two "endochondral components (scapula and coracoid)" in this species, while in others they refer to a "scapulocoracoid" (e.g., in their fig. 5). As their figures do not allow us to discern this feature, I prefer to prudently code †*Tiktaalik roseae* as '?'. Some adult *Acipenser* and *Psephurus* have a configuration such as CS-1, while others have a configuration such as CS-0; these two taxa are thus coded as '?'.

138. *Mesial limb of coracoids (or scapulo-coracoids, or cleithro-scapulo-coracoids) broad and anteroposteriorly elongated (ordered multistate character) (inspired from, e.g., Arratia, 1997, 1999; Diogo, 2004a).* In taxa of CS-0 the mesial limb of the coracoid (or scapulo-coracoid, or cleithro-scapulo-coracoid), whether ossified or not, is a somewhat slender and/or anteroposteriorly short structure (e.g., Fig. 8B) [State 0]. In taxa of CS-1 it is a broad and anteroposteriorly elongated structure (e.g., Fig. 33) [State 1]. In taxa of CS-2 this median limb is still broader and meets its counterpart in a strong median interdigitation (e.g., Fig. 3-56) [State 2]. Since some *Galaxias* specimens exhibit a configuration such as CS-0, while others seemingly exhibit a configuration such as CS-1 (e.g., Swinnerton, 1903; McDowall, 1969), this genus is coded as '?'.
139. *Posteroventral process on ventral surface of coracoid (or scapulo-coracoids, or cleithro-scapulo-coracoids) (inspired from, e.g., Gosline, 1969; Markle and Merrett, 1980; Gayet, 1981; Begle, 1992; Johnson and Patterson, 1996; Diogo, 2004a).* Unlike taxa of CS-0 [State 0], taxa of CS-1 display this feature (e.g., Fig. 30) [State 1] (which should not be confused with the presence of the posterior process that is found on the posteroventral surface of the coracoid, or scapulo-coracoid, of numerous osteichthyan taxa: e.g., Fig. 44). Although the configuration of these posteroventral processes in taxa of CS-1 is somewhat variable, I prefer not to exclude, *a priori*, the hypothesis that these processes could eventually constitute a primary homology; this primary homology hypothesis will be tested by the cladistic analysis also including all the other characters listed in this Section.
140. *Presence of peculiar "coracoid bridge"*. Unlike taxa of CS-0 [State 0], taxa of CS-1 display this feature (e.g., Diogo, 2004a: fig. 3-56) [State 1].
141. *Absence of distinct mesocoracoid arch (ossified or not) (inspired from, e.g., Marshall, 1962; Begle, 1992; Forey et al., 1996; Johnson and Patterson, 1996; Johanson et al., 2004).* According to, for example, Johanson et al.

(2004), the present of a distinct mesocoracoid arch may well be plesiomorphic for the osteichthyan taxa included in the present cladistic analysis; such an arch is found in numerous actinopterygians and seemingly in at least some sarcopterygians. I am thus tentatively coding here the presence of a distinct mesocoracoid arch (e.g., Fig. 30B) as CS- 0 [State 0]. Those cases in which this arch is either undifferentiated or completely fused anteriorly with the posterior margin of the coracoid are thus tentatively coded as CS-1 (e.g., Fig. 95) [State 1]. Johanson et al. (2004) stated that the "supraglenoid buttress" of †*Acanthostega gunnari* might eventually correspond to the mesocoracoid arch of taxa of CS-0. As I could not analyze the pectoral girdle of specimens of †*Acanthostega gunnari* in order to check whether they exhibit a mesocoracoid arch such as that I found in taxa of CS-0, I prefer to prudently code this fossil species as '?'. I also code †*Tiktaalik roseae* as '?', since the descriptions and figures provided by Shubin et al. (2006) do not allow us to discern whether a mesocoracoid arch such as that of taxa of CS-0 may eventually be present in this taxon. I found no distinct mesocoracoid arch in the specimens examined of *Timon* and *Ambystoma*, and the descriptions and illustrations of Millot and Anthony (1958) seem to indicate that such an arch is also missing in *Latimeria*.

142. *Mesocoracoid arch ossified*. Among those taxa examined exhibiting a distinct mesocoracoid arch (see above), the plesiomorphic condition seems to be that in which this structure is not ossified (e.g., fig. 17) [State 0]. In taxa of CS-1 this arch is ossified (e.g., Fig. 30) [State 1]. In the adult specimens examined of the gymnotiform genera *Gymnotus* and *Brachyhypopomus* the mesocoracoid arch is at least partly ossified; in the adult specimens of these two genera described by, for example, De la Hoz and Chardon (1984), this structure is seemingly completely unossified: these genera are thus coded as '?'. Some adult members of the genera *Acipenser* and *Psephurus* have a configuration such as CS-1, while others have a configuration such as CS-0 (e.g., Hilton and Bemis, 1999; Hilton, pers. comm.; this work); these acipenseriform genera are therefore coded as '?'.

143. *Mesocoracoid arch firmly and rigidly attached, trough suture or complete fusion, to coracoid and/or scapula (or scapulo-coracoid)*. Among those taxa examined with an distinct, ossified mesocoracoid arch, the plesiomorphic condition is seemingly that in which the mesocoracoid arch is not firmly and rigidly attached, through suture or complete fusion, to the coracoid and/or scapula (it is worth noting that this condition is found in some specimens of acipenseriform genera such as *Acipenser* in which the mesocoracoid arch ossifies: Hilton, pers. comm.). This is the case, for example, in elopomorph taxa such as

Elops, *Megalops* and *Albula* and in osteoglossomorph taxa such as *Hiodon* and *Mormyrus*, in which the mesocoracoid arch articulates ventrally with the coracoid and/or scapula and, thus, in which this arch has some mobility in relation to these latter bones (e.g., Figs. 30, 33) [State 0]. Those taxa of CS-1 have an ossified mesocoracoid arch that is firmly and rigidly associated, often through suture or complete fusion, with the coracoid and/or scapula (or scapulo-coracoid) (e.g., Fig. 51) [State 1].

144. *Mesocoracoid arch (either ossified or not) markedly elongated dorsoventrally (inspired from, e.g., Greenwood and Thompson, 1960; Le Danois, 1967; Taverne, 1978).* Unlike taxa of CS-0 [State 0], specimens of *Pantodon* exhibit this feature (e.g., Taverne, 1978: figs. 30, 44) [State 1].
145. *Mesocoracoid arch (either ossified or not) markedly enlarged.* Unlike taxa of CS-0 [State 0], taxa of CS-1 exhibit this feature (e.g., Findeis, 1997: fig. 9A) [State 1]. Some *Mormyrus* specimens display a configuration such as CS-1 (e.g., those examined in the present work), while others display a configuration such as CS-0 (e.g., Taverne, 1972: fig. 1); this genus is thus coded as '?'.
146. *Single pectoral fin endoskeletal element ("humerus") articulating with pectoral girdle (inspired from, e.g., Jarvik, 1944, 1965; Daget, 1950; Bjerring, 1973; Rosen et al., 1981; Shubin and Alberch, 1986; Coates, 1994, 1996; Nelson and Tabin, 1995; Shubin et al., 1997, 2004, 2006; Coates and Cohn, 1999; Joss and Longhurst, 2001; Hinchliffe et al., 2001; Zhu and Schultze, 2001; Coates et al., 2002; Cohn et al., 2002; Ruta et al., 2003; Davis et al., 2004; Mabee and Noordsy, 2004; Garvey et al., 2005; Ahlberg and Clack, 2006; Daeschler et al., 2006).* Unlike taxa of CS-0 (e.g., Fig. 9) [State 0], taxa of CS-1 exhibit this feature (e.g., Figs. 95, 114) [State 1].
147. *Presence of sternum, either ossified or cartilaginous (inspired from, e.g., Jarvik, 1944, 1965; Romer, 1944; Lecuru, 1968a; Kardong, 2002; Walthall and Ashley-Ross, 2006).* Unlike in taxa of CS-0 [State 0], a cartilaginous sternum is found in the specimens examined of *Ambystoma ordinarium* and an ossified sternum is found in the *Timon* specimens analyzed (e.g., Lecuru, 1968a: fig. 10) [State 1]. Since it is difficult to discern whether or not a cartilaginous sternum is absent in †*Tiktaalik roseae* and †*Acanthostega gunnari*, these fossils are coded as '?' (as were the other fossils included in the cladistic analysis).
148. *Presence of ossified, broad extracleithrum (inspired from, e.g., Millot and Anthony, 1958; Cloutier, 1991; Forey, 1991, 1998).* Unlike in taxa of CS-0 [State 0], an ossified, broad extracleithrum is found in *Latimeria* (e.g., Fig. 87) [State 1].
149. *Marked robusticity and mobility of distal elements of pectoral fins/forelimbs (inspired from, e.g., Daeschler et al., 2006; Shubin et al., 2006).* As noted by

these authors, the robusticity and mobility of the distal elements of the pectoral fins/forelimbs of †*Tiktaalik roseae* and †*Acanthostega gunnari*, and of extant tetrapods such as *Timon* and *Ambystoma* (e.g., Shubin et al., 2006: figs. 2, 7) [State 1], are greater than those of the distal elements of the pectoral fins of the other taxa included in the present analysis [State 0].

150. *Glenoid orientation with lateral component (inspired from, e.g., Daeschler et al., 2006; Shubin et al., 2006).* As also stated by these authors, the glenoid articulatory surfaces between the pectoral girdle and the pectoral fins/forelimbs of taxa of CS-1 (e.g., Shubin et al., 2006: figs. 1, 7) [State 1] are more lateral than those of the other taxa included in the present cladistic analysis (e.g., Fig. 95) [State 0].
151. *Presence of digits on forelimbs (inspired from, e.g., Rosen et al., 1981; Mabee, 2000; Joss and Longhurst, 2001; Hinchliffe et al., 2001; Zhu and Schultze, 2001; Clack, 2002).* Unlike taxa of CS-0 [State 0], taxa of CS-1 have digits on the forelimbs (e.g., Figs. 97, 103, 112) [State 1].
152. *Absence of ossified clavicle (inspired from, e.g., Romer, 1924; Jarvik, 1944; Lecuru, 1968a,b; Liem and Woods, 1973; Carroll, 1977; Patterson, 1977b; Rieppel, 1992; Bels et al., 1993; Zhu and Schultze, 2001; Schultze and Cumbaa, 2001; Cloutier and Arratia, 2004; Mabee and Noordsy, 2004).* Unlike taxa of CS-0 (e.g., Fig. 97) [State 0], taxa of CS-1 do not exhibit an ossified clavicle [State 1]. Jarvik (1944), Liem and Woods (1973), Gardiner et al. (1996), and Cloutier and Arratia (2004) stated that *Amia* and/or *Lepisosteus* may exhibit structures that are homologous to clavicles and/or interclavicles or that are at least "vestiges" of these elements (e.g., Fig. 19B: '?'; Jarvik, 1944: fig. 1E; Patterson, 1977b: fig. 6). This view was contradicted by, for example, Arratia and Schultze (1990). I consider that one should not completely exclude the hypothesis that some of the structures found in the specimens examined of *Amia* and *Lepisosteus* might eventually be homologous to clavicles and/or interclavicles. However, until more data is available on this subject, I prefer to prudently code these two genera as '?' for both this character and the character below.
153. *Absence of characteristic, large ossified interclavicle (inspired from, e.g., Romer, 1924; Jarvik, 1944; Millot and Anthony, 1958; Liem and Woods, 1973; Zhu and Schultze, 2001; Schultze and Cumbaa, 2001; Cloutier and Arratia, 2004; Mabee and Noordsy, 2004).* Unlike taxa of CS-0 (e.g., Fig. 97) [State 0], taxa of CS-1 do not exhibit a characteristic, large ossified interclavicle [State 1] (see above). In Millot and Anthony's (1958) descriptions of *Latimeria* these authors refer to an interclavicle that is small and not ossified (see Fig. 87); this taxon is therefore coded as CS-1.
154. *Presence of "pectoral splints" (inspired from, e.g., Taverne, 1974; Gosline, 1980; Forey et al., 1996).* As noted by Taverne (1974), Gosline (1980), Forey et

al. (1996) and others, unlike in taxa of CS-0 [State 0], "pectoral splints" are found in *Lepisosteus* and *Amia*, as well as in extant teleosts such as *Elops*, *Megalops*, *Notacanthus* and *Albula* (e.g., Figs. 22, 30) [State 1].

155. *First pectoral ray, if present, articulating directly with scapula and/or eventually with coracoid (inspired from, e.g., Jessen, 1972; Patterson, 1977a; Gosline, 1980; Arratia, 1999).* Unlike taxa of CS-0 [State 0], taxa of CS-1 display this feature (e.g., Fig. 30) [State 1]. In taxa such as *Acipenser* the propterygium is firmly attached to, but not completely incorporated into, the first pectoral ray; therefore, this latter ray does not articulate directly with the scapula and/or coracoid (e.g., Findeis, 1997: fig. 2).
156. *Peculiar, thin, somewhat barbel-like pectoral fin (inspired from, e.g., Bischoff, 1840; Owen, 1841; Greenwood, 1986).* Unlike taxa of CS-0 [State 0], specimens of *Lepidosiren* exhibit this feature [State 1].

Lateral Cephalic Musculature

157. *Most lateral bundles of adductor mandibulae not exclusively inserted on mandible and/or on primordial ligament, near the insertion of this ligament on the mandible.* The plesiomorphic condition for the osteichthyan taxa included in the cladistic analysis is seemingly that in which the most external bundles of the adductor mandibulae are exclusively attached to the mandible (and/or eventually to the primordial ligament, near the insertion of this ligament on the mandible) and, thus, do not insert directly on other bony structures (e.g., Figs. 7, 46, 92) [State 0]. In taxa of CS-1 part of the external bundles of the adductor mandibulae attaches also, or even exclusively, on other bony structures such as the maxilla, the premaxilla and/or other bones (e.g., lacrimal) (e.g., Figs. 53, 53, 74, 83) [State 1]. It should be noted that the muscle retractor tentaculi apomorphically present in some catfish taxa is the result of the differentiation of a mesial, not a lateral, bundle of the adductor mandibulae; catfishes with a retractor tentaculi are thus not coded as CS-1 (e.g., McMurrich, 1884; Lightoller, 1939; Eaton, 1948; Alexander, 1965; Howes, 1983a; Adriaens and Verraes, 1996; Diogo and Chardon, 2000; Diogo and Vandewalle, 2003; Diogo, 2004a).
158. *When most lateral bundles of adductor mandibulae attach also, or exclusively, on bony structures other than the mandible, the non-mandibular insertions include bones such as those of the infraorbital series or the dermopalatine/autopalatine.* In those taxa examined in which the most lateral bundles of the adductor mandibulae attach also, or exclusively, on bony structures other than the mandible, their non-mandibular insertions are usually on the maxilla and/or premaxilla (e.g., Fig. 74) [State 0]. In taxa of CS-1, however, these non-mandibular insertions

include bones such as those of the infraorbital series (e.g., lacrimal) or the dermopalatine/autopalatine (e.g., Fig. 83) [State 1].

159. *Presence of adductor mandibulae A1-OST (inspired from, e.g., Fink and Fink, 1981, 1996; Gosline, 1989; Diogo and Chardon, 2000)*. Unlike taxa of CS-0 [State 0], taxa of CS-1 present a peculiar, large ventrolateral bundle of the adductor mandibulae usually attaching on the posterior and/or posterolateral surface of the mandible, which was named adductor mandibulae A1-OST by Diogo and Chardon, 2000 (e.g., Fig. 80) [State 1].

160. *Presence of adductor mandibulae A1 (inspired from, e.g., Winterbottom, 1974; Fink and Weitzman, 1982; Gosline, 1989; Diogo and Chardon, 2000; Sato and Nakabo, 2002; Wu and Shen, 2004)*. Unlike taxa of CS-0 [State 0], taxa of CS-1 exhibit a well-developed dorsolateral bundle of the adductor mandibulae (adductor mandibulae A1 of Diogo and Chardon, 2000; not to be confused with the bundle named adductor mandibulae A1-OST by these authors: see character above) usually attaching on the maxilla (e.g., Fig. 43) [State 1]. Sato and Nakabo (2002) and Wu and Shen (2004: 726) stated that in some specimens of the aulopiform genus *Chlorophthalmus* the A1 is present but may eventually be fused to the A2. In the specimens of this genus examined in the present work the A1 and A2 are well separated. Nevertheless, even if the A1 and A2 are fused in certain members of this genus, they are still coded as CS-1, as they do have an A1.

161. *Presence of a well-developed, distinct adductor mandibulae A3-MAX (inspired from, e.g., Bishai, 1967; Rosen, 1973; Greenwood, 1977; Winterbottom, 1974; Kershaw, 1976; Wu and Shen, 2004)*. The name "adductor mandibulae A3-MAX" has not been used in the literature before. However, to create a new name for this adductor mandibulae section seems to me the best option, since there has been much confusion in the literature concerning its nomenclature and homologies of this section. The adductor mandibulae A3-MAX, present in taxa of CS-1 [State 1] and absent in taxa of CS-0 [State 0], corresponds to the "A1-b" of, for example, Greenwood (1977), Winterbottom (1974) and Wu and Shen (2004) (see, e.g., Greenwood, 1977: figs. 3, 10, 11, 12; Wu and Shen, 2004: fig. 2). However, as recognized by these authors, various other names have been used to designate this section, such as "pterygo-maxillaire", "pterygo-maxillaris", or "levator maxillae superior". In my opinion the reason many authors do not use the name "A1-b" or a similar designation for this section is that the section is almost always mesial, and not lateral, to the A2. In reality, the position and origin of this section (mesially to the main body of the A2) is somewhat similar to that of the A3'/A3" of other taxa examined in the present work, with the difference that in

taxa of CS-1 the section attaches on the maxilla and not on the mandible. That is why, in order to differentiate this section from the A3'/A3'' of other taxa, I choose the name A3-MAX. I consider this name appropriate because it pertains to an adductor mandibulae section (and not really a new, completely distinct muscle—that is why I do not agree with names such as "pterygo-maxillaire", "pterygo-maxillaris", or "levator maxillae superior"; the A3-MAX should also not be confused with the muscle retractor tentaculi of certain catfishes: e.g., Diogo, 2004a) that is mesial to the main body of the A2 (thus, an "A3") and attaches on the maxilla (thus, an "A3-MAX").

162. *Presence of sections palatomandibularis minor and palatomandibularis major of the adductor mandibulae (inspired from, e.g., Edgeworth, 1935; Lauder, 1980a).* Unlike other taxa included in the cladistic analysis [State 0], these two sections are found in *Lepisosteus* (e.g., Figs. 15, 16) [State 1]. In Lauder's (1980a) table II it is suggested that the sections palatomandibularis minor and palatomandibularis major of *Lepisosteus* are likely homologous to the levator maxillae superioris 3 and 4 of *Amia* (see below), since all these structures represent an "anterior division" of the adductor mandibulae. However, the overall configuration, position and attachments of the palatomandibularis minor and major of *Lepisosteus* are markedly different from those of the levator maxillae superioris 3 and 4 of *Amia*. Just to give an example, the palatomandibularis minor and major of *Lepisosteus* originate dorsally on the ectopterygoid/entopterygoid and insert ventrally on the mandible, while, for example, the Section 3 of the levator maxillae superioris of *Amia* originates dorsally on the neurocranium and orbital bones and inserts ventrally mainly on the autopalatine (e.g., Fig. 19A). Thus, I prefer not to code the presence of these bundles as a primary homology between *Amia* and *Lepisosteus*. However, if the cladistic analysis comprising also all the other numerous characters listed in this Section does support the position that *Amia* is more closely related to *Lepisosteus* than to the other taxa included in the analysis, the hypothesis that these bundles may be homologous should, of course, be reconsidered (see Chapter 4, Section 4.2).

163. *Presence of sections levator maxillae superioris 3 and 4 of the adductor mandibulae (inspired from, e.g., Allis, 1897; Edgeworth, 1935; Kesteven, 1942-1945; Lauder, 1980a).* As explained above, unlike taxa of CS-0 [State 0], *Amia* exhibits this feature (e.g., Fig. 19A) [State 1].

164. *Presence of retractor anguli oris of the adductor mandibulae (inspired from, e.g., Luther, 1913; Edgeworth, 1935; Bemis, 1986; Bemis and Lauder, 1986; Moro and Abdala, 2000; Montero et al., 2002; Abdala and Moro, 2003).* Unlike in taxa of CS-0 [State 0], a retractor anguli oris is found in *Lepidosiren* (e.g., Fig. 91) [State 1]. In the *Timon* specimens examined, I

found a muscle that may correspond to the retractor anguli oris of dipnoans such as *Lepidosiren* (and *Protopterus,* not included in the present cladistic analysis: see, e.g., Bemis, 1986; Bemis and Lauder, 1986), and that is often named, among researchers working mainly with lepidosaurs, "levator anguli oris" (e.g., Fig. 109). I thus tentatively code *Timon lepidus* as CS-1; this primary homology hypothesis will be tested in face of the results obtained in this work and will be discussed in Chapter 4, Section 4.2.

165. *Presence of adductor mandibulae A2-PVM.* In the specimens examined of *Timon, Ambystoma, Neoceratodus* and *Lepidosiren* I found a peculiar section of the adductor mandibulae that has some fibers associated with those of the adductor mandibulae A2 but is distinct from it. As in those taxa this small bundle is somewhat posterior, ventral, and medial to the main body of the A2, I have decided to call it "adductor mandibulae A2-PVM" (PVM thus meaning posteroventromesial) (e.g., Figs. 92, 100). This bundle is seemingly present in many amphibians other than *Ambystoma,* being often named by researchers working with amphibians as "adductor mandibulae posterior" (e.g., Carroll and Holmes, 1980; Iordansky, 1992). It is also seemingly present in many amniotes other than *Timon,* being also named "adductor mandibulae posterior" by researchers working with, for example, lepidosaurs (e.g., Moro and Abdala, 2000; Montero et al., 2002; Abdala and Moro, 2003). Bemis and Lauder (1986) and Miyake et al. (1992) have designated the assemblage formed by the A2-PVM and the A2 of dipnoans (sensu this work) as an "adductor mandibulae posterior". Unlike in *Timon, Ambystoma, Neoceratodus* and *Lepidosiren,* coded as CS-1 [State 1], I did not find such an adductor mandibulae A2-PVM in the other taxa included in the cladistic analysis. The textual descriptions and illustrations of the adductor mandibulae of *Latimeria* provided by, for example, Millot and Anthony (1958) and Adamicka and Ahnelt (1992) indicate that such an A2-PVM is also seemingly absent in this taxon [State 0]. It is important to stress that the A2-PVM should not be confused with the retractor anguli oris (see above): the retractor anguli oris is usually situated posteroventrolaterally to the A2, being, thus, mainly *superficial* to the A2, while the A2-PVM situates posteroventromesially to the A2, being thus mainly *mesial* to this latter bundle. In fact, the two bundles are often found in the same taxon, as is the case in dipnoans such as *Lepidosiren* (e.g., Figs. 91, 92).

166. *Presence of depressor mandibulae (inspired from, e.g., Luther, 1913; Edgeworth, 1935; Frazzetta, 1962; Gasc, 1968; Gans et al., 1985; Bemis, 1986; Bemis and Lauder, 1986; Bauer, 1992; Iordansky, 1992; Miyake et al., 1992; Moro and Abdala, 2000; Haas, 2001; Montero et al., 2002; Abdala and*

Moro, 2003; Ericsson and Olsson, 2004; Ericsson et al., 2004). Unlike taxa of CS-0 [State 0], taxa of CS-1 exhibit this feature (e.g., Figs. 91, 99, 109) [State 1]. Although the depressor mandibulae inserts on the mandible and, as its name indicates, its contraction is seemingly often associated with mandibular depression, it is not derived from the mandibular muscle plate, but from the hyoid muscle plate (e.g., Edgeworth, 1935; Bemis, 1986; Bemis and Lauder, 1986; Miyake et al., 1992; see Chapter 4, Section 4.2).

167. *Presence of a levator hyoideus in at least some developmental stages (inspired from, e.g., Luther, 1913; Edgeworth, 1935; Bemis, 1986; Bemis and Lauder, 1986; Miyake et al., 1992)*. According to Edgeworth (1935), Forey (1986), Bauer (1997), and others, unlike in taxa of CS-0 [State 0] a levator hyoideus is found in at least some developmental stages of taxa of CS-1 (e.g., Fig. 93; Edgeworth, 1935: figs. 28B, 31, 32; Bartsch, 1994: fig. 2B) [State 1]. In contrast with most other characters used in the cladistic analysis, this character does not refer exclusively to the condition found in adults; it also refers to the condition found in other developmental stages. Thus, for instance, *Neoceratodus* is coded as CS-1, since in young developmental stages (e.g., larvae of 28.5 mm TL: e.g.. Bartsch, 1994) there is a recognizable levator hyoideus; in older stages (e.g., adults) the muscle becomes completely mixed with the complex formed by the interhyoideus plus the hyohyoideus and the constrictor operculi (see below) (e.g., Bartsch, 1994: fig. 14). Juveniles and adults of *Lepidosiren* (and also of *Protopterus*, not included in the present cladistic analysis) continue to exhibit a recognizable levator hyoideus (e.g., Fig. 93). Edgeworth (1935) and Bauer (1997) suggested that part of the depressor mandibulae found in numerous tetrapods, such as *Ambystoma* and *Timon*, corresponds to the levator hyoideus of dipnoans (e.g., Edgeworth, 1935: figs. 313, 327). I tentatively code *Ambystoma* and *Timon*, together with *Neoceratodus* and *Lepidosiren*, as CS-1. It should be noted that in *Latimeria* there is no depressor mandibulae, nor any other hyoid muscle that seemingly corresponds to, or includes, the levator hyoideus of taxa of CS-1 (e.g., Millot and Anthony, 1958; Miyake et al., 1992). See Chapter 4, Section 4.2.

168. *Presence of abductor mandibulae (inspired from, e.g., Tchernavin, 1947a,b, 1953)*. Unlike in taxa of CS-0 [State 0], part of the adductor mandibulae of *Eurypharynx* has differentiated into an abductor mandibulae (e.g., Tchernavin, 1947a: figs. 4, 5) [State 1]. This latter structure originates on the neurocranium and inserts on the posterior end of the mandible, behind the quadrato-mandibular articulation, its contraction thus seemingly resulting in the opening of the mouth.

169. *Presence of cervicomandibularis (inspired from, e.g., Edgeworth, 1935; Frazzetta, 1962; Levet, 1987; Moro and Abdala 2000; Montero et al., 2002;*

Abdala and Moro, 2003; Deufel and Cundall, 2003). Unlike taxa of CS-0 [State 0], the specimens of *Timon* exhibit this feature (e.g., Fig. 109) [State 1]. As stated by Edgeworth (1935), Levet (1987), and others, although the cervicomandibularis inserts on, or near, the mandible, this muscle is seemingly derived from the hyoid muscle plate (see Chapter 4, Section 4.2).

170. *Presence of peculiar bundle of adductor mandibulae extending far anteriorly in order to attach to anterodorsal surface of mandible (inspired from, e.g., Bishai, 1967)*. Unlike taxa of CS-0 [State 0], the specimens of *Mormyrus* exhibit this feature (e.g., Bishai, 1967: figs. 1, 2, 3) [State 1].
171. *Quite peculiar configuration of adductor mandibulae A1-OST (inspired from, e.g., Howes, 1985a)*. Unlike in other taxa examined having an adductor mandibulae A1-OST [State 0], in *Phractolaemus* this section exhibits a quite peculiar configuration in which its anterior portion is almost perpendicular to its posterior portion (e.g., Fig. 63) [State 1].
172. *Presence of several small, peculiar tendons branching off from adductor mandibulae A2 (inspired from Howes, 1985a)*. Unlike taxa of CS-0 [State 0], *Phractolaemus* exhibits this feature (e.g., Fig. 63) [State 1].
173. *Origin of adductor mandibulae A1-OST on neurocranium (inspired from, e.g., Diogo, 2004a)*. The plesiomorphic condition for those taxa having an adductor mandibulae A1-OST is seemingly that in which this section originates on the suspensorium [State 0]; among the taxa included in the cladistic analysis this section originates on the neurocranium only in the catfish *Cetopsis* (e.g., Diogo, 2004a: fig. 3-43) [State 1].
174. *Adductor mandibulae A2 essentially lateral to A1-OST (inspired from, e.g., Diogo, 2004a)*. Unlike taxa of CS-0 [State 0], taxa of CS-1 exhibit this feature (e.g., Diogo, 2004a) [State 1].
175. *Presence of peculiar, distinct bundle A1-OST-M running from anteroventral surface of quadrate to maxilla (inspired from, e.g., Howes, 1985a)*. Unlike taxa of CS-0 [State 0], taxa of CS-1 exhibit this feature (e.g., Figs. 63, 68) [State 1].
176. *Attachment of mainly undivided A2 on mesial surface of mandible accomplished by means of two well-distinguished, thick tendons*. Unlike taxa of CS-0 [State 0], the specimens examined of taxa of CS-1 exhibit a peculiar configuration of the A2: although this section is constituted by a single, mainly undivided mass of fibers, its attachment on the mesial surface of the mandible is accomplished by means of two (and not one) well-distinguished, thick tendons (e.g., Fig. 43) [State 1]. The most lateral of these tendons usually attaches on the coronomeckelian; the most mesial one usually attaches on the back of the Aw (when the Aw is absent, it often attaches to the mesial mandibular region in which

this bundle is usually lodged) (e.g., Fig. 43). The members of taxa that exhibit a distinct, separated section A3' of the adductor mandibulae were coded as "Inapplicable", since this latter section may well be the result of a division of the main body of the adductor mandibulae plus the incorporation of one of the two ventral tendons mentioned above (see below; note that, exceptionally, *Megalops* and *Albula* have both a "mainly undivided A2 attaching on the mesial surface of the mandible by means of two well-distinguished tendons" and a distinct, separated section A3', and that, despite having an A3', these genera are thus exceptionally coded as CS-1).

177. *Direct insertion of adductor mandibulae A2 far anteriorly on the anteromesial surface of dentary*. Unlike taxa of CS-0 [State 0], taxa of CS-1 exhibit this feature (e.g., Fig. 62) [State 1].
178. *Presence of recognizable dilatator operculi (inspired from, e.g., Danforth, 1913; Tchernavin, 1947a,b, 1953; Lauder, 1980a; Gardiner, 1984; Miyake et al., 1992; Mallat, 1997; Carroll and Wainwright, 2003)*. Unlike taxa of CS-0 [State 0], taxa of CS-1 exhibit this feature (e.g., Figs. 6, 15, 19) [State 1]. According to Danforth (1913), Edgeworth (1935), Carroll and Wainwright (2003), and others, the protractor hyomandibulae of extant acipenseriforms (see below) corresponds to the levator arcus palatini and eventually also to the dilatator operculi of other actinopterygians (e.g., Fig. 12). Be that as it may, in extant acipenseriforms there is no recognizable, separate dilatator operculi; *Acipenser* and *Psephurus* are thus coded as CS-0. Some authors (e.g., Edgeworth, 1935) have stated that in *Polypterus* there is a muscle "spiracularis" lying near the dilatator operculi. No such muscle was, however, described in *Polypterus* by Lauder (1980a), nor was it found in the specimens of this genus examined in the present work.
179. *Absence of recognizable adductor operculi (inspired from, e.g., Tchernavin, 1947a,b, 1953)*. Unlike taxa of CS-0 [State 0], in taxa of CS-1 there is no recognizable adductor operculi (e.g., Fig. 109) [State 1]. The "opercularis" of extant acipenseriforms (e.g., Danforth, 1913; Carroll and Wainwright, 2003) seems to be homologous to the adductor operculi of other actinopterygians (Miyake et al., 1992; see Fig. 12); *Acipenser* and *Psephurus* are thus coded as CS-0. As stressed by, for example, Edgeworth (1935) and Miyake et al. (1992) the muscles derived from the hyoid plate in extant dipnoans are hardly comparable to those muscles of other osteichthyans. The dipnoan interhyoideus seems to correspond to the interhyoideus (sensu this work) of other osteichthyans. The dipnoan "constrictor hyoideus ventralis" (sensu, e.g., Miyake et al., 1992), is deeply mixed with the interhyoideus and may be derived from the portion of the hyoid muscle plate giving rise to the hyohyoideus of actinopterygians (see

Chapter 4, Section 4.2). However, the other three hyoid dipnoan muscles are rather peculiar. The levator hyoideus (e.g., Fig. 93), for example, runs peculiarly from the neurocranium to the ceratohyal (see above). The portion of the hyoid muscle plate from which this muscle originates may eventually correspond to that from which the adductor arcus palatini of other bony fishes originates. In fact, in extant dipnoans the hyomandibula is considerably reduced in size or even completely missing, and the palatoquadrate is fused to the neurocranium, being thus much less mobile than that of most other bony fishes. Therefore, the portion of the hyoid muscle plate from which the adductor arcus palatini originates may have lost its usual attachments on the hyomandibula and/or palatoquadrate in dipnoans and become attached on the ceratohyal; the dorsal surface of the ceratohyal lies in fact more dorsally than in most other bony fishes, occupying a position similar to that occupied by the hyomandibula in most other bony fishes (see Fig. 93). My observations of *Lepidosiren* revealed that in this taxon the levator hyoideus actually attaches on the dorsal surface of the ceratohyal, but also on part of its dorsomesial margin (e.g., Fig. 93). Thus, as the adductor arcus palatini of other bony fishes usually attaches on the dorsomesial margin of the hyomandibula/palatoquadrate in order to adduct these structures, it seems that the levator hyoideus might not only elevate, but also adduct, the dorsal surface of the ceratohyal. Interestingly, Edgeworth (1935: 102) stated that the levator hyoideus of dipnoans originates from the same region of the constrictor hyoideus from which the retractor hyomandibulae of extant acipenseriforms derives, which seems to correspond precisely to the adductor arcus palatini of other bony fishes (see below). Regarding the muscle "constrictor operculi" ("constrictor hyoideus dorsalis" sensu, e.g., Miyake et al., 1992) of extant dipnoans, it could perhaps originate from the same portion of the constrictor hyoideus from which the adductor operculi of other osteichthyans originates (see Fig. 91). In fact, according to, for example, Campbell and Barwick (1986), some basal fossil dipnoans exhibit well-defined scars on the mesial margins of the opercular bones for the attachment of muscles that seem to have been somewhat like the adductores operculae of other bony fishes. Be that as it may, in adult extant dipnoans the "constrictor operculi" is completely mixed with other hyoid muscles such as the "interhyoideus" and the "constrictor hyoideus ventralis" (sensu Miyake et al., 1992) (e.g., Fig. 91). Therefore, even if this "constrictor operculi" is developed from the same portion of the constrictor hyoideus from which the adductor operculi of other bony fishes develops, in adult extant dipnoans there is no recognizable, separate adductor operculi; *Neoceratodus* and

Lepidosiren are thus coded as CS-1. The other hyoid muscle found in extant dipnoans as *Lepidosiren* and *Protopterus*, the depressor mandibulae, has been discussed above.

180. *Adductor operculi inserting exclusively on subopercle (inspired from, e.g., Carroll and Wainwright, 2003).* Unlike taxa of CS-0 [State 0], taxa of CS-1 exhibit this feature (e.g., Fig. 12) [State 1].
181. *Separation between adductor operculi and adductor arcus palatini.* Among those taxa with an adductor operculi (see above), the plesiomorphic condition seems to be that in which the anterior fibers of this muscle are significantly mixed with the posterior fibers of the adductor arcus palatini, that is, in which the separation of these two muscles is not complete (e.g., Edgeworth, 1935) (see, e.g., Fig. 12; Lauder, 1980ab: fig. 8A,B) [State 0]. In taxa of CS-1 the separation between these muscles is far more marked than it is in taxa of CS-0 (e.g., Fig. 63) [State 1]. It should be noted that the adductor arcus palatini of this work corresponds to the adductor arcus palatini of Winterbottom (1974) and Diogo (2004a) and, thus, should not be confused with the "adductor hyomandibulae" muscles found in certain osteichthyans (see below).
182. *Dilatator operculi markedly lateral to A2.* In taxa of CS-0 the dilatator operculi is mainly mesial and/or dorsal to the adductor mandibulae A2 (e.g., Fig. 38) [State 0], but in taxa of CS-1 it is clearly lateral to the latter muscle (e.g., Fig. 80; Greenwood, 1977: fig. 10) [State 1].
183. *Adductor mandibulae A2 exhibiting thick tendon that is perpendicular to its main body and that connects this section to the anteroventral surface of the quadrate (inspired from, e.g., Howes, 1985a).* Unlike taxa of CS-0 [State 0], specimens of *Cromeria* exhibit this feature (e.g., Fig. 68B) [State 1].
184. *Levator arcus palatini lateral to all bundles of adductor mandibulae (inspired from, e.g., De la Hoz, 1974; Chardon and De la Hoz,1973; De la Hoz and Chardon, 1984; Aguilera, 1988).* Unlike in taxa of CS-0, in which the levator arcus palatini is essentially mesial (and/or eventually dorsal) to the most external bundles of the adductor mandibulae [State 0], in the specimens examined of taxa of CS-1 it lies laterally to all adductor mandibulae sections (e.g., Fig. 83) [State 1].
185. *Absence of distinct section A3' of adductor mandibulae.* As explained by Lauder (1980a: table II), the plesiomorphic condition for osteichthyans is seemingly that in which there are two "mesial adductor mandibulae divisions", as is the case in *Polypterus*, *Amia* and *Lepisosteus* and many other actinopterygian and sarcopterygian taxa included in this cladistic analysis (e.g., Figs. 6, 7, 15, 16, 19, 21, 100). The two mesial adductor mandibulae divisions found in, for example, *Polypterus*, *Amia* and *Lepisosteus* correspond to the A3' and

A3″ of Diogo and Chardon (2000) and of the present work. Therefore, the presence of an A3' and an A3″ is coded here as CS-0 [State 0] (see also character below). In taxa of CS-1 there is no recognizable, separated section A3' (e.g., Fig. 43). Sanford (2000: fig. 94) reported a distinct A3' in specimens of the genus *Galaxias*. In the *Galaxias* specimens examined in the present work part of the dorsal fibers of the A2 lie mesial to the levator arcus palatini, but ventrally these fibers meet and deeply mix with the remaining fibers of the A2; in these specimens there is thus no distinct, separated A3'. *Galaxias* is consequently coded as '?'. Contrary to Greenwood (1977), I consider that one of the adductor mandibulae bundles attaching on the coronomeckelian bone in *Albula* corresponds to the A3'; I thus tentatively code this genus as CS-0.

186. *Absence of distinct, separated section A3″ of adductor mandibulae*. Unlike in the taxa of CS-0 [State 0], in taxa of CS-1 there is no distinct, separated section A3″ of the adductor mandibulae (see above) [State 1]. It should be noted that the taxa coded as CS-1 in the character above are coded here as "Inapplicable", since if the section A3' is missing there is obviously no section mesial to it, that is, there is no A3″.

187. *Absence of adductor mandibulae A*w. The presence of an adductor Aw (e.g., Figs. 7, 21) is seemingly plesiomorphic for osteichthyans (e.g., Edgeworth, 1935; Lauder, 1980a,b) [State 0]. In the specimens examined of taxa of CS-1 this bundle is missing (e.g., fig. 16) [State 1]. The bundle named "Aw" in Vari's (1979) fig. 42 of *Distichodus* seems to be part of the A1-OST sensu this work and not a true Aw. In fact, in the *Distichodus* specimens examined the Aw is missing, a condition that is very likely related to the peculiar articulation and high mobility between the angulo-articular and the dentary bone found in these specimens. An adductor mandibulae Aw is seemingly found in *Latimeria* (e.g., Millot and Anthony, 1958: fig. 19; Lauder, 1980b). In *Timon* the adductor mandibulae has a large anteroventral portion that is lodged in the "adductor fossa" (sensu Lauder, 1980b) and that seems in fact very similar to the Aw of other osteichthyans. A similar anteroventral portion of the adductor mandibulae lodging in the "adductor fossa" was also described in other extant amniotes, such as *Crocodilus* (e.g., Edgeworth, 1935). Therefore, I prefer not to completely exclude the hypothesis that the portion of the adductor mandibulae of *Timon* is eventually homologous to the Aw of taxa of CS-0. Thus, I prefer to tentatively code *Timon* as CS-0 and to check whether or not this primary homology hypothesis stands in light of the phylogenetic results obtained in the present work (see Chapter 4, Section 4.2).

188. *A*w *divided into well-developed, distinct A*w*-D and A*w*-V bundles, the A*w*-V attaching anteriorly on the suspensorium and/or opercular series (inspired*

from, e.g., Lauder and Liem, 1983; Sato and Nakabo, 2002). Unlike taxa of CS-0 [State 0], specimens of *Aulopus* and *Chlorophthalmus* exhibit this feature (e.g., Fig. 43) [State 1]. Gosline (1986, 1989) described the Aw of *Aulopus japonicus* as an undivided section that does not attach on the suspensorium and/or on the opercular series. The *Aulopus* specimens I analyzed do have an Aw divided into a well-developed Aw-V (e.g., Fig. 43) attaching posteriorly on the opercular series and a well-developed Aw-D (e.g., Fig. 43). This is also the case of the aulopiform specimens examined of the genus *Chlorophthalmus*, as well as of several other aulopiform and non-aulopiform eurypterygians described in the literature, which exhibit a configuration strikingly similar to that found in the *Aulopus* specimens analyzed in the present study (e.g., Winterbottom, 1974; Gosline, 1986; Sato and Nakabo, 2002; Wu and Shen, 2004). I thus consider that it is probable that Gosline (1986, 1989) failed to detect the Aw-V in the *Aulopus* specimens he observed. In fact, in the first observations of the specimens of this genus I also failed to detect the Aw-V. Not because this bundle was small or really absent in those specimens, but because when one separates the mandible from the other head structures, this bundle often remains attached to the opercular series: when one then analyzes the separated mandible and the structures attached to it, one can thus easily fail to detect the Aw-V. Only the subsequent examination of the opercular series of those specimens drew my attention to this adductor mandibulae bundle, which I confirmed to be, in dissections of other specimens, actually part of the Aw.

189. *Presence of small bundle of adductor mandibulae attaching to lateral ethmoid by means of a thin, long tendon.* Unlike taxa of CS-0 [State 0], specimens of *Cobitis* exhibit this feature [State 1].
190. *Presence of adductor arcus palatini (inspired from, e.g., Edgeworth, 1935; Winterbottom, 1974; Miyake et al., 1992).* The presence of an adductor arcus palatini (sensu Winterbottom, 1974, and the present work: e.g., Fig. 38) running from the neurocranium to the mesial surface of the suspensorium/palatoquadrate seems to be plesiomorphic for osteichthyans (e.g., Edgeworth, 1935; Miyake et al., 1992) [State 0]. In taxa of CS-1 there is no recognizable adductor arcus palatini (e.g., Fig. 109) [State 1]. Contrary to what was stated by Greenwood (1977), the specimens examined of the genera *Albula* and *Notacanthus* do display a well-developed adductor arcus palatini. Although peculiarly modified, a muscle corresponding to the adductor arcus palatini of other bony fishes is seemingly present in extant acipenseriforms ("retractor" of hyomandibula: e.g., Fig. 12; see character below). As explained above, the levator hyoideus/depressor mandibulae found in at least some developmental stages of, for example, *Neoceratodus*,

Lepidosiren, *Ambystoma* and *Timon* may eventually be derived from the same portion of the hyoid muscle plate from which the adductor arcus palatini of other bony fishes originates. Be that as it may, the adults of these latter four genera do not have an adductor arcus palatini like that found in taxa of CS-0; these four genera are therefore coded as CS-1.

191. *Adductor arcus palatini peculiarly modified in a "retractor" of the hyomandibula (inspired from, e.g., Danforth, 1913; Sewertzoff, 1928; Edgeworth, 1935; Carroll and Wainwright, 2003).* Unlike in taxa of CS-0 [State 0], in specimens of *Acipenser* and *Psephurus* the adductor arcus palatini is peculiarly modified in a "retractor" of the hyomandibula (e.g., Fig. 12; Carroll and Wainwright, 2003: fig. 3A) [State 1].
192. *Levator arcus palatini (and eventually also a portion of the constrictor dorsalis corresponding to the dilatator operculi) peculiarly modified into a "protractor" of the hyomandibula (inspired from, e.g., Danforth, 1913; Sewertzoff, 1928; Edgeworth, 1935; Carroll and Wainwright, 2003).* Unlike in taxa of CS-0 [State 0], in specimens of *Acipenser* and *Psephurus* the levator arcus palatini (and eventually also a portion of the constrictor dorsalis corresponding to the dilatator operculi: e.g., Danforth, 1913; Carroll and Wainwright, 2003) is peculiarly modified in a "protractor" of the hyomandibula (e.g., Fig. 12; Carroll and Wainwright, 2003: fig. 3A) [State 1].
193. *Adductor arcus palatini, or muscle differentiated from it, inserting on autopalatine (inspired from, e.g., Takahasi, 1925; Alexander, 1965; Winterbottom, 1974; Diogo, 2004a).* Unlike taxa of CS-0 [State 0], taxa of CS-1 exhibit this feature (e.g., Diogo, 2004a: fig. 3-39) [State 1].
194. *Separation between extensor tentaculi and adductor arcus palatini (ordered multistate character) (inspired from, e.g., Takahasi, 1925; Alexander, 1965; Winterbottom, 1974; Diogo, 2004a).* Among those taxa coded as CS-1 in the character above the plesiomorphic condition seems to be that in which there is no differentiated muscle extensor tentaculi (e.g., Diogo, 2004a) [State 0]. This muscle, present in all catfishes included in the cladistic analysis, results from a differentiation of the anterior fibers of the adductor arcus palatini. Taxa of CS-1 [State 1] have an extensor tentaculi, but the fibers of this muscle and those of the adductor arcus palatini are somewhat mixed; in taxa of CS-2 [State 2] these two muscles are well separated (e.g., Diogo, 2004a: fig. 3-39).
195. *Neither adductor arcus palatini nor muscle differentiated from it inserting on hyomandibula (inspired from, e.g., Diogo, 2004a).* Unlike taxa of CS-0 [State 0], taxa of CS-1 exhibit this feature (e.g., Diogo, 2004a) [State 1].
196. *Adductor arcus palatini inserting on preopercle (inspired from, e.g., Howes, 1985a).* Unlike taxa of CS-0 [State 0], taxa of CS-1 exhibit this feature (e.g., Howes, 1985a) [State 1].

197. *Significant part of fibers of adductor arcus palatini inserting on lateral surface of suspensorium (inspired from, e.g., Oliveira et al., 2002; Diogo, 2004a).* Unlike taxa of CS-0 [State 0], taxa of CS-1 exhibit this feature (e.g., Fig. 46) [State 1].
198. *Presence of both a peculiar protractor pterygoidei and a peculiar levator pterygoidei (inspired from, e.g., Edgeworth, 1935; Frazzetta, 1962; Moro and Abdala, 2000; Montero et al., 2002; Abdala and Moro, 2003).* Unlike in taxa of CS-0 [State 0], both these muscles (e.g., Frazzetta, 1962: fig. 5) are found in the *Timon* specimens examined [State 1]. In these latter specimens the muscles are somewhat mixed with each other, and their position is somewhat similar to that of the levator arcus palatini found in many osteichthyan groups. In fact, according to, for example, Edgeworth (1935), both the protractor pterygoidei and the levator pterygoidei are derived from the dorsal portion of the mandibular plate from which the levator arcus palatini of other osteichthyans is derived (see Chapter 4, Section 4.2).
199. *Presence of pterygomandibularis (inspired from, e.g., Edgeworth, 1935; Frazzetta, 1962; Gans et al., 1985; Moro and Abdala, 2000; Montero et al., 2002; Abdala and Moro, 2003).* Unlike taxa of CS-0 [State 0], the *Timon* specimens examined exhibit this muscle (e.g., Fig. 111) [State 1]. According to, for example, Edgeworth (1935), this pterygomandibularis is derived from the mesial portion of the adductor mandibulae.
200. *Presence of peculiar, massive retractor bulbi (inspired from, e.g., Edgeworth, 1935; Frazzetta, 1962; Larsen and Guthrie, 1975; Bjerring, 1993).* Unlike in taxa of CS-0 [State 0], I have found such a peculiar, massive retractor bulbi (e.g., Larsen and Guthrie, 1975: fig. 3) in specimens of *Timon* and *Ambystoma* [State 1].
201. *Presence of peculiar basicranial muscle (inspired from, e.g., Millot and Anthony, 1958; Bemis and Northcutt, 1991; Bjerring, 1993; Northcutt and Bemis, 1993).* Unlike taxa of CS-0 [State 0], the members of *Latimeria* exhibit a peculiar basicranial muscle connecting the anterior and posterior mobile halves of their neurocranium (e.g., Fig. 85) [State 1]. Bemis and Northcutt (1991) and Northcutt and Bemis (1993) suggested that this muscle may eventually correspond to the peculiar retractor bulbi of tetrapods (see character above). Bjerring (1993: 296) and other authors considered that such a suggestion "has no embryological or morphological support whatsoever". According to Bjerring (1993), both a basicranial muscle and a retractor bulbi were present in certain sarcopterygian fossils; this would thus raise serious doubts about the homology between these muscles. Be that as it may, a peculiar basicranial muscle connecting the anterior and posterior mobile halves of the neurocranium such as that found in *Latimeria* is

absent in all the other extant taxa included in the present cladistic analysis, and, thus, *Latimeria* is the only taxon coded as CS-1 here (the fossils included in the analysis were coded as '?').

202. *Levator arcus palatini*. Unlike in taxa of CS-0 [State 0], a recognizable levator arcus palatini is absent in taxa of CS-1 (e.g., Fig. 91) [State 1]. As explained above, although peculiarly modified, the levator arcus palatini is seemingly present in *Psephurus*, *Acipenser* and *Timon*.
203. *Differentiation of levator arcus palatini into two well-differentiated bundles (inspired from, e.g., Diogo, 2004a; Greenwood, 1968, 1977)*. Unlike taxa of CS-0 [State 0], taxa of CS-1 exhibit this feature (e.g., Fig. 38) [State 1].
204. *Levator arcus palatini not inserting on metapterygoid*. The plesiomorphic condition for osteichthyans is seemingly that in which the levator adductor arcus palatini is at least partly inserted on the metapterygoid. This is the case in *Polypterus*, *Lepisosteus*, *Amia* and in numerous others actinopterygians, as well as in sarcopterygians such as *Latimeria* (e.g., Fig. 38) [State 0]. In taxa of CS-1 this muscle does not insert on the metapterygoid [State 1]. It is often mentioned in the literature that in most osteichthyans the levator arcus palatini attaches only on the hyomandibula; as can be seen by the numerous taxa coded as CS-0 in this character, this is far from being the case in the groups analyzed in the present work. Just to give an example, Kershaw (1976) stated that in the members of *Pantodon buchholzi* the levator arcus palatini "inserts on the hyomandibula", but in the specimens of this species examined in the present work this muscle clearly also inserts on the metapterygoid.
205. *Origin of levator arcus palatini on dorsal surface of cranial roof (inspired from, e.g., De Pinna and Vari, 1995; Diogo, 2004a)*. Unlike taxa of CS-0 [State 0], taxa of CS-1 exhibit this feature (e.g., Diogo, 2004a) [State 1].
206. *Origin of dilatator operculi on dorsal surface of cranial roof*. Unlike taxa of CS-0 [State 0], taxa of CS-1 exhibit this feature (e.g., Fig. 80) [State 1].
207. *Dilatator operculi exhibiting peculiar configuration, being almost completely covered in lateral view by dorsal surface of preopercle*. Unlike in taxa of CS-0 [State 0], in the *Denticeps* specimens examined the dilatator operculi is almost completely covered in lateral view by the dorsal surface of the preopercle (e.g., Fig. 46) [State 1]. This has led Greenwood (1968) and other authors to incorrectly state that the dilatator operculi is absent in the members of this genus.
208. *Differentiation of dilatator operculi into different bundles (inspired from, e.g., Diogo, 2004a)* . Unlike taxa of CS-0 [State 0], taxa of CS-1 exhibit this feature (e.g., Diogo, 2004a) [State 1].
209. *Presence of distinct levator operculi (inspired from, e.g., Edgeworth, 1935; Schaeffer and Rosen, 1961; Lauder, 1980a)*. The plesiomorphic condition

for osteichthyans is seemingly that in which there is no distinct levator operculi (Fig. 6; e.g., Lauder, 1980a) [State 0]. A distinct levator operculi is found in taxa of CS-1 (e.g., Figs. 19, 38) [State 1]. Kardong (2002: fig. 19.39B) suggested that a levator operculi is present in dipnoans such as *Neoceratodus*. However, as explained above, these taxa have a "constrictor operculi" that may eventually correspond to the adductor operculi of most other bony fishes, but clearly not with the characteristic levator operculi of taxa of CS-1. Millot and Anthony (1958) stated that, apart from a muscle adductor operculi, *Latimeria* has also a muscle "levator operculi". As can be seen in their descriptions and their figures (e.g., their plate VII), and as explicitly recognized in their page 61, the fibers of their "levator operculi" are deeply mixed with those of the adductor operculi. This has led Lauder (1980c) and other authors to be very skeptical about the presence, in *Latimeria,* of a distinct levator operculi such as that of taxa of CS-1. However, Adamicka and Ahnelt (1992: 108) have reaffirmed, seemingly on the basis of their own direct observations of *Latimeria*, that this taxon "does have a levator operculi muscle differentiated out of the adductor (operculi)". As I could not personally examine this feature in members of the genus *Latimeria,* I will follow the statement of Adamicka and Ahnelt (1992) and, thus, tentatively code *Latimeria chalumnae* as CS-1. A discussion of this feature, taking into account the results of the present cladistic analysis, is given in Chapter 4, Section 4.2.

210. *Insertion of levator operculi on significant part of lateral surface of opercle (inspired from Howes, 1983a; Diogo, 2004a).* Unlike taxa of CS-0 [State 0], taxa of CS-1 exhibit this feature (e.g., Diogo, 2004a: fig. 3-88) [State 1].
211. *Hypertrophy of levator arcus branchialis V (ordered multistate character) (inspired from, e.g., Takahasi, 1925; Matthes, 1963; Winterbottom, 1974; Vandewalle, 1975).* Unlike in taxa of CS-0, in which the levator arcus branchialis is moderately developed or eventually missing [State 0], in specimens of taxa of CS-1 [State 1], and particularly of taxa of CS-2 [State 2], there is a remarkable hypertrophy of this muscle (e.g., Plates 1, 2, 3, 4; Matthes, 1963: plate 9C).
212. *Levator operculi peculiarly divided into an anterior, mesial bundle and a posterior, lateral bundle (inspired from, e.g., De la Hoz and Chardon, 1984; Aguilera, 1988).* Unlike taxa of CS-0 [State 0], taxa of CS-1 exhibit this feature (e.g., Fig. 83) [State 1].
213. *Presence of drumming muscle of swimbladder (inspired from Ladich, 2001; Diogo, 2004a).* Unlike taxa of CS-0 [State 0], specimens of *Pimelodus* exhibit this feature (e.g., Diogo, 2004a: fig. 3-99) [State 1].
214. *Fibers of hypaxialis and/or epaxialis covering great part of neurocranial floor (inspired from, e.g., Gosline, 1969; Fink and Fink, 1986).* Unlike in

taxa of CS-0 [State 0], in taxa of CS-1 the fibers of the hypaxialis and/or epaxialis peculiarly cover a great part of the ventral surface of the neurocranial floor (e.g., Fig. 38; Gosline, 1969: fig. 8; Günther and Deckert, 1959: figs. 11, 12), inclusively covering, in certain cases, the ventral surface of the prevomer/vomer (e.g., in some *Stomias* specimens examined) [State 1].

Splanchnocranium

215. *Maxilla fused with infraorbitals, forming peculiar, long-toothed compound structure (inspired from, e.g., Patterson, 1973; Wiley, 1976; Jollie, 1984a; Arratia and Schultze, 1991; Arratia, 1999).* Unlike taxa of CS-0 [State 0], taxa of CS-1 exhibit this feature (e.g., Fig. 10) [State 1].
216. *Absence of maxillary teeth (inspired from, e.g., Regan 1911a,b; Alexander, 1965; Gosline, 1975; Gosline, 1980; Fink and Fink, 1981, 1996; Begle, 1992; Johnson and Patterson, 1996; Poyato-Ariza, 1996; Belouze, 2002).* Unlike taxa of CS-0 [State 0], taxa of CS-1 lack maxillary teeth (e.g., Fig. 53) [State 1]. I agree with, for example, Belouze (2002) in that the upper jaw of *Saccopharynx* and *Eurypharynx* is at least partly constituted by the maxilla. The presence of a well-defined, proximal head of the toothed element constituting the upper jaw in the *Eurypharynx* specimens examined, the presence of a strong ligament between the distal surface of this element and the mandible, and the overall position and shape of this element, conjugated with the developmental data of Orton (1963), seem to support this interpretation. Gosline (1969) and Sanford (2000) stated that adult *Alepocephalus* analyzed by them have maxillary teeth. However, the adult specimens of this genus that I have examined, as well as those examined by, for example, Begle (1992), lack maxillary teeth. It thus seems that a few, very small maxillary teeth can eventually be found in some adult *Alepocephalus* while in others such teeth are missing; this genus is thus coded as '?'. As stressed by, for example, Greenwood (1968), *Denticeps* specimens exhibit numerous odontodes in various bones of the skull, and it is somewhat difficult, in certain cases, to discern whether the "teeth" present in some bones are odontodes or "real teeth"; therefore, I prefer to prudently code *Denticeps* as '?' here. The presence/absence of maxillae in extant acipenseriforms has been, and continues to be, a subject of controversy. For instance, Bridge (1878), Jollie (1980), Arratia and Schultze (1991), and Grande and Bemis (1991) consider that these fishes lack maxillae (see, e.g., Grande and Bemis, 1991: fig. 35); Findeis (1997), Bemis et al. (1997), and Carroll and Wainwright (2003) maintain that they do have maxillae (see, e.g., Bemis et al., 1997: fig. 18). The observation of the *Acipenser*

and *Psephurus* specimens examined did not make it possible to strongly support either of these hypotheses; therefore, these genera are coded as '?' in the present character, as well as in those other characters below concerning the configuration of the maxillae.

217. *Maxilla not contacting or articulating indirectly with mesethmoidal/prevomerine/vomerine region (inspired from, e.g., Fink and Fink, 1981, 1996).* Unlike in taxa of CS-0 [State 0], in the few taxa of CS-1 the maxilla does not contact or articulate indirectly (e.g., by means of small cartilages eventually situated between the maxilla, autopalatine and/or neurocranium: see below) with the mesethmoidal/prevomerine/vomerine region (e.g., Diogo, 2004a: fig. 3-39) [State 1]. Fink and Fink (1996) suggested that the gonorynchiforms *Kneria*, *Parakneria* and *Cromeria* may display a configuration such as CS-1. However, in the specimens of these three genera examined by the author, as well as in the specimens of *Kneria* and *Parakneria* described by Lenglet (1974: e.g., his figs. 3 and 5) and in the specimens of *Cromeria* described by D'Aubenton (1961: e.g., the descriptions given in his page 146), the maxilla does contact directly at least one structure of the mesethmoidal/prevomerine region (note: it was not possible to appropriately discern this feature in the *Grasseichthys* specimens examined). Fink and Fink's (1981) fig. 3F might give the idea that the members of the gymnotiform *Sternopygus* have a configuration such as CS-1. However, this is very likely due to the fact that the figure does not illustrate the well-developed "cartilage-type 2" (sensu this work) lying between the proximal margin of the maxilla and the ethmoid region. In fact, as noted by, for example, De la Hoz (1974) and De la Hoz and Chardon (1975, 1984), in the *Sternopygus* specimens they analyzed there is an indirect articulation between the maxilla and the ethmoid region by means of a this "cartilage-type 2" (= one of their "submaxillary cartilages"; e.g., De la Hoz and Chardon, 1975: fig. 5A). Such an indirect articulation between the maxilla and the ethmoid region by means of a "cartilage-type 2" is also found in the *Sternopygus* specimens examined in the present work. As suggested by, for example, De la Hoz (1974), in the specimens examined of the gymnotiforms *Brachyhypopomus* and *Gymnotus* there is no such indirect articulation. In the *Brachyhypopomus* specimens examined there is a "cartilage-type 2", but it does not contact the maxilla, lying instead between the autopalatine and/or the "cartilage-type 1" (see below) and the ethmoid region (see, e.g., fig. 5D of De la Hoz and Chardon, 1975, which illustrates a configuration similar to that found in these specimens). In the *Gymnotus* specimens observed there is no "cartilage-type 2" (e.g., De la Hoz, 1974: figs. 121, 133).

218. *Maxilla, if present, exhibiting deep, lateral fossa on distal margin of maxilla (inspired from Howes, 1985a).* Unlike in taxa of CS-0 [State 0], in *Phractolaemus* there is a deep, lateral fossa on the distal margin of the maxilla, in which part of the adductor mandibulae attaches (e.g., Fig. 63) [State 1].
219. *Some bones of ethmoid region forming, together with premaxillae and prevomer/vomers, a peculiar compound structure (inspired from, e.g., Tchernavin, 1947a; Forey et al., 1996; Smith, 1989a,b; Belouze, 2002).* Unlike taxa of CS-0 [State 0], this feature is found in taxa of CS-1 [State 1] (e.g., Belouze, 2002: fig. 2A).
220. *Supramaxillae present as independent elements (inspired from, e.g., Taverne, 1972, 1977b, 1978; Fink and Fink, 1981, 1996; Jollie, 1986; Grande and Bemis, 1998; Hilton, 2003; Cloutier and Arratia, 2004).* Unlike taxa of CS-0 [State 0], taxa of CS-1 exhibit this feature (e.g., Fig. 38) [State 1]. Since some adult *Astronesthes* have independent supramaxillae and others do not (e.g., Weitzman, 1967b), this genus is coded as '?'.
221. *Maxillae not markedly ankylosed with neurocranium (inspired from, e.g., Patterson, 1973; Lauder, 1980a; Grande and Bemis, 1998).* Although in a few taxa of CS-1 the maxillae may eventually be somewhat firmly attached to the neurocranium (e.g., Figs. 19, 25) [State 1], they are not as firmly attached/ankylosed to this latter structure as in taxa of CS-0 (e.g., Fig. 108) [State 0].
222. *Presence of distinct, strong, relatively elongated ligaments connecting anterior surface/anterior cartilage of the autopalatines and/or dermopalatines and the maxillae and/or premaxillae.* Unlike taxa of CS-0 [State 0], taxa of CS-1 exhibit this feature (e.g., Fig. 73) [State 1].
223. *Presence of well-developed "rostral" cartilaginous or cartilaginous-like structures associated with posterior surface of well-developed premaxillary dorsomedial processes attaching to/articulating with ethmoid region (inspired from, e.g., Fink and Weitzman, 1982; Fink, 1984; Rosen, 1985; Hartel and Stiassny, 1986; Stiassny, 1986; Johnson, 1992).* The presence/absence and homologies of the "rostral" premaxillary cartilages have been the subject of controversy. For instance, Fink and Weitzman (1982) and Fink (1984) maintained that certain members of the orders Stomiiformes and Aulopiformes, which are usually seen as basal Neoteleostei orders (see above), have "rostral" premaxillary cartilages similar to those found in members of more derived neoteleostean groups. However, Stiassny (1996: 455) argued that some of the structures certain authors consider to be "rostral" cartilages are in fact "not composed of hyaline cartilage" but, instead, "of an essentially fibrous composition of minimal matrix secretion". I thus prefer to define CS1 [State 1] in a way that includes all those cases in which I have found "well-developed rostral cartilaginous or cartilaginous-

like structures associated with the posterior surface of well-developed premaxillary dorsomesial processes attaching to/articulating with the ethmoid region". In fact, whether these structures are paired or not, or whether they are really completely cartilaginous or not, among all the adult specimens examined I only found such well-developed, peculiar structures associated with the posterior surface of well-developed premaxillary dorsomedial processes (e.g., Rosen, 1985: figs. 40A,C, 41A,B,C) in the few taxa coded as CS-1. Fink's (1984: 204) examination of small juvenile cichlids (derived Neoteleostei) revealed "that the rostral cartilage appears to develop ontogenetically from bilateral cartilage bodies which fuse at midline". According to this author "this is suggestive of corroboration of Fink and Weitzman's (1982) hypothesis that the rostral cartilage (of derived Neoteleostei) evolved from paired cartilages anterior to the ethmoid region", like those found in the adult specimens examined in the present work of, for example, *Salmo* and *Thymallus*. I thus think that one should not completely exclude the hypothesis that the peculiar configuration seen in those taxa coded as CS-1 might be homologous. The *Astronesthes* specimens examined seem to have a configuration somewhat similar to those of the genus *Stomias*. However, my observations of these *Astronesthes* specimens did not allow me to be completely confident about their coding in this character; therefore, I prudently code *Astronesthes* as '?'. *Coregonus* is coded as '?' because it is also very difficult to appropriately discern this feature in the specimens of this genus I have observed.

224. *Premaxillae peculiarly fused into a single, median structure (inspired from, e.g., Ridewood, 1904c; 1905a; Taverne, 1972, 1977b, 1978; Bels et al., 1993; Lee, 1998; Belouze, 2002; Hilton, 2003)*. Unlike taxa of CS-0 [State 0], taxa of CS-1 exhibit this feature (e.g., Fig. 108) [State 1].
225. *Premaxillae not markedly ankylosed with neurocranium (inspired from, e.g., Ridewood, 1904c, 1905a; Taverne, 1972, 1977b, 1978; Patterson, 1973; Grande and Bemis, 1998; Arratia, 1999)*. Although in a few taxa of CS-1 the premaxillae may eventually be somewhat firmly attached to the neurocranium [State 1], the ankylosis (but not complete fusion: see above) between these structures is much more marked in taxa of CS-0 (e.g., Figs. 6, 108) [State 0].
226. *Maxilla exhibiting peculiar, somewhat $\wedge$-shaped overall configuration*. Unlike taxa of CS-0 [State 0], taxa of CS-1 exhibit this feature (e.g., Figs. 63, 64; D'Aubenton, 1961: fig. 8; Lenglet, 1974: figs. 12, 13) [State 1].
227. *Presence of peculiar medial interdigitations between the premaxillae (inspired from, e.g., Weitzman, 1962; Vari, 1979)*. Unlike taxa of CS-0 [State 0], taxa of CS-1 [State 1] exhibit this feature (e.g., Weitzman, 1962: figs. 2, 4).

228. *Presence of prominent, well-defined, roundish anterior process of maxilla for articulation with posterior/mesial surface of premaxilla (inspired from, e.g., Pasleau, 1974; Rosen, 1985; Grande and Poyato-Ariza, 1999).* Unlike taxa of CS-0 [State 0], taxa of CS-1 exhibit this feature (e.g., Fig. 53; Sulak, 1977: figs. 3A, 7A) [State 1].
229. *Markedly short maxilla (inspired from, e.g., Bornbusch, 1991, 1995; Diogo, 2004a).* Unlike in taxa of CS-0 [State 0], in taxa of CS-1 the maxilla is peculiarly reduced to a very short, small structure (e.g., Diogo, 2004a: fig. 3-43) [State 1].
230. *Mesial surface of distal portion of maxillae/supramaxillae firmly attached to lateral surface of mandibles (inspired from, e.g., Jollie, 1984c; Diogo, 2004a).* Unlike in taxa of CS-0 [State 0], in taxa of CS-1 the mesial surface of the distal portion of the maxillae/supramaxillae is firmly attached to the lateral surface of the mandibles (e.g., Figs. 38, 53, 54), usually by short, strong ligamentous tissue [State 1].
231. *Attachment of primordial ligament on posterolateral surface of mandible.* The plesiomorphic condition for those taxa with a primordial ligament is seemingly that in which this ligament connects the maxilla to the dorsal surface of the coronoid process, as seen in the *Polypterus*, *Lepisosteus* and *Amia* specimens examined as well as in many other osteichthyans observed (e.g., Figs. 7, 16, 19) [State 0]. In the specimens examined of taxa of CS-1 the primordial ligament connects the maxilla to the posterolateral surface of the mandible, somewhat near its articulation with the quadrate (e.g., Fig. 80) [State 1].
232. *Presence of strong, well-defined ligament between premaxilla and proximal surface of maxilla (inspired from, e.g., Gosline, 1975; Sanford, 2000; Diogo, 2004a).* Unlike taxa of CS-0 [State 0], taxa of CS-1 exhibit this feature (e.g., Diogo, 2004a: fig. 3-55) [State 1]. Since some *Albula* specimens do not have such a ligament (e.g., those examined in this work), while others seemingly do (e.g., Greenwood, 1977), this genus is coded as '?'.
233. *Each premaxilla presenting well-developed, roughly circular articulatory surface to articulate mesially with its counterpart.* Unlike in taxa of CS-0 [State 0], in specimens of *Gonorynchus* each premaxilla presents a well-developed, roughly circular articulatory surface to articulate mesially with its counterpart ("bouton articulaire median inter-premaxillaires"; Monod, 1963: fig. 11) [State 1].
234. *Presence of prominent, roughly triangular anterolateral processes of premaxillae (inspired from, e.g., Greenwood and Rosen, 1971).* Unlike taxa of CS-0 [State 0], specimens of *Searsia* exhibit this feature [State 1] (e.g., Greenwood and Rosen, 1971: fig. 24).
235. *"Premaxillae syndesmotically attached to proximal head of maxillae" (inspired from, e.g., Chapman, 1941; Howes and Sanford, 1987b; Begle,*

1992; Johnson and Patterson, 1996). Unlike taxa of CS-0 [State 0], specimens of *Plecoglossus* exhibit this feature (e.g., Howes and Sanford, 1987b: fig. 2) [State 1].

236. *"Peculiar lateral excavation of upper and lower jaws, in which are anchored numerous outer, epithelially implanted comb-teeth" (inspired from, e.g., Chapman, 1941; Howes and Sanford, 1987b)*. Unlike taxa of CS-0 [State 0], specimens of *Plecoglossus* exhibit this feature (e.g., Howes and Sanford, 1987b: figs. 2, 16, 17) [State 1].

237. *"Presence of large, peculiar tooth-bearing interpremaxillary pad between premaxillae in adults" (inspired from, e.g., Howes and Sanford, 1987b)*. Unlike taxa of CS-0 [State 0], specimens of *Plecoglossus* exhibit this feature (e.g., Howes and Sanford, 1987b: fig. 2) [State 1].

238. *Absence of premaxillary teeth (inspired from, e.g., Regan, 1911a,b; Alexander, 1965; Greenwood, 1968; Gosline, 1975; Fink and Fink, 1981, 1996; Poyato-Ariza, 1996; Johnson and Patterson, 1996; Grande and Poyato-Ariza, 1999)*. Unlike taxa of CS-0 [State 0], taxa of CS-1 lack premaxillary teeth (e.g., Fig. 53) [State 1]. The "premaxillary teeth" reported by Monod (1963) in *Gonorynchus* are peculiar, non-ossified structures (see below); this genus is thus coded as CS-1. As explained above, the specimens of *Denticeps* present numerous odontodes in many bones of the skull, and it is somewhat difficult to discern, in certain cases, whether the "teeth" present in some bones are odontodes or are "real teeth"; therefore, *Denticeps* is thus prudently coded as '?'. Howes and Sanford (1987b) stated that in adult *Plecoglossus* there are "real premaxillary teeth", since they considered that, although the teeth of the peculiar tooth-bearing interpremaxillary pad described in the character above are not directly in contact with the premaxillae, these teeth are in fact homologous with the premaxillary teeth of other osteichthyans. However, I prefer to prudently code *Plecoglossus* as '?', since the configuration in the adult taxa does not exactly correspond to the configuration found in other adult taxa coded here as CS-0, in which the premaxillary directly supports the premaxillary teeth, nor does it correspond to the configuration found in other adult taxa coded here as CS-1, in which there are no premaxillary teeth at all. Unlike in *Eurypharynx*, coded here as '?' because it is difficult to discern whether the premaxillae are present (e.g., fused with mesethmoid and/or prevomer/vomer) or not (in this case, this character would be inapplicable), there is a good amount of evidence to support the view that in *Anguilla* and *Conger* the premaxillae are fused to the mesethmoid and prevomer/vomer, and thus that these two taxa do have premaxillary teeth (e.g., Belouze, 2002: fig. 2D).

239. *Presence of long, strong ligament between premaxilla and anteromesial surface of mandible (inspired from Thys van den Audenaerde, 1961).* Unlike taxa of CS-0 [State 0], members of *Phractolaemus* exhibit this feature (e.g., Thys van den Audenaerde, 1961: fig. 18) [State 1].
240. *Presence of "rictal cartilages" between upper jaws and between lower jaws (inspired from, e.g., Thys van den Audenaerde, 1961; Howes, 1985a).* Unlike taxa of CS-0 [State 0], members of *Phractolaemus* exhibit this feature (e.g., Figs. 63, 64) [State 1].
241. *Presence of "gingival teeth" on the upper jaw (inspired from Poyato-Ariza, 1996; Grande and Poyato-Ariza, 1999).* As explained above, the "gingival teeth" reported by Monod (1963) in *Gonorynchus* [State 1], which are present in other taxa included in the analysis [State 0], are non-osseous structures, being described by Grande and Poyato-Ariza (1999: 210) as "fringes on the soft issue of the premaxillae".
242. *Presence of peculiar ridged tooth plates on the jaws (inspired from, e.g., Miles, 1977; Lauder and Liem, 1983; Rosen et al., 1981; Bemis, 1986; Schultze, 1986; Campbell and Barwick, 1986; Cavin et al., in press).* Unlike taxa of CS-0 [State 0], taxa of CS-1 exhibit this feature (e.g., Figs. 89, 90) [State 1].
243. *"Autostylic suspension of mandibular arch" (inspired from, e.g., Rosen et al., 1981; Forey, 1986; Bartsch, 1992, 1994; Cloutier and Ahlberg, 1996; Clack, 2002; Shubin et al., 2006).* As noted by these and other authors, unlike taxa of CS-0 [State 0], the adult specimens of CS-1 exhibit an "autostylic suspension of the mandibular arch" (e.g., Figs. 89, 101, 108) [State 1]. Some of these authors consider that the way in which this feature is realized in some of the groups included in the present analysis is not completely similar and, thus, may not be homologous. However, I prefer to tentatively code the "autostylic suspension" found in taxa coded as CS-1 as a primary homology and to test this hypothesis against the phylogenetic results of the cladistic analysis also including all the other numerous characters listed here, rather than to exclude, based on *a priori* assumptions, the hypothesis that this feature is eventually homologous within at least some of these taxa. I have explained throughout this work the reasons for doing so, and I will return to this subject further below. Clack (2002: 294) stated that "reassessments suggested that the skulls of the earliest tetrapods were not after all fully autostylic but that their stapes seems still to have been acting as a brace between braincase and palate"; "*Acanthostega* is now though to have functional internal gills, and the stapes may still have been involved with breathing". I agree that the skulls of taxa such as †*Acanthostega* may well not be "fully autostylic". However, examination of the descriptions and illustrations provided in the literature does seem to indicate that the condition found in this

fossil taxon is more similar to that found in extant tetrapods such as *Ambystoma* and *Timon* (i.e., nearer a "full autostyly") than to that found in taxa of CS-0. I therefore tentatively code †*Acanthostega* here as CS-1. Again, it should be kept in mind that this is just a primary homology hypothesis, which can be contradicted (or supported) by the results of the present cladistic analysis. I code †*Tiktaalik roseae* as '?' because the paper by Daeschler et al. (2006) did not provide much information on this feature; it is hoped that more information will be available soon (Shubin, pers. comm.).

244. *Pterygoids/dermopalatines/pterygopalatines, if present, peculiarly attached, or fused, to their counterparts in the midline (inspired from, e.g., Rosen et al., 1981; Forey, 1986; Schultze, 1986; Schultze and Campbell, 1986; Clack, 2002; Shubin et al., 2006).* Unlike in taxa of CS-0 [State 0], in taxa of CS-1 the "pterygoids/dermopalatines/pterygopalatines are peculiarly attached, or fused, to their counterparts in the midline" (e.g., Fig. 90; Kesteven, 1942-1945: fig. 108; Schultze, 1986: fig. 7C) [State 1] (like Bartsch, 1994, and Kemp, 1999, I prefer to use the name "pterygopalatines" for the elements found in the dipnoans *Lepidosiren* and *Neoceratodus*, rather than names such as "pterygoids" or "entopterygoids", since, as shown by these authors, the dipnoan pterygopalatine is seemingly "more than a pterygoid and more than a palatine bone, and referring to the element as an entopterygoid, pterygoid or palatal is inadequate"—Kemp, 1999: 133). Some of the authors mentioned above consider that one should not use a character such as the present one, since the way in which CS1 is realized in the different groups coded as CS-1 is not completely similar and, thus, may not be homologous. For the reasons explained above, I prefer to tentatively code the condition of the taxa coded as CS-1 as a primary homology and to test this hypothesis against the phylogenetic results of the cladistic analysis rather than to completely exclude, based on *a priori* assumptions, the hypothesis that this feature is eventually homologous within at least some of these taxa.

245. *Absence of toothed bony dermopalatines and/or "predermopalatines" (inspired from, e.g., Ridewood, 1904c; Taverne, 1972; Fink and Fink, 1981, 1996; Arratia and Schultze, 1991; Arratia, 1992, 1999; Johnson and Patterson, 1996; Ruta et al., 2003).* Taxa of CS-0 have toothed bony dermopalatines and/or "predermopalatines" (e.g., Fig. 20), which may be eventually fused to other bony elements (e.g., ectopterygoids) [State 0]. In taxa of CS-1 toothed bony dermopalatines and "predermopalatines" are missing (e.g., Fig. 55) [State 1]. I agree with, for example, Johnson and Patterson (1996) in that in taxa such as *Retropinna* and *Stokellia* the toothed structures situated ventrally and posteroventrally to the autopalatines are very likely the result of the

fusion of the dermopalatines with the ectopterygoids. Therefore, these two genera are tentatively coded as CS-0. As explained above, the "pterygopalatines" of *Lepidosiren* and *Neoceratodus*, which are associated with peculiar ridged tooth plates, may eventually include some dermal "palatine" elements; however, the precise homologies of these "pterygopalatines" are still not clear, and these two genera are prudently coded here as '?'.

246. *Presence of compound dermopalatine-ectopterygoid toothed structure (inspired from, e.g., Taverne, 1978; Johnson and Patterson, 1996).* As explained above, I agree that in taxa such as *Retropinna* and *Stokellia* the toothed structures situated ventrally and posteroventrally to the autopalatines are very likely the result of fusion between the dermopalatine and the ectopterygoids. Thus, unlike the other taxa included in the cladistic analysis [State 0], these two genera are tentatively coded as CS-1 [State 1] (e.g., Johnson and Patterson, 1996: fig. 4A). Taverne (1978) and Hilton (2003) stated that the osteoglossomorph *Xenomystus*, as well as many other osteoglossomorphs, also exhibits compound, dermopalatine-ectopterygoid toothed structures (e.g., Taverne, 1978: fig. 104). The observations of the present work support this statement, and thus *Xenomystus* is coded as CS-1. Ridewood (1905a), Kershaw (1970), Taverne (1978), Hilton (2003), and Moritz and Britz (2005) documented that one of those other osteoglossomorphs exhibiting a compound, dermopalatine-ectopterygoid structure is *Pantodon*. The observations of the present work also support this latter statement, but Arratia and Schultze's (1991) fig. 20A seems to show that in certain *Pantodon* specimens the dermopalatine and the ectopterygoid may not be completely fused: this genus is thus prudently coded as '?'. *Lepidosiren* and *Neoceratodus* are coded as CS-0 since, although the precise homologies of their "pterygopalatines" are not clear (see above), it is commonly accepted that these structures include at least part of the entopterygoids (sensu this work); thus, they do not seem to correspond to the compounds found in taxa of CS-1, which are exclusively formed by dermopalatines plus ectopterygoids.

247. *Autopalatines missing or remaining almost completely, or completely, cartilaginous in adults (inspired from, e.g., Millot and Anthony, 1958; Taverne, 1972, 1977b, 1978; Fink and Fink, 1981, 1996; De la Hoz and Chardon, 1984; Arratia and Schultze, 1991; Grande and Bemis, 1991; Arratia, 1992; Albert and Campos-da-Paz, 1998; Albert, 2001; Hilton, 2003).* Unlike taxa of CS-0 [State 0], taxa of CS-1 exhibit this feature (e.g., Figs. 101, 108) [State 1]. The homologies of the "pterygopalatines" of *Neoceratodus* and *Lepidosiren* are not clear and these structures may eventually include dermal "palatine" elements;

however, they seem to not include elements corresponding to the ossified autopalatines of taxa of CS-0 (e.g., Figs. 89, 90) (e.g., Arratia and Schultze, 1991; Kemp, 1999). Since some adult *Acipenser* and *Psephurus* exhibit a configuration such as that of CS-0, while others exhibit a configuration such as that of CS-1, these two genera are coded as '?'.

248. *Mobile articulation, either direct or indirect, between the autopalatine and/or dermopalatine and the maxilla.* Unlike in taxa of CS-0 [State 0], in taxa of CS-1 there is a mobile articulation, either direct or indirect (e.g., by means of a pre-ethmoid, as in some cyprinids, or by means of small cartilaginous structures such as those mentioned above), between the maxilla and the autopalatine and/or dermopalatine (e.g., Fig. 38) [State 1]. Despite its rather small size, there is a maxillary articulatory facet for the autopalatine and/or dermopalatine in the osteoglossomorph specimens examined of the genera *Hiodon*, *Pantodon* and *Xenomystus* (however, it was not possible to discern whether or not this is also the case in the *Mormyrus* specimens analyzed).
249. *Entopterygoid exhibiting a peculiar, large, deep anterodorsal concavity for articulation with a well-developed ventral articulatory facet of lateral ethmoid, when this latter bone is present (inspired from, e.g., Vandewalle, 1975).* Unlike taxa of CS-0 [State 0], taxa of CS-1 exhibit this feature (e.g., Vandewalle, 1975: fig. 51; see also character above) [State 1]. Note that the entopterygoid of this work corresponds to the bone that is often named "pterygoid" in amniotes such as *Timon* and in amphibians such as *Ambystoma* (e.g., Figs. 101, 108).
250. *Pars autopalatina, if present, markedly separated from pars pterygoquadrata, allowing a peculiar "palatine-maxillary" system (inspired from, e.g., Fink and Fink, 1981, 1996; Arratia, 1987, 1992; Arratia and Schultze, 1991; Diogo, 2004a).* Unlike taxa of CS-0 [State 0], taxa of CS-1 exhibit this feature (e.g., Diogo, 2004a: fig. 3-39) [State 1].
251. *Autopalatine (ossified or not), if present, with large anteroventral expansion of laminar bone (inspired from, e.g., Ridewood, 1905b; Monod, 1963; Pasleau, 1974).* Unlike taxa of CS-0 [State 0], specimens of *Gonorynchus* exhibit this feature (e.g., Monod, 1963: fig. 30) [State 1].
252. *Anterior portion and/or anterior cartilage of autopalatine (or pars autopalatina) forming peculiar "broad hook" covering a great portion of the proximal portion of maxilla in lateral view (inspired from, e.g., Rosen, 1985; Stiassny, 1986; Baldwin and Johnson, 1996; Sanford, 2000).* As noted by, for example, Sanford (2000), although in certain taxa of CS-0 [CS-0] the anterior portion and/or cartilage of the autopalatine (or pars autopalatina) may be significantly elongated anteroposteriorly and/ or eventually form a "small hook", a peculiar "broad hook" covering

a great portion of the proximal portion of the maxilla in lateral view is only found, within the taxa included in the analysis, in taxa of CS-1 [State 1] (e.g., Sanford, 2000: fig. 32; Stiassny, 1986: fig. 5A).

253. *Prominent, peculiar "processus dorsomedialis" of autopalatine (inspired from Fink and Fink, 1981, 1996; Arratia, 1992).* Unlike in taxa of CS-0 [State 0: e.g., Fig. III.1.13.6], such a process is found in taxa of CS-1 (e.g., Fig. 74; Fink and Fink, 1981: figs. 3B, 9) [State 1].

254. *Semimovable articulation between posterior portion and/or posterior cartilage of autopalatine and a concave facet of entopterygoid (inspired from, e.g., Fink and Fink, 1981, 1996; Arratia, 1992).* As explained by Fink and Fink (1981, 1996), although some taxa coded here as CS-0 [State 0] may exhibit an articulation between the pterygoid bones and the autopalatine (e.g., some gonorynchiforms) that is somewhat similar to the articulation found in taxa of CS-1, only in these latter taxa is the articulation made between a characteristic, concave facet of the entopterygoid and the posterior portion/posterior cartilage of the autopalatine [State 1].

255. *Autopalatine, if present, with posterior portion markedly enlarged dorsoventrally.* Unlike taxa of CS-0 [State 0], taxa of CS-1 exhibit this feature (e.g., Fig. 38) [State 1].

256. *Strong, long ligament connecting anterior surface of ectopterygoid and/or ectopterygoid to maxilla (inspired from, e.g., Diogo, 2004a).* Unlike in taxa of CS-0 [State 0], in taxa of CS-1 there is such a strong, long ligament (e.g., Diogo, 2004a: fig. 3-39C) [State 1].

257. *Autopalatine, when present, being highly mobile in relation to the rest of the suspensorium/palatoquadrate (inspired from, e.g., Fink and Fink, 1981, 1996; Grande and Poyato-Ariza, 1999).* Among those taxa with an autopalatine, the plesiomorphic condition seems to be that in which there is a relatively low mobility between this element and the rest of the suspensorium [State 0]. Such a mobility is increased in siluriforms, in which there is a peculiar separation between the pars quadrata and the pars autopalatina (see above). But such a peculiar embryological separation is not a necessary condition for an increased mobility of the autopalatine, as this latter is also found in some non-siluriform taxa examined, as can be seen in the list of taxa coded as CS-1 [State 1] (see Table 1). As pointed out by Patterson (1984), it seems possible to discern this character in certain fossils: it is, for instance, very likely that in a taxon such as †*Chanoides macropoma* the firm, wide connection between the autopalatine and the ectopterygoid prevented a high mobility of the autopalatine in relation to the rest of the suspensorium (e.g., Patterson, 1984: fig. 5).

258. *Ossification of ligaments connecting anterior margin of suspensoria to ethmoid region.* Unlike in taxa of CS-0 [State 0], in taxa of CS-1 there is a

partial or complete ossification of these ligaments [State 1]. It should be noted that the "sesamoid bones 2 and 3 of the suspensorium" found in various siluriforms are seemingly the result of ossification of the ligaments connecting the pterygoids to the autopalatines, and not of the ligaments connecting the pterygoids to the ethmoid region (e.g., Diogo et al., 2001b; Diogo, 2004a).

259. *Presence of large "metapterygoid-quadrate fenestra" (inspired from, e.g., Fink and Fink, 1981, 1996; Arratia, 1992).* Unlike in taxa of CS-0 [State 0], in taxa of CS-1 there is a large "metapterygoid-quadrate fenestra" (sensu, e.g., Fink and Fink, 1981, 1996) [State 1].

260. *Ectopterygoid missing (inspired from, e.g., McDowall, 1969; Begle, 1992; Johnson and Patterson, 1996; Grande and Poyato-Ariza, 1999; Diogo, 2004a).* Unlike in taxa of CS-0 [State 0], in taxa of CS-1 [State 1] the ectopterygoid is seemingly missing (e.g., Fig. 101). There are some taxa, other than those listed as CS-1, in which the ectopterygoid is not present as an independent element but in which this structure is seemingly not missing: it is very likely fused with other structures, such as the entopterygoid or the dermopalatine (see above). According to, for example, Diogo et al. (2001b), Diogo and Chardon (2003) and Diogo (2004a), the structure often called "metapterygoid" in catfish is seemingly an ento-ectopterygoid. Chardon and De la Hoz (1973), De la Hoz (1974) and others pointed out that the "entopterygoid" of most gymnotiforms has "features typical of the entopterygoid (e.g., ligamentous connection with the neurocranium; relation with the adductor arcus palatini), ectopterygoid (e.g., antero-dorsal relation with the autopalatine) and entopterygoid + ectopterygoid (e.g., spatial position) of other ostariophysan taxa". However, Fink and Fink (1981) and Arratia (1992) consider that the gymnotiform "entopterygoid" corresponds to the entopterygoid of other teleosts. There is also controversy about the identity of the "pterygoid" of elopomorph taxa such as *Conger* and *Anguilla*; some authors consider it an entopterygoid, while others interpret it as an ento-ectopterygoid (see, e.g., Belouze, 2002 for a discussion of this subject). I am thus prudently coding the catfish and gymnotiform taxa included in the cladistic analysis, as well as taxa such as *Conger* and *Anguilla*, as '?' here. As explained above, in light of the data available one cannot completely exclude the hypothesis that the "pterygopalatines" of *Lepidosiren* and *Neoceratodus* eventually include elements corresponding to the ectopterygoids of taxa of CS-0; these two genera are thus also coded as '?'.

261. *Entopterygoid missing (inspired from, e.g., Weitzman, 1967b; Taverne, 1972; Hilton, 2003).* Unlike in taxa of CS-0 [State 0], in taxa of CS-1 the entopterygoid is seemingly missing (e.g., Taverne, 1972: fig. 7). In some

adult *Astronesthes* the entopterygoid is present, while in others it is seemingly missing (e.g., Weitzman, 1967b); this genus is coded as '?'. The homologies of the "pterygopalatines" of *Lepidosiren* and *Neoceratodus* are not clear, but there is some consensus that these structures include elements corresponding to the entopterygoids of the present work (see above). As also explained above, the entopterygoid of this work corresponds to the bone that is often named "pterygoid" in tetrapod taxa such as †*Acanthostega*, *Timon* and *Ambystoma* (e.g., Fig. 101); these three latter taxa are thus coded as CS-0.

262. *Peculiar configuration of anterior portion of suspensorium in which the entopterygoid and ectopterygoid are widely separated by the quadrate, resulting in almost no contact, or no contact at all, between these two pterygoid structures (inspired from, e.g., Lekander, 1949; Ramaswami, 1953)*. Unlike taxa of CS-0 [State 0], specimens of *Cobitis* exhibit this feature (e.g., Lekander, 1949: fig. 67) [State 1].
263. *No teeth on, or clearly associated with, pterygoid bones (inspired from, e.g., Fink and Fink, 1981, 1996; Arratia, 1992; Johnson and Patterson, 1996; Sato and Nakabo, 2002; Diogo, 2004a)*. The specimens of taxa of CS-0 have teeth on, or clearly associated with, the entopterygoid and/or ectopterygoid (and/or eventually the metapterygoid) [State 0]. In taxa of CS-1 there are no teeth on, or clearly associated with, these bones (e.g., Fig. 55). As stressed by Johnson and Patterson (1996), some adult *Alepocephalus* have teeth on the ectopterygoid and, thus, on the pterygoid region. However, certain adults of this genus analyzed in the present work do not have such teeth; the genus is thus coded as '?'. There is seemingly also a variation of this feature in adults of the genus *Distichodus*: in the specimens examined in this work there are no teeth on the pterygoid region, but Buckup (1998) mentioned that in the specimens he observed there were teeth on both the entopterygoid and the ectopterygoid; this genus is thus also coded as '?'. As explained above, the specimens of *Denticeps* present numerous odontodes in many bones of the skull, and it is difficult to discern, in certain cases, whether the "teeth" present in some bones are odontodes or "real teeth"; this genus is coded as '?'. The adult specimens of *Timon lepidus* and the adult, metamorphosed specimens of *Ambystoma ordinarium* examined in the present work do not have teeth on the pterygoid region; these taxa are thus coded as CS-0. †*Tiktaalik roseae* is coded as '?' because Daeschler et al. (2006) did not provide much detail on this character.
264. *Ectopterygoid with stout, prominent posteromesial process for articulation with ethmoid region (inspired from Thys van den Audenaerde, 1961).*

Unlike taxa of CS-0 [State 0], *Phractolaemus* exhibits this feature (e.g., Thys van den Audenaerde, 1961: fig. 17) [State 1].

265. *Ectopterygoid with "prominent dorsal process abutting in infraorbitals" (inspired from, e.g., Ridewood, 1904b; Forey et al., 1996).* Unlike taxa of CS-0 [State 0], *Albula* exhibits this feature [State 1].
266. *Presence of prominent "hyomandibular lateral spur at or below the level of the opercular process, projecting caudally to contact the preopercle" (inspired from, e.g., Fink, 1984; Howes and Sanford, 1987a; Begle, 1991, 1992; Johnson and Patterson, 1996; Sanford, 2000).* Unlike taxa of CS-0 [State 0], taxa of CS-1 exhibit this feature (e.g., Johnson and Patterson, 1996: fig. 4B) [State 1]. In Johnson and Patterson's (1996: 326) appendix 1, they code this character as unordered, *Plecoglossus* being, for example, coded as CS-1 ("short vertical crest fitting against preopercle"), *Osmerus* as CS-2 ("triangular spur"), and *Galaxias* as CS-3 ("obliquely orientated spurlike crest"). In contrast with what happens for most of the characters discussed by Johnson and Patterson (1996), in which I mostly agree with the criticism of these authors with respect to Begle's (1992) coding, regarding this specific character I prefer to follow Begle's (1992) coding. In fact, I do not consider that the peculiar "lateral hyomandibular spur at or below the level of the opercular process projecting caudally to contact the preopercle" found in these taxa being, for example, a little bit more vertical (e.g., in *Plecoglossus*) or a little less vertical (e.g., in *Galaxias*) is enough to exclude, *a priori*, the possibility that this feature eventually constitutes a primary homology within at least some taxa of CS-1.
267. *No hyomandibula, bony or cartilaginous (inspired from, e.g., Bemis, 1986; Bemis and Lauder, 1986).* As explained by these and other authors, unlike taxa of CS-0 [State 0], adults of the genus *Lepidosiren* do not seem to exhibit any structure, bony or cartilaginous, homologous with the hyomandibula of other osteichthyans (e.g., Bemis, 1986: fig. 4; Bemis and Lauder, 1986: fig. 1).
268. *Presence of prominent, long, thin and posteroventrally directed posterior process of hyomandibula.* Unlike taxa of CS-0 [State 0], *Gonorynchus* exhibits this feature (e.g., Monod, 1963: figs. 36 and 37) [State 1].
269. *Quadrate (ossified or not) with prominent anteroventromesial process (inspired from, e.g., Lenglet, 1974).* Unlike taxa of CS-0 [State], *Parakneria* and *Kneria* exhibit this feature (e.g., Lenglet, 1974: fig. 12) [State 1].
270. *Quadrate (ossified or not) with prominent, thin, anterodorsally directed anteroventrolateral process (inspired from, e.g., Thys van den Audenaerde, 1961; Howes, 1985a).* Unlike taxa of CS-0 [State 0], *Phractolaemus* exhibits this feature (e.g., Thys van den Audenaerde, 1961: fig. 17) [State 1].

271. *Metapterygoids not present as independent ossifications (inspired from, e.g., Monod, 1963; Pasleau, 1974; Arratia and Schultze, 1991; Begle, 1992; Bemis, 1986; Bemis and Lauder, 1986; Bartsch, 1994; Kemp, 1999; Johnson and Patterson, 1996; Grande and Poyato-Ariza, 1999).* Unlike in taxa of CS-0 [State 0], in taxa of CS-1 the metapterygoids are not present as independent ossifications (e.g., Fig. 101) [State 1]. †*Tiktaalik roseae* is coded as '?' because Daeschler et al. (2006) did not provide much detail on this character. Although more information is available for †*Acanthostega gunnari*, this taxon is also coded as '?' because some authors refer to an incomplete suture between the "epipterygoids" (which are usually considered to correspond to at least part of the metapterygoid of the present work) and the "pterygoids" (which correspond to the entopterygoids of the present work: see above), while others refer to seemingly independent "epipterygoids" (e.g., Clack, 1998, 2002; Kardong, 2002; Brazeau and Ahlberg, 2006). In the *Timon* specimens examined I found a complete separation between the "pterygoids" and the "epipterygoids"; thus, tentatively accepting the homology of these "epipterygoids" with at least part of the metapterygoids of the present work, I code this taxon as CS-0. Since in some adults of *Bathylagus* the metapterygoids, although small, are present, while in others they are missing, this genus is coded as '?'.
272. *Symplectic present as independent ossification (inspired from, e.g., Taverne, 1972, 1977b, 1978; Swinnerton, 1903; D'Aubenton, 1961; Thys van den Audenaerde, 1961; Howes, 1985a; Patterson, 1982; Jollie, 1986; Arratia and Schultze, 1991; Poyato-Ariza, 1996; Grande and Poyato-Ariza, 1999; Gardiner et al., 2005).* The plesiomorphic condition for the osteichthyan taxa included in the cladistic analysis is seemingly that in which the symplectic is not present as an independent ossification (e.g., Figs. 10, 89; Jollie, 1986; Gardiner et al., 2005) [State 0]. In taxa of CS-1 the symplectic is present as an independent ossification [State 1] (e.g., Fig. 55; see discussion below concerning the presence/absence of an ossified interhyal). Patterson (1982) argued that the "symplectics" of coelacanths such as *Latimeria* (e.g., Fig. 85) are seemingly not homologous to those found in other taxa coded as CS-1. However, I prefer not to completely exclude, *a priori*, the hypothesis that such structures might eventually be homologous. I thus prefer to tentatively code *Latimeria* as CS-1 and to check whether or not the results of the cladistic analysis support this primary homology hypothesis.
273. *Quadratojugals not present as independent ossifications (inspired from, e.g., Patterson, 1973; Jollie, 1980, 1984a,b,c; Benton, 1985; Arratia and Schultze, 1991; Cloutier, 1991; Grande and Bemis, 1991, Bartsch, 1994; Kemp, 1999; Clack, 2002; Daeschler et al., 2006).* Unlike in taxa of CS-0 (e.g., Fig. 10)

[State 0], in taxa of CS-1 the quadratojugals are not present as independent ossifications (e.g., Fig. 55) [State 1].

274. *More than one well-defined articulatory facet of suspensorium/ palatoquadrate for mandible (inspired from, e.g., Patterson, 1973; Miles, 1977; Jollie, 1984a,b; Bemis, 1986; Bemis and Lauder, 1986; Campbell and Barwick, 1986; Arratia and Schultze, 1991; Bartsch, 1994; Kemp, 1999).* Unlike in taxa of CS-0 [State 0], in taxa of CS-1 there is more than one well-defined articulatory facet of the suspensorium/palatoquadrate for the mandible (e.g., Figs. 21, 90) [State 1].
275. *Presence of peculiar toothed "dermometapterygoids" (inspired from, e.g., Arratia and Schultze, 1991; Grande and Bemis, 1998).* Unlike taxa of CS-0 [State 0], taxa of CS-1 exhibit peculiar toothed "dermometapterygoids" (e.g., Arratia and Schultze, 1991: figs. 22, 42B) [State 1].
276. *Quadrate (ossified or not) with large fossa or foramen on its anteroventral surface (inspired from, e.g., Diogo, 2004a).* Unlike taxa of CS-0 [State 0], taxa of CS-1 exhibit this feature (e.g., Diogo, 2004a) [State 1].
277. *Two hyomandibular articulatory heads for neurocranium (inspired from, e.g., Arratia and Schultze, 1991; Arratia, 1992, 1997, 1999; Cloutier and Ahlberg, 1996; Grande and Poyato-Ariza, 1999; Clack, 2001; Zhu and Schultze, 2001; Schultze and Cumbaa, 2001; Sato and Nakabo, 2002; Hilton, 2003).* The plesiomorphic condition for the osteichthyan taxa included in the present cladistic analysis is seemingly that in which the hyomandibula does not articulate with the neurocranium by means of two well-defined articulatory heads (e.g., Fig. 20) (although in some taxa coded as CS-0 such as *Megalops* it might seem that there are two separate hyomandibular cartilaginous articulatory heads for the neurocranium, these cartilaginous articulatory structures are in fact continuous) [State 0]. In taxa of CS-1 [State 1] the hyomandibula presents two well-defined, separate articular facets for the neurocranium (e.g., Fig. 48). The description of Filleul and Maisey (2004) regarding the articulations of the hyomandibula of †*Santanichthys diasii* is somewhat superficial: it states that, as in many other teleosts, there are two main articulations, one with the neurocranium and one with the opercle. This is also the case of the description of †*Clupavus maroccanus* provided by Taverne (1977a, 1995). The illustrations provided by these authors do not make it possible to discern whether in these two fossil species the articulation between the hyomandibula and the neurocranium is in fact realized by one or two separate hyomandibular articulatory heads. Therefore, until more information becomes available, I prefer to prudently code this character as '?' for these two species. In the *Neoceratodus* specimens examined I did not find two well-defined hyomandibular articulatory heads for the neurocranium such as those found in taxa

of CS-1; I am thus coding this taxon as CS-0; *Lepidosiren* is coded as "Inapplicable" since it seemingly lacks a hyomandibula (see above). †*Acanthostega* and *Latimeria* are coded as CS-1 (e.g., Millot and Anthony, 1958; Cloutier and Ahlberg, 1996; Clack, 1998, 2002). †*Tiktaalik roseae* is coded as '?', because no detailed information about this character was given by Daeschler et al. (2006). I prudently code *Timon* and *Ambystoma* as '?', since the specimens of these two taxa analyzed in the present work exhibit rather peculiar types of association between the hyomandibula (which is often called "stapes" in these taxa) and the neurocranium. For instance, in the *Ambystoma* specimens analyzed the hyomandibula is a peculiarly small element that does not really articulate directly with the neurocranium by means of one or more well-defined articulatory heads such as those found in taxa of CS-0 and CS-1, but rather attaches to the neurocranium mainly by means of short ligamentous tissue (e.g., Fig. 102).

278. *Presence of prominent posterodorsal projection of hyomandibula (or hyomandibulo-metapterygoid) firmly attached to neurocranium by strong connective tissue (inspired from, e.g., Diogo, 2004a).* Unlike taxa of CS-0 [State 0], taxa of CS-1 exhibit this feature (e.g., Diogo, 2004a: fig. 3-89) [State 1].
279. *Presence of peculiar, broad "cartilaginous palatal complex" (inspired from, e.g., Findeis, 1997).* Unlike taxa of CS-0 [State 0], *Acipenser* exhibits this feature (e.g., Findeis, 1997: fig. 10) [State 1].
280. *Prominent articulatory facet on posterolateral margin of suspensorium/palatoquadrate for mesial surface of preopercle.* The taxa of CS-1 [State 1] exhibit a prominent articulatory facet on the posterolateral margin of the suspensorium/palatoquadrate (often at, or near, the region where the hyomandibula, the metapterygoid and/or the symplectic meet) for the mesial surface of the preopercle. Since in many of these taxa there is some space between the anterior margin of the preopercle and the posterior margin of the suspensorium/palatoquadrate (e.g., Jollie, 1975: fig. 5), the preopercle can, by means of this well-developed articulation, articulate with, and thus be somewhat mobile in relation to, the suspensorium/palatoquadrate. This is not the case in the specimens examined of taxa of CS-0 [State 0].
281. *Absence of subopercle as an independent element (inspired from, e.g., Tchernavin, 1947b; Taverne, 1972, 1977b, 1978; Larsen and Guthrie, 1975; Fink and Fink, 1981, 1996; Forey, 1991; Clack, 2002; Bels et al., 1993; Hilton, 2003; Daeschler et al., 2006).* Unlike taxa of CS-0 [State 0], taxa of CS-1 exhibit this feature (e.g., Fig. 101) [State 1].
282. *Presence of interopercle as an independent element (inspired from, e.g., Tchernavin, 1947a; Kershaw, 1970; Patterson, 1973; Taverne, 1972, 1977b,*

1978; Jollie, 1980, 1984c, 1986; Grande and Bemis, 1991; Findeis, 1997). Unlike taxa of CS-0 [State 0], taxa of CS-1 exhibit this feature (e.g., Fig. 101) [State 1]. According to Kershaw (1970) an independent interopercle is missing in most specimens of the genus *Pantodon* but may eventually be present in others; this genus is thus coded as '?'.

283. *Subopercle articulating directly with hyoid arch by a distinguishable, prominent anterior spine (inspired from, e.g., Ridewood, 1904a).* Unlike taxa of CS-0 [State 0], taxa of CS-1 exhibit this feature (e.g., Ridewood, 1904a: fig. 135C) [State 1].
284. *No independent opercular bone (inspired from, e.g., Tchernavin, 1947a; Grande and Bemis, 1991, 1996; Bemis et al., 1997; Zhu and Schultze, 2001; Cloutier and Arratia, 2004).* Unlike taxa of CS-0 [State 0], taxa of CS-1 exhibit this feature (e.g., Fig. 108) [State 1].
285. *Opercle exhibiting a peculiar, deep, dorsoventrally elongated mesial crest.* Unlike taxa of CS-0 [State 0], *Gonorynchus* displays this feature (e.g., Monod, 1963: fig. 16) [State 1].
286. *Presence of distinct fenestra on anterodorsal surface of opercle (inspired from Vari, 1979).* Unlike taxa of CS-0 [State 0], *Distichodus* exhibits this feature (e.g., Vari, 1979: fig. 19B) [State 1].
287. *Preopercle not present as an independent ossified element (inspired from, e.g., Tchernavin, 1947a; Grande and Bemis, 1991, 1996; Bemis et al., 1997; Findeis, 1997).* Unlike taxa of CS-0 [State 0], taxa of CS-1 exhibit this feature (e.g., Fig. 101) [State 1].
288. *Opercle with peculiar, roughly triangular shape (inspired from, e.g., Fink and Fink, 1981, 1996).* Unlike taxa of CS-0 [State 0], taxa of CS-1 exhibit this feature (e.g., Fig. 83) [State 1].
289. *Posterior and/or posteroventral portions of opercle exhibiting several peculiar thin spines (inspired from, e.g., Gegenbaur, 1878; Gosline, 1969; Greenwood and Rosen, 1971).* Unlike in taxa of CS-0 [State 0], in taxa of CS-1 the posterior and/or posteroventral portions of the opercle, and often also of the subopercle, interopercle and/or preopercle, exhibit several peculiar thin spines (e.g., Fig. 38) [State 1]. Some specimens of the genera *Elops*, *Megalops*, *Galaxias* and *Esox* exhibit such spines, while others seemingly do not (e.g., Ridewood, 1904b; Vrba, 1968; Taverne, 1974; Winterbottom, 1974; this work); these genera are thus coded as '?'.
290. *Presence of prominent, thin, dorsally oriented anterodorsal projection of opercle (inspired from, e.g., Diogo, 2004a).* Unlike taxa of CS-0 [State 0], taxa of CS-1 exhibit this feature (e.g., Figs. 53, 55) [State 1].
291. *Posterior margin of preopercle exhibiting prominent, enlarged "spine-like process" (inspired from, e.g., Greenwood, 1968).* Unlike taxa of CS-0 [State 0], *Denticeps* exhibits this feature (e.g., Fig. 46) [State 1].

292. *Preopercles markedly expanded ventrally, one preopercle overlapping the other along ventral midline (inspired from, e.g., Ridewood, 1905a; Thys van den Audenaerde, 1961; Howes, 1985a; Poyato-Ariza, 1996; Grande and Poyato-Ariza, 1999).* Unlike taxa of CS-0 [State 0], *Phractolaemus* exhibits this feature (e.g., Grande and Poyato-Ariza, 1999) [State 1].
293. *Absence of ligamentous connection between interopercle and mandible (inspired from, e.g., Diogo, 2004a).* Among those taxa included in the cladistic analysis having interopercles, the plesiomorphic condition seems to be that in which these structures are ligamentously connected to the mandibles (e.g., Fig. 46) [State 0]. In taxa of CS-1 there is no ligamentous connection between the interopercles and the mandibles (e.g., Diogo, 2004a) [State 1].
294. *Presence of peculiar, broad anterolateral articulatory facet of interopercle for articulation with quadrate.* Unlike taxa of CS-0 [State 0], taxa of CS-1 exhibit this feature (e.g., Fig. 66B) [State 1].
295. *Presence of well-defined, long, strong ligament running from anterodorsal surface of interopercle to posterodorsal surface of preopercle (inspired from, e.g., Diogo, 2004a).* Unlike taxa of CS-0 [State 0], taxa of CS-1 exhibit this feature (e.g., Diogo, 2004a) [State 1].
296. *Interopercle reduced to a long, very thin spine (inspired from, e.g., Thys van den Audenaerde, 1961; Howes, 1985a; Grande and Poyato-Ariza, 1999).* Unlike taxa of CS-0 [State 0], taxa of CS-1 exhibit this feature (e.g., Grande and Poyato-Ariza, 1999) [State 1].
297. *Absence of contact between opercle/subopercle and interopercle (inspired from, e.g., Gosline, 1969).* Unlike taxa of CS-0 [State 0], *Xenodermichthys* exhibits this feature [State 1]. (e.g., Gosline, 1969: fig. 3A).
298. *No ossified ceratohyals (inspired from, e.g., Patterson, 1982; Jollie, 1986).* Unlike in taxa of CS-0 [State 0], in taxa of CS-1 [State 1] there are no ossified ceratohyals (e.g., Fig. 99).
299. *Presence of two independent, ossified ceratohyals (inspired from, e.g., Patterson, 1982; Bemis, 1986; Bemis and Lauder, 1986; Bartsch, 1994; Kemp, 1999; Gardiner and Schaeffer, 1989; Gardiner et al., 2005).* Among those taxa with ossified ceratohyals (see character above) the plesiomorphic condition is seemingly that in which in each side of the body there is a single ossified ceratohyal element [State 0]. In taxa of CS-1 [State 1] there are two independent, ossified ceratohyals in each side of the body (e.g., Fig. 61). This feature has been the subject of much controversy. For instance, Jollie (1984c), Lauder (1980a), and Arratia and Schultze (1990) consider that *Polypterus* has two independent, ossified ceratohyals in each side of the body, which would thus correspond to the anterior and posterior ceratohyals of the present work (e.g., Arratia and Schultze, 1990: fig. 1). However, Edgeworth

(1935), Jessen (1968), Jarvik (1963), Patterson (1982), Gardiner and Schaeffer (1989), Gardiner et al. (1996, 2005) and other authors have interpreted the more posterior of these two bones as an ossified interhyal (e.g., Patterson, 1982: fig. 1C). The observations of the present work strongly support the view of these latter authors (see Fig. 6). In fact, in *Polypterus* the posterior bone is significantly more mobile in relation to the anterior one than is the ossified posterior ceratohyal of the taxa of CS-1 in relation to the anterior ceratohyal; such mobility is, in fact, much like that commonly seen between the interhyal and the hyoid arch in most taxa analyzed. Also, in *Polypterus* there is short ligamentous tissue between the anterior bone and the posterior one; in other taxa examined ligamentous tissue such as this is usually found between the interhyal and the posterior ceratohyal, and not between the anterior ceratohyal and the posterior ceratohyal. Moreover, in *Polypterus* the mandibulohyoid ligament runs from the mandible to the anterior bone, and not to the posterior one (e.g., Fig. 6). In the vast majority of osteichthyans with two ossified ceratohyals and with a distinct mandibulohyoid ligament this latter ligament inserts on the posterior, and not on the anterior, ceratohyal. These arguments thus do seem to indicate that the more anterior bone in *Polypterus* includes both the anterior and posterior ceratohyals (or, eventually, that the posterior ceratohyal is missing or remains unossified in the adult members of this genus) and that the more posterior one corresponds to the interhyal of other osteichthyans (see, e.g., Fig. 6; Patterson, 1982: fig. 1C).

300. *Presence of independent, ossified dermohyal (inspired from, e.g., Patterson, 1982; Jollie, 1984c; Zhu and Schultze, 2001; Schultze and Cumbaa, 2001; Cloutier and Arratia, 2004; Gardiner et al., 2005).* Unlike taxa of CS-0 [State 0], specimens of *Polypterus* have an independent, ossified dermohyal (e.g., Patterson, 1982: fig. 1C) [State 1]. Gardiner et al. (2005) suggested (or at least this is my interpretation of their text) that *Lepisosteus* has a dermohyal. In the *Lepisosteus* specimens examined I could not find a structure corresponding to the dermohyal I observed in the specimens of *Polypterus*. It is not clear if the statement of Gardiner et al. (2005) is thus due to a real variation of this character in the members of the genus *Lepisosteus* or if it is due to an error. I prudently code *Lepisosteus* as '?'.
301. *"Posterodorsal portion of ceratohyal reaching far dorsally" (inspired from, e.g., Rosen et al., 1981; Forey, 1986).* In the specimens examined of taxa of CS-1 (e.g., Fig. 93) [State 1] the posterodorsal portion of the ceratohyal extends far more dorsally than the posterodorsal portion of the ceratohyal (or of the posterior ceratohyal, in those taxa with both

anterior and posterior ceratohyals) of taxa of CS-0 (e.g., Fig. 85) [State 0].

302. *Absence of ossified interhyal in adults (inspired from, e.g., Ridewood, 1904c, 1905a; Taverne, 1972, 1977b, 1978; Patterson, 1982; Jollie, 1984a,b, 1986; Mo, 1991; Gayet, 1993; Grande and Poyato-Ariza, 1999).* Unlike taxa of CS-0 (e.g., Figs. 6, 85) [State 0], taxa of CS-1 do not have an independent, ossified interhyal (e.g., Fig. 77) [State 1]. Some members of the genera *Chanos, Gonorynchus, Phractolaemus, Nematogenys* and *Catostomus* do not have an ossified interhyal, while others seemingly exhibit such a structure (e.g., Weisel, 1960; Arratia, 1992; Grande and Poyato-Ariza, 1999; Diogo, 2004a; this work): these genera are coded as '?'.

303. *Interhyal (ossified or not) thin and markedly elongated dorsoventrally (inspired from, e.g., Tchernavin, 1953; Günther and Deckert, 1959; Weitzman, 1967a,b).* Unlike taxa of CS-0 [State 0], *Stomias* exhibits this feature [State 1] (e.g., Günther and Deckert, 1959: fig. 33).

304. *Interhyal (ossified or not) connected by ligaments to, and thus not articulating directly with, hyoid arch and suspensorium (inspired from, e.g., Arratia, 1992; Diogo, 2004a).* Ossified or not, the interhyal of taxa of CS-0 usually articulates directly with the hyoid arch and/or the suspensorium [State 0]. In taxa of CS-1 the interhyal is connected by strong, somewhat long ligaments to the hyoid arch and to the suspensorium and thus does not articulate directly with these structures (e.g., Diogo, 2004a: fig. 3-66) [State 1]. Arratia (1992) described this derived in catfishes. However, the gymnotiform specimens examined of the genera *Sternopygus, Gymnotus* and *Brachyhypopomus* also have distinct ligaments between the hyoid arch and the interhyal and between this latter structure and the suspensorium; this is also the case in the specimens of these genera examined by other authors, such as De la Hoz and Chardon (1984: e.g., their fig. 8).

305. *Interhyal (ossified or not) with peculiar, somewhat dumbbell shape (inspired from, e.g., Begle, 1992).* Unlike taxa of CS-0 [State 0], taxa of CS-1 exhibit this feature [State 1]. The original definition of this character by Begle (1992) was "interhyal short, dumbbell-shaped". However, as stressed by, for example, Johnson and Patterson (1996), in taxa such as *Bathylagus* the interhyal is not significantly shorter than in certain other taxa coded as CS-0. Therefore, I prefer to define the character as "interhyal with peculiar, somewhat dumbbell shape": such a description of the interhyal does apply to the condition found in the adult specimens examined in the present work of those taxa coded as CS-1.

306. *Absence of ligamentous connection between hyoid arch and mandible.* Unlike in taxa of CS-0 (e.g., Fig. 6) [State 0], in the specimens examined of taxa of CS-1 there is no distinct ligament between the hyoid arch and the mandible [State 1]. I did not find a distinct mandibulohyoid ligament in the *Timon* specimens examined; I am thus coding this taxon as CS-1.
307. *Mandibulohyoid and mandibulointeropercular ligaments not well separated from each other.* Within those taxa having mandibulohyoid and mandibulointeropercular ligaments (see above), the plesiomorphic condition is seemingly that in which these ligaments are mainly well separated from each other (e.g., Fig. 19), the former running from the mandible to the hyoid arch and the latter running from the mandible to the interopercle (e.g., Fig. 19) [State 0]. This is not the case in the specimens examined of the taxa of CS-1 [State 1], in which one of these conditions, or a combination of them, occurs: the two ligaments are deeply mixed anteriorly, giving the appearance that there is a single ligament that bifurcates posteriorly to attach on the posterior ceratohyal and on the interopercle (e.g., Fig. 43); a significant portion of the mandibulointeropercular ligament attaches also on the posterior ceratohyal; a significant portion of the mandibulohyoid ligament attaches also on the interopercle. Some authors referred to a "shift in insertion of the mandibulohyoid ligament to the interopercle" as a potential synapomorphy of the Eurypterygii (e.g., Lauder and Liem, 1983; Johnson, 1992) or of the Neoteleostei (e.g., Rosen, 1985). Thus, according to Rosen (1985) taxa such as the basal neoteleostean stomiiforms and the basal eurypterygian aulopiforms seemingly exhibit such a feature, while according to Lauder and Liem (1983) and Johnson (1992) the stomiiforms do not exhibit such a feature, but the aulopiforms do. Stiassny (1996) and Sato and Nakabo (2002), however, argued that such a feature is not present in Stomiiformes nor in many Aulopiformes. The observations of the present work support the statement of these latter authors. This is because the stomiiform and aulopiform taxa analyzed exhibit a condition similar to that found in many non-neoteleostean taxa coded as CS-1: the mandibulohyoid and mandibulointeropercular ligaments are both present but are deeply mixed anteriorly, giving the appearance that there is a single ligament bifurcating posteriorly to attach on the posterior ceratohyal and on the interopercle (e.g., Fig. 43).
308. *Presence of well-developed, posterodorsally pointed dorsal process of posterior ceratohyal (inspired from Thys van den Audenaerde, 1961).* Unlike taxa of CS-0 [State 0], *Phractolaemus* exhibits this feature (e.g., Thys van den Audenaerde, 1961: fig. 17) [State 1].

309. *Presence of well-developed ventrolateral laminar expansion of anterior ceratohyal (inspired from, e.g., Diogo, 2004a).* Unlike taxa of CS-0 [State 0], *Nematogenys* exhibits this feature (e.g., Diogo, 2004a) [State 1].
310. *Presence of prominent, broad anteroventral lamina of anterior ceratohyal (inspired from, e.g., Diogo, 2004a).* Unlike taxa of CS-0 [State 0], *Nematogenys* exhibits this feature (e.g., Diogo, 2004a) [State 1].
311. *Presence of tooth plates associated with ceratohyals (inspired from, e.g., Taverne, 1974; Arratia and Schultze, 1990).* Unlike taxa of CS-0 [State 0], taxa of CS-1 have peculiar tooth plates associated with the ceratohyals (e.g., Arratia and Schultze, 1990: fig. 2C,D) [State 1]. It should be noted that, like members of the genus *Elops*, the *Megalops* specimens examined exhibit such tooth plates (*Megalops* was not listed among the taxa examined by Arratia and Schultze, 1990). The descriptions provided by Millot and Anthony (1958) regarding the presence/absence of such tooth plates in *Latimeria* are not very clear: in some instances they give the idea that *Latimeria* may eventually exhibit tooth plates associated with the ceratohyals, but in other instances they seem to indicate that this is not the case. Since I could not personally examine the hyoid arch of members of this taxon, I prefer to prudently code it here as '?'.
312. *Peculiar articulation between prominent anteromesial process of anterior ceratohyal and broad, deep, circular concavity formed by the lateral margins of both the urohyal and basihyal (inspired from, e.g., Ridewood, 1904c; Taverne, 1972).* Unlike taxa of CS-0 [State 0], *Mormyrus* exhibits this feature (e.g., Taverne, 1972: figs. 9, 10) [State 1].
313. *Main body of urohyal/"urohyal"/parahyoid/"tendon urohyal"/parurohyal markedly elongated anteroposteriorly, being inclusively longer than the whole hyoid arch.* Unlike in taxa of CS-0 (e.g., Fig. 99) [State 0], in taxa of CS-1 the main body of the urohyal/"urohyal"/parahyoid/"tendon urohyal"/parurohyal is markedly thin and long, being longer than the whole hyoid arch (i.e., hypohyals, anterior ceratohyal, and/or posterior ceratohyals, when these elements are present) (e.g., Fig. 61) [State 1].
314. *Urohyal exhibiting one broad, circular, deep concavity in each of its anterodorsolateral margins for lodging anteroventral portions of anterior ceratohyals (inspired from, e.g., Belouze, 2002).* Unlike taxa of CS-0 [State 0], taxa of CS-1 exhibit this feature (e.g., Belouze, 2002: fig. 37) [State 1].
315. *No branchiostegal rays (inspired from, e.g., Tchernavin, 1947a; Jollie, 1984c; Arratia and Schultze, 1990; Cloutier and Arratia, 2004).* According to, for example, Cloutier and Arratia (2004) the branchiostegal rays are plesiomorphically present in osteichthyans [State 0]; I am thus tentatively coding the presence of these structures as CS-0 [State 0]

and their absence as CS-1 (e.g., Fig. 99) [State 1]. In the extant sarcopterygians examined I did not find any structures seemingly corresponding to the branchiostegal rays found in taxa of CS-0. Branchiostegal rays such as those found in taxa of CS-0 also seem to be lacking in †*Acanthostega gunnari* and †*Tiktaalik roseae* (e.g., Cloutier and Ahlberg, 1996; Daeschler et al., 2006).

316. *Few, highly modified branchiostegal rays (inspired from, e.g., Grande and Bemis, 1991; Findeis, 1997)*. Unlike taxa of CS-0 [State 0], taxa of CS-1 exhibit this feature (e.g., Fig. 12) [State 1].
317. *Presence of ossified urohyal/"urohyal"/parahyoid/"tendon urohyal"/ parurohyal (inspired from, e.g., Forey, 1980; Arratia and Schultze, 1990)*. In contrast with the condition seemingly found in acanthodians and found in many osteichthyans included in the cladistic analysis [State 0], taxa of CS-1 have an ossified urohyal/"urohyal"/parahyoid/ "tendon urohyal"/parurohyal (e.g., Figs. 61, 86, 99) [State 1]. Although Arratia and Schultze (1990) have shown that some of these "different types" or urohyals are seemingly developed in different ways, I prefer not to completely exclude, *a priori*, the hypothesis that at least some of them may be phylogenetically homologous (for example, new discoveries in the field of evolutionary developmental biology are revealing more and more cases of structures that are considered phylogenetically homologous but that may show different types of ontogenetic development: e.g., Gould, 2002; West-Eberhard, 2003; Carroll et al., 2005; Kirschner and Gerhart, 2005).
318. *Branchiostegal rays exhibiting peculiar long and thin cartilage for articulation with hyoid arch (inspired from Reis, 1998)*. Unlike taxa of CS-0 [State 0], *Callichthys* exhibits this feature [State 1] (e.g., Reis, 1998).
319. *Presence of several peculiar branchiostegal photophores (inspired from, e.g., Harold and Weitzman, 1996)*. Unlike taxa of CS-0 [State 0], taxa of CS-1 exhibit this feature [State 1].
320. *Absence of ossified hypohyals (inspired from, e.g., Millot and Anthony, 1958; Taverne, 1972; Jollie, 1980; Grande and Bemis, 1991; Janvier, 1996; Hilton, 2003)*. Unlike in taxa of CS-0 (e.g., Fig. 61) [State 0], which have ossified hypohyals (both ventral and dorsal ossified hypohyals may eventually be present), in taxa of CS-1 there are no such ossified structures (e.g., Fig. 99). Although very small, hypohyals are seemingly present in *Mormyrus*, as explained by, for example, Taverne (1972) and Hilton (2003). Some authors consider that the "hypohyals" found in, for example, tetrapod taxa such as the lizard *Timon*, in sarcopterygian fishes such as the coelacanth *Latimeria*, and in actinopterygian fishes such as *Polypterus* may be non-homologous; however, for the reasons explained above, I prefer not to completely

exclude the hypothesis that at least some of these "hypohyals" are eventually homologous. Therefore, I prefer to tentatively code here the presence of ossified "hypohyals" as the seemingly plesiomorphic condition for osteichthyans and to check whether or not the results of the cladistic analysis support this primary homology hypothesis. There is some confusion concerning this character in the descriptions and illustrations of †*Acanthostega* provided in the literature; this taxon is thus prudently coded here as '?'. *Timon* and *Psephurus* are also coded as '?' since in the members of these species there is seemingly a variation of this character: in some cases the hypohyals are completely cartilaginous, in others they may be partly ossified (e.g., Edgeworth, 1935; Smith, 1988; Herrel et al., 2005; this study).

321. *Some branchiostegal rays articulating with hypohyals (inspired from, e.g., McAllister, 1968; Fink and Weitzman, 1982; Baldwin and Johnson, 1996; Harold and Weitzman, 1996).* Unlike taxa of CS-0 [State 0], taxa of CS-1 exhibit this feature [State 1] (e.g., Weitzman, 1967b: fig. 11). It should be noted that, as observed by, for example, Baldwin and Johnson (1996), certain aulopiforms not included in the present study as well as certain other teleosts such as some myctophiforms also have some branchiostegal rays articulating with the hypohyals.
322. *Absence of ossified gular plates (inspired from, e.g., Greenwood et al., 1966; Jessen, 1968; Taverne, 1974; Forey, 1980; Arratia and Schultze, 1991; Forey et al., 1996; Arratia, 1999; Taverne, 1999; Hilton, 2003; Cloutier and Arratia, 2004; Daeschler et al., 2006; Cavin et al., in press).* Unlike taxa of CS-0 (e.g., Figs. 19A, 25) [State 0], taxa of CS-1 lack ossified gular plates (e.g., Fig. 99) [State 1]. Forey et al. (1996) coded *Albula* as having ossified gular plates, but in the specimens examined of this genus this was not so clear; this taxon is thus prudently coded as '?'.
323. *Absence of mandibular teeth and tooth plates (inspired from, e.g., Allis, 1922; Regan 1911a,b; Alexander, 1965; Schaeffer, 1973; Gosline, 1975; Howes and Sanford, 1987b; Fink and Fink, 1981, 1996; Poyato-Ariza, 1996; Arratia 1997, 1999).* Unlike taxa of CS-0 [State 0], taxa of CS-1 exhibit this feature (e.g., Fig. 53) [State 1]. The "mandibular teeth" described by Monod (1963) in specimens of *Gonorynchus* are non-osseous structures (see above); this genus is thus coded as CS-1. As explained above, the specimens of *Denticeps* present numerous odontodes on many bones of the skull, and it is somewhat difficult to discern, in certain cases, whether the "teeth" present on some bones are odontodes or "true teeth"; *Denticeps* is therefore prudently coded as '?' here. Since in some adults of *Coregonus*, *Psephurus* and *Acipenser* there are small mandibular teeth while in others such teeth are seemingly missing (e.g., Shaposhnikova, 1967; Nelson, 1973; Grande and Bemis, 1991; Sanford, 2000; this work), these three genera are coded as '?'. It is

not possible to discern this character in †*Lusitanichthys characiformis* because mandibular teeth are shown in Gayet's (1981) fig. 4 but are not shown in Gayet's (1985) fig. 18 of the same species. In a paper published after Gayet's (1981, 1985) works, Cavin (1999) described a new species of †*Lusitanichthys*, †*L. africanus*, which, according to that author, does not have mandibular teeth. Until more information is available, I prefer to prudently code †*Lusitanichthys characiformis* as '?'.

324. *Surangulars present as independent ossifications (inspired from, e.g., Patterson, 1973; Gardiner and Schaeffer, 1989; Arratia, 1999; Kemp, 1999; Clack, 2002; Cloutier and Arratia, 2004).* Unlike in taxa of CS-0 [State 0], in taxa of CS-1 the surangulars (= "supraangulars" of, e.g., Cloutier and Arratia, 2004) are present as independent ossifications (e.g., Fig. 97) [State 1].

325. *Coronoid bones not present as independent ossifications (inspired from, e.g., Nelson, 1973; Patterson, 1973; Ahlberg and Clack, 1998; Forey, 1998; Arratia, 1999; Kemp, 1999; Clack, 2002; Cloutier and Arratia, 2004; Daeschler et al., 2006).* Unlike in taxa of CS-0 [State 0], in taxa of CS-1 the coronoid bones are not present as independent ossifications (e.g., Fig. 54) [State 1]. *Ambystoma ordinarium* is coded as '?' because, as in other species of *Ambystoma*, there is seemingly a variation of this character in the adult members of this species: in some instances the coronoid bones seem to be present as very small ossifications, while in others they are seemingly completely missing (e.g., Bonebrake and Brandon, 1971; Larsen and Guthrie, 1975; Lauder and Shaffer, 1985; this work).

326. *Prearticulars not present as independent ossifications (inspired from, e.g., Nelson, 1973; Patterson, 1973; Bemis, 1986; Grande and Bemis, 1991; Findeis, 1997; Arratia, 1999; Kemp, 1999; Cloutier and Arratia, 2004).* Unlike in taxa of CS-0 [State 0], in taxa of CS-1 the prearticulars are not present as independent ossifications (e.g., Fig. 54) [State 1]. *Acipenser* is coded as '?' because some adults of this genus exhibit a configuration such as CS-1, while others exhibit a configuration such as CS-0 (e.g., Nelson, 1973; Jollie, 1980; Grande and Bemis, 1991; this work). *Ambystoma ordinarium* is also coded as '?' since, as explained by, for example, Bonebrake and Brandon (1971), the so-called "prearticular" of the members of this and other *Ambystoma* species may actually be a prearticulo-angular, but this is far from being clear (e.g., Fig. 100). *Timon* is coded as CS-1 because in the specimens examined of this taxon the prearticular, the articular, and possibly the retroarticular of this work are fused in a single element, as is the case in various other lizards (e.g., Fig. 108; Frazzetta, 1962: fig. 1).

327. *Presence of "strong, immobile bony mandibular symphysis including exclusively the prearticulars" (inspired from, e.g., Bemis, 1986; Grande and*

Bemis, 1986). Unlike taxa of CS-0 [State 0], *Lepidosiren* exhibits this feature (e.g., Bemis, 1986: fig. 6E) [State 1].

328. *Mentomeckelian bones not present as independent ossifications (inspired from, e.g., Nelson, 1973; Jollie, 1984c; Grande and Bemis, 1991; Findeis, 1997; Ahlberg and Clack, 1998; Clack, 2002; Jeffery, 2003).* Unlike in taxa of CS-0 [State 0], in taxa of CS-1 mentomeckelian bones are not present as independent ossifications (e.g., Fig. 54) [State 1]. There is some confusion in the literature, particularly in the literature referring to sarcopterygians, on the use of the term "mentomeckelian bones". Therefore, it is sometimes difficult to appraise whether mentomeckelian bones sensu this work (= sensu, e.g., Nelson, 1973) are or are not present as independent ossifications in certain taxa included in the cladistic analysis that I could not personally examine, such as †*Acanthostega gunnari*. I will, however, tentatively code this latter species as CS-0, because in the review by Jeffery (2003: 271), in which this author provides some useful comparisons between the condition found in certain sarcopterygian fishes with that found in certain tetrapods and certain actinopterygians, this author explicitly states that in †*Acanthostega* "at least part of the mentomeckelian ossification remains". *Acipenser* is coded as '?' because some adult members of this taxon exhibit a configuration such as CS-1, while others exhibit a configuration such as CS-0 (e.g., Nelson, 1973; Jollie, 1980; Grande and Bemis, 1991; this work).
329. *No separate bones corresponding to splenials, to postsplenials or to splenialpostsplenials (inspired from, e.g., Kemp, 1999; Ahlberg and Clack, 1998; Clack, 2002; Jeffery, 2003; Cloutier and Arratia, 2004).* The plesiomorphic condition for the osteichthyan taxa included in the cladistic analysis is seemingly that found in most sarcopterygians and in basal actinopterygians such as †*Dialipina*, in which splenial, postsplenial and/or splenialpostsplenial bones (sensu, e.g., Kemp, 1999: fig. 4) are present (e.g., Fig. 97; Cloutier and Arratia, 2004) [State 0]. In taxa of CS-1 [State 1] there are no separate bones corresponding to the splenials, to the postsplenials or to the splenialpostsplenials of taxa of CS-0 (e.g., Fig. 54).
330. *Articular mainly fused with angular (and/or retroarticular) (inspired from, e.g., Greenwood, 1968; Nelson, 1973; Taverne, 1974; Grande 1985a,b; Arratia, 1997, 1999; Hilton, 2003).* Unlike in taxa of CS-0 (e.g., Fig. 16) [State 0], in taxa of CS-1 the articular is mainly fused with the angular (and/or retroarticular) (e.g., Fig. 54) [State 1]. According to, for example, Nelson (1973), Taverne (1974) and Hilton (2003) the adult members of *Megalops* and *Elops* exhibit an articular seemingly not fused with the angulo-retroarticular. However, Arratia (1999: 275, 289) coded *Megalops* as having an articular "fused with angular and

retroarticular bones". As the observations of the present work did not allow us to discern this feature in *Megalops*, this genus is prudently coded as '?'. Arratia (1999: 275, 289) coded *Elops* as having an articular "partially fused with anguloretroarticular late in ontogeny". *Elops* is coded here as CS-0, since, even if in some adult specimens of this genus the angular is "partially" fused with the anguloretroarticular, as stated by Arratia (1999), it is not "mainly" fused with this latter structure. As stressed by, for example, Bartsch (1994), although many researchers use the name "angular" to designate one of the mandibular bony elements of *Neoceratodus* and *Lepidosiren* (a procedure that I followed in Fig. 92), it is not really known whether this element includes only the angular or also includes other structures. However, what seems to be clear is that none of the bony elements constituting the mandibles of the adult members of these two dipnoan genera include structures corresponding to the articulars and/or retroarticulars of other osteichthyans; these two genera are thus coded as CS-0. The so-called "articulars" of tetrapods such as *Timon lepidus, Ambystoma ordinarium* and †*Acanthostega gunnari*, as well as of taxa such as †*Tiktaalik roseae*, seem to correspond to both the articulars and retroarticulars of the present work; the so-called "retroarticular process" often found in these sarcopterygian taxa seemingly corresponds, as its name indicates, to part of the retroarticular. According to, for example, Nelson (1973), Forey (1991) and Kardong (2002), the presence of an articulo-retroarticular compound, either resulting from a single ossification center or from two centers followed by an ulterior fusion, seems to be a rather common feature within sarcopterygians. Therefore, I am tentatively coding *Timon, Ambystoma ordinarium*, †*Acanthostega gunnari* and †*Tiktaalik roseae* as CS-1.

331. *Retroarticular not present as independent ossification (inspired from, e.g., Frazzetta, 1962; Nelson, 1973; Arratia, 1992, 1997; Ahlberg and Clack, 1998; Clack, 2002; Daeschler et al., 2006).* Unlike in taxa of CS-0 (e.g., Fig. 16) [State 0], in taxa of CS-1 the retroarticular is not present as an independent element (e.g., Fig. 100) [State 1]. According to, e.g., Nelson (1973) and Hilton (2003), in *Hiodon* the retroarticular is fused with the angular. However, the *Hiodon* specimens illustrated by Arratia (1997: fig. 85B) and Taverne (1977b: fig. 15) appear to have an independent retroarticular. This genus is thus prudently coded as '?'.
332. *Retroarticular not included in quadrate-mandibular joint (inspired from, e.g., Nelson, 1973; Lauder and Liem, 1983; Arratia, 1992, 1999; Hilton, 2003).* In taxa of CS-0 the retroarticular is included in the quadrate-mandibular joint (e.g., Fig. 16) [State 0]. In taxa of CS-1 the retroarticular is excluded from this joint (e.g., Fig. 54) [State 1]. Arratia

(1992) stated that adult gymnotiforms display a condition such as CS0. However, although some adult gymnotiforms do indeed exhibit this condition, in others, such as the adult *Sternopygus* specimens examined in the present work, as well as in studies such as De la Hoz and Chardon (1984), the retroarticular is not included in the quadrate-mandibular joint (e.g., De la Hoz and Chardon, 1984: figs. 2, 8). Nelson (1973: 340) documented that in the notopterid osteoglossomorphs (which include *Xenomystus*), as well as in the osteoglossomorph *Pantodon*, the retroarticular is not included in the quadrato-mandibular joint. But Hilton (2003) questioned this statement, since according to him in *Pantodon* the retroarticular is included in this joint. Concerning *Xenomystus*, Hilton (2003) stated: "of the notopterids examined here, in *Xenomystus* the retroarticular comes closest to being in contact with the jaw joint". In the specimens of *Xenomystus* and *Pantodon* analyzed in the present work, the retroarticular is included in the quadrate-mandibular joint. So, there is seemingly a variation of this character within the members of the genus *Xenomystus* and of the genus *Pantodon*. These two genera are accordingly coded as '?'. It should be noted that those taxa coded as CS-1 in the character above were coded as "Inapplicable" in the present character, since in those taxa the retroarticular is not present as an independent ossification.

333. *Mandible highly modified, presenting a series of peculiar, prominent, lateral, mesial and dorsal processes (inspired from, e.g., Lenglet, 1974; Howes, 1985a; Grande and Poyato-Ariza, 1999)*. Unlike taxa of CS-0 [State 0], taxa of CS-1 exhibit this feature (e.g., Howes, 1985a: fig. 13) [State 1].
334. *Right and left halves of lower jaw loosely attached to each other at midline (inspired from, e.g., Howes and Sanford, 1987b)*. Unlike taxa of CS-0 [State 0], taxa of CS-1 exhibit this feature (e.g., Howes and Sanford, 1987b) [State 1].
335. *Dorsal tip of coronoid process markedly curved mesially (inspired from, e.g., Diogo, 2004a)*. Unlike taxa of CS-0 [State 0], taxa of CS-1 exhibit this feature (e.g., Diogo, 2004a: fig. 3-91) [State 1].
336. *Presence of prominent, posterodorsally pointed anterolateral spine of dentary bone in adults*. Unlike taxa of CS-0 [State 0], *Danio* exhibits this feature (e.g., Fig. 74) [State 1].
337. *Angulo-articular, if present, presenting prominent, thin, dorsoventrally elongated posterolateral spine (inspired from, e.g., De la Hoz and Chardon, 1984)*. Unlike taxa of CS-0 [State 0], *Sternopygus* exhibits this feature [State 1] (e.g., De la Hoz and Chardon, 1984: fig. 2).
338. *Presence of broad dorsolateral lamina of dentary bone (inspired from, e.g., Diogo, 2004a)*. Unlike taxa of CS-0 [State 0], *Cetopsis* exhibits this feature [State 1].

339. *Angulo-articular, if present, exhibiting broad, dorsally directed posterodorsal process (inspired from, e.g., Monod, 1963; Howes, 1985a).* Unlike taxa of CS-0 [State 0], *Gonorynchus* exhibits this feature [State 1] (e.g., Fig. 62).

340. *Presence of "interdentary" (inspired from, e.g., Howes and Sanford, 1987b; Johnson and Patterson, 1996).* Unlike taxa of CS-0 [State 0], the specimens of *Plecoglossus* [State 1] exhibit a "mandibular postsymphysial element with ossified tips", named "interdentary" by Howes and Sanford (1987b: fig. 3). It should, however, be noted that Howes and Sanford (1987b) and Johnson and Patterson (1996) found a spherical, partly ossified structure lying posterior to the mandibular symphysis in a specimen of *Osmerus mordax* of 83mm SL, which, according to Howes and Sanford (1987b), appears to be somewhat similar to the "interdentary" of *Plecoglossus. Osmerus* is thus coded here as '?'.

341. *Presence of characteristic ascending portion of Meckel's cartilage (inspired from, e.g., Mo, 1991; Arratia, 1992; Diogo, 2004a).* Unlike taxa of CS-0 [State 0], taxa of CS-1 exhibit this feature [State 1] (e.g., Diogo, 2004a: fig. 3-35).

342. *Presence of independent, ossified coronomeckelian bone (inspired from, e.g., Nelson, 1973; Jollie, 1984a,b; Mo, 1991; De Pinna, 1993).* Unlike in taxa of CS-0 [State 0], an independent, ossified coronomeckelian bone is present in taxa of CS-1 (e.g., Fig. 62) [State 1].

343. *Coronomeckelian bone lying dorsally to dorsal margin of other mandibular bones (inspired from, e.g., Lenglet, 1974; Howes, 1985a).* Unlike taxa of CS-0 [State 0], taxa of CS-1 exhibit this feature (e.g., Howes, 1985a: fig. 13) [State 1]. Since *Mormyrus* specimens seemingly exhibit a condition such as CS-1, while others exhibit a condition such as CS-0, this genus is coded as '?'.

344. *Angulo-articular, if present, exhibiting prominent, roughly circular articulatory surface on its dorsomesial margin for articulation with quadrate and/or ectopterygoid.* Unlike taxa of CS-0 [State 0], *Argentina* exhibits this feature [State 1].

Miscellaneous

345. *Presence of peculiar "saclike bioluminescent organ" on posterior margin of pectoral fin (inspired from, e.g., Parr, 1951, 1960; Matsui and Rosenblatt 1987; Begle, 1992; Johnson and Patterson, 1996).* Unlike taxa of CS-0 [State 0], *Searsia* exhibits this feature [State 1].

346. *Presence of peculiar "tongue-bite mechanism with dorsal teeth on parasphenoid" (and eventually also other bones) (inspired from, e.g., Li and Wilson, 1996; Arratia, 1999; Sanford and Lauder, 1989, 1990; Sanford,*

2000, 2001a,b). Unlike taxa of CS-0 [State 0], taxa of CS-1 exhibit this feature (e.g., Sanford and Lauder, 1989: fig. 2) [State 1]. As stressed by, for example, Sanford (2000, 2001a,b), a few taxa coded as CS-0, such as some salmoniforms (sensu this work), may exhibit a mechanism that is somewhat similar to that found in taxa of CS-1, but this mechanism does not involve the presence of dorsal teeth on the parasphenoid.

347. *Presence of peculiar muscle retractor dorsalis (inspired from, e.g., Nelson, 1967, 1969; Winterbottom, 1974; Rosen, 1973, 1985; Fink and Weitzman, 1982; Lauder and Liem, 1983; Johnson, 1992)*. In taxa of CS-1 there is a peculiar, distinct muscle retractor dorsalis extending from the anterior vertebrae to the dorsal gill arch elements (e.g., Winterbottom, 1974) [State 1]. As noted by, for example, Johnson (1992: 11), "although musculature between the vertebral column and dorsal gill arch elements occurs in a few lower euteleosts", this musculature has been interpreted by most authors as "nonhomologous with the retractor dorsalis of neoteleosts". According to this author, the "homology of the neoteleostean retractor dorsalis is corroborated by an associated modification of the dorsal gill arch muscles" of taxa such as stomiiforms and aulopiforms: the "insertion of the third internal levator on the fifth upper pharyngeal toothplate". This latter feature is found in the stomiiform and aulopiform specimens examined in the present work. As also explained by Johnson (1992: 11), "Fink (1984) noted the presence of a 'retractor dorsalis' in *Lepidogalaxias*", a genus not included in the present cladistic analysis that is nowadays commonly placed in the superfamily Galaxioidea (e.g., Johnson and Patterson, 1996). According to Johnson (1992) the configuration found in *Lepidogalaxias* is unlike that found in taxa such as stomiiforms and aulopiforms, since the former "retain insertion of the third internal levator on the fourth pharyngobranchial cartilage", this "lack of the associated neoteleostean modification" being "at least consistent with an independent origin of the 'retractor dorsalis' of *Lepidogalaxias*". This latter statement cannot be supported/contradicted by the observations of the present study since, as explained above, *Lepidogalaxias* was not included in the present work.

348. *Presence of a pair of well-developed "nasal tubes" on anterolateral surfaces of head (inspired from, e.g., Thys van den Audenaerde, 1961; Howes, 1985a)*. Unlike taxa of CS-0 [State 0], *Phractolaemus* exhibits this feature (e.g., Thys van den Audenaerde, 1961: fig. 1) [State 1].

349. *Presence of peculiar "accessory cartilage of the fifth ceratobranchial" (inspired from, e.g., Nelson, 1967, 1969, 1970; Greenwood and Rosen, 1971; Begle, 1992, Johnson and Patterson, 1996)*. Unlike taxa of CS-0 [State 0], taxa of CS-1 exhibit a peculiar "accessory cartilage of the fifth

ceratobranchial" (sensu Nelson, 1967), which forms part of the "crumeral organ" of these taxa (e.g., Greenwood and Rosen, 1971: figs. 1A, 2A, 4A, 6B) [State 1].

350. *Presence of peculiar multicuspid teeth (inspired from, e.g., Weitzman, 1962; Fink and Fink, 1981, 1996).* Unlike taxa of CS-0 [State 0: e.g.. Fig. ???], taxa of CS-1 have peculiar multicuspid teeth (e.g., Fig. 80; e.g., Weitzman, 1962: fig. 10) [State 1].
351. *Presence of characteristic, paired "maxillary barbels" associated with peculiar "palatine-maxillary system" (inspired from, e.g., Regan, 1911a,b; Alexander, 1965; Gosline, 1975; Fink and Fink, 1981, 1996; Diogo and Chardon, 2001; Diogo et al., 2000, 2003).* Unlike taxa of CS-0 [State 0], taxa of CS-1 exhibit this feature (e.g., Diogo, 2004a: fig. 3-88) [State 1].
352. *Presence of characteristic, paired "mandibular barbels" associated with broad, peculiar basal cartilage (inspired from, e.g., Alexander, 1965; Diogo, 2004a).* Unlike taxa of CS-0 [State 0], taxa of CS-1 exhibit this feature (e.g., Diogo, 2004a: fig. 3-54) [State 1].
353. *"Leptocephalus larvae" (inspired from, e.g., Greenwood et al., 1966; Lauder and Liem, 1983; Forey et al., 1996; Arratia, 1999; Inoue et al., 2004).* Unlike taxa of CS-0 [State 0], taxa of CS-1 have peculiar "leptocephalus larvae" (e.g., Inoue et al., 2004: figs. 1, 5) [State 1]. Gosline (1973a) and others stated that the presence of a "leptocephalus larva" is also seen in some taxa other than those usually included in the Elopomorpha, and that the character state described as CS-1 might in fact be a plesiomorphic feature for teleosts. However, Forey (1973b) and other authors have pointed out that the "leptocephalus larvae" of "elopomorphs" exhibit in fact several peculiar, derived features not seen in the larvae of other teleosts (see, e.g., Forey, 1973b, for more details on these peculiar features). In the present work the use of the term "leptocephalus larva" follows the more restrictive definition of, for example, Forey (1973b).
354. *Presence of "symphysial barbel" (inspired from, e.g., Howes, 1985a).* Unlike taxa of CS-0 [State 0], taxa of CS-1 present a peculiar median "symphysial barbel" on the anteroventral surface of the cephalic region, which is connected to each premaxilla by a bifurcated ligament (see Howes, 1985a: fig. 7) [State 1].
355. *Presence of "luminous chin barbel" (inspired from, e.g., Greenwood et al., 1966; Fink, 1985; Harold and Weitzman, 1996).* Unlike taxa of CS-0 [State 0], taxa of CS-1 have a peculiar "luminous chin barbel" [State 1] (e.g., Weitzman, 1967b: fig. 1).
356. *One or more abdominal scutes, each of a single element which crosses the ventral midline of the body (inspired from, e.g., Grande, 1985a,b; Arratia, 1997, 1999; Chang and Maisey, 2003).* Unlike taxa of CS-0 [State 0], taxa

of CS-1 exhibit this feature [State 1] (e.g., Grande, 1985a,b). As noted by, for example, Hilton (2003: 68), some osteoglossomorphs, such as the notopterid *Xenomystus*, have "abdominal scutes", but the abdominal scutes found in clupeiforms are unique because they are "formed as single median elements, whereas those of notopterids are paired, and therefore do not pass the test of similarity in the establishment of (primary) homology". *Xenomystus* is thus tentatively coded here as CS-0.

Chapter 4

Comparative Anatomy, Higher-level Phylogeny and Macroevolution of Osteichthyans—A Discussion

In this chapter, I will discuss some key topics regarding the comparative anatomy, higher-level phylogeny and macroevolution of osteichthyans. As explained in Chapter 1, the vast majority of the discussions provided in the literature on osteichthyan comparative anatomy, higher-level phylogeny and/or macroevolution are mainly focused on osteological and/or external features. In the present work I will therefore pay special attention to features that have not often been discussed, such as those concerning the configuration of the muscles of the cephalic region and of the pectoral girdle. Thus, Sections 4.2 and 4.3 are mainly focused on the comparative anatomy, development, homologies and evolution of the cephalic muscles within the major osteichthyan groups examined in this work. Section 4.4 is mainly concerned with the comparative anatomy, homologies and evolution of the pectoral girdle muscles. Section 4.5 concerns the origin and evolution of one of the most remarkable and enigmatic structural complexes of osteichthyans, the Weberian apparatus. In fact, as explained in the previous chapters, one of the main reasons to include various extant and fossil otocephalan representatives in the cladistic analysis of the present work is precisely to try to shed light on the puzzling evolution of this structural complex. As will be seen in Section 4.5, the Weberian apparatus does constitute a precious case study to illustrate a crucial point emphasized throughout this book: that the origin and evolution of a structural complex can be adequately understood only when all its various constituents—its bones as well as its cartilage, ligaments, and other parts—are analyzed in a global, integrative, functional perspective. Section 4.6 concerns the use of myological versus osteological structures in phylogenetic reconstructions.

Since some readers might eventually be interested in reading this Chapter 4 without looking in detail at the phylogenetic analysis of Chapter 3, before passing to Sections 4.2, 4.3, 4.4, 4.5 and 4.6, I consider it appropriate to provide here a very brief summary of the main results obtained in that analysis.

4.1 BRIEF SUMMARY OF THE PHYLOGENETIC RESULTS OBTAINED IN THE CLADISTIC ANALYSIS

As explained in Chapter 1, despite the progress achieved on osteichthyan phylogeny in the last decades (see Figs. 1, 2), some crucial questions remain highly controversial:

- Which are the closest extant relatives of teleosts, the amiids of the genus *Amia*, or both these fishes and the lepisosteids of the genera *Lepisosteus* and *Atractosteus*, that is, the members of the three extant genera of an eventual clade Halecomorphi + Ginglymodi?
- Is the Elopomorpha really a monophyletic clade?
- Which are the most basal extant teleosts, the osteoglossomorphs or the elopomorphs (if these latter do in fact constitute a monophyletic group)?
- Are the Ostariophysi and the Otocephala really monophyletic groups, or are some members of these groups deeply related to fishes that are usually included in the Euteleostei, such as the Alepocephaloidea?
- Which are the closest living relatives of tetrapods, the members of the actinistian genus *Latimeria* or the members of the dipnoan genera *Protopterus*, *Lepidosiren* and *Neoceratodus*?

Concerning these five crucial questions, the phylogenetic results of Chapter 3:

- Strongly corroborated the hypothesis that the closest living relatives of teleosts are the members of the genus *Amia*, as is accepted in various general textbooks (see Figs. 3, 4; compare, e.g., with Fig. 1). This hypothesis is supported in the cladistic analysis of Chapter 3 by numerous osteological and myological features, including some features that appear as homoplasy-free in the analysis.
- Provided evidence to support the hypothesis that the elopomorphs do constitute a monophyletic unit. As explained in Section 3.1, although this evidence is not particularly strong, this is the first published cladistic morphological analysis supporting the monophyly of the Elopomorpha as a whole (see Figs. 3. 4; compare, e.g., with Figs. 1, 2).

- Provided evidence (although not particularly strong) to support the hypothesis that the most basal extant teleosts are the elopomorphs, and not the osteoglossomorphs, as is accepted in various general textbooks (see Figs. 3, 4; compare, e.g., with Fig. 2).
- Provided strong evidence to support the monophyly of the Ostariophysi and of the Otocephala, as well as to support the hypothesis that the Alepocephaloidea are the sister-group of the Argentinoidea, as is usually defended in most morphological studies (see Figs. 3, 4; compare, e.g., with Figs. 1, 2).
- Strongly corroborated the hypothesis that the closest living relatives of tetrapods are extant dipnoans, as is accepted in some general textbooks (see Figs. 3, 4). This hypothesis is supported in the cladistic analysis of Chapter 3 by numerous osteological and myological features, including some features that appear as homoplasy-free in the analysis; some of these features have not been proposed in previous works as potential synapomorphies supporting a close relationship between dipnoans and tetrapods.

Thus, in a very brief, overall summary of the phylogenetic results obtained in Chapter 3 (see Fig. 4), it can be said that *Polypterus* (Cladistia) appears as the most basal extant actinopterygian, the acipenseriforms (Chondrostei) appear as the most basal living actinopterans, and *Lepisosteus* (Ginglymodi) appears as the most plesiomorphic extant neopterygian, as expected. *Amia* (Halecomorphi) is placed as the closest living relative of teleosts, this hypothesis being strongly supported by the results of Chapter 3. These results also provide evidence to support the monophyly of elopomorphs, which appear as the most basal extant teleosts, although this evidence is not particularly strong. The osteoglossomorphs included in the cladistic analysis are also placed in a monophyletic clade, which appears as the sister-group of the remaining non-elopomorph teleostean taxa included in the analysis. The present work strongly supports the monophyly of the Ostariophysi and of the Otocephala, thus contradicting the results of some recent molecular cladistic analyses placing the Alepocephaloidea inside the Otocephala. The monophyly of the clade Alepocephaloidea + Argentinoidea is strongly supported; evidence is also provided to support the clade Salmoniformes + Neoteleostei and the clade Esociformes + (Galaxioidea + Osmeroidea), although this evidence is not very strong. These three latter major euteleostean clades are placed in an unresolved trichotomy. Concerning the sarcopterygians, the phylogenetic results of this work strongly support the hypothesis that tetrapods are more closely related to extant dipnoans (Dipnoi) than to *Latimeria* (Actinistia). As expected, †*Tiktaalik* is placed near the tetrapods. As also expected, the amphibian salamander *Ambystoma* and the amniote lizard *Timon* appear more closely related to each other than to the early tetrapod fossil †*Acanthostega*.

4.2 COMPARATIVE ANATOMY, HOMOLOGIES, AND EVOLUTION OF OSTEICHTHYAN CRANIAL MUSCLES

As explained in Chapter 1, one of the aims of this book is to provide a fresh look at the comparative anatomy, homologies and evolution of osteichthyan muscular structures, which have been much less studied and discussed than the osteological bones. As also mentioned in Chapter 1, the most extensive, detailed comparative analyses that have been done on osteichthyan muscles, and that were actually based on a direct observation of these muscles in osteichthyan taxa as varied as, for example, Teleostei, Halecomorphi, Ginglymodi, Chondrostei, Cladistia, Dipnoi, Amphibia and Amniotes, and not mainly on a recompilation from the literature, were provided some decades ago in works such as Luther (1913-1914), Kesteven (1942-1945), and principally the impressive volume of Edgeworth (1935). Despite the qualities of these works, their authors had no access to much information now available, for example, on the muscles of the coelacanth *Latimeria chalumnae* or on the data being provided by evolutionary developmental biology supporting, for instance, the essential role of neural crest cells in the development and patterning of vertebrate cranial muscles (e.g., Le Lièvre and Le Douarin, 1975; Noden, 1983, 1984, 1986; Couly et al., 1992; Köntges and Lumsden, 1996; Schilling and Kimmel, 1997; Olsson et al., 2001; Ericsson and Olsson, 2004). It is also important to note that some of the evolutionary hypotheses proposed by those authors were based on phylogenetic scenarios that been contradicted by numerous studies (see Chapter 1).

About 15 years ago, Miyake et al. (1992) published a paper that was mainly focused on the cranial muscles of chondrichthyan batoids, but in which they reexamined and discussed some of the general hypotheses proposed by Edgeworth (1935). For instance, they noted that "Noden (1983, 1984, 1986) elegantly demonstrated with quail-chick chimeras that cranial muscles are embriologically of somitic origin, and not as commonly thought, of lateral plate origin, and in doing so corroborated the nearly forgotten work of Edgeworth" (Miyake et al., 1992: 214). They also pointed out that molecular developmental studies such as Hatta et al. (1990, 1991) "have corroborated one of Edgeworth's findings: the existence of one premyogenic condensation (the constrictor dorsalis) in the cranial region of teleost fish" (Miyake et al., 1992: 214). As explained by Miyake et al. (1992), Edgeworth recognized various presumptive premyogenic condensations, including them in five different main cranial muscle plates: mandibular, hyoid, branchial, epibranchial, and hypobranchial. Edgeworth viewed development of cranial muscles in the light of developmental pathways leading from these condensations to different states in each cranial arch. According to him these developmental pathways involve migration of premyogenic cells, differentiation of myofibers, directional growth of myofibers and possibly interactions with surrounding structures. These events occur in very specific locations, for example, dorsal, medial or ventral areas of each cranial arch, as shown in Fig. 115. This diagram, based on Edgeworth's studies, is similar to that proposed by Mallat (1997). The differences between the two diagrams are, in fact, mainly nomenclatural ones (e.g., the "hyoidean and mandibular superficial constrictors" of Edgeworth correspond to the "hyoidean and mandibular interbranchial muscles" of Mallat—see, e.g., Mallat's table 2). According to Edgeworth, although exceptions may occur (see below), the mandibular muscles are generally innervated by the Vth nerve, the hyoid muscles by the VIIth nerve and the branchial muscles by the IXth and Xth nerves. Also according to this author, the epibranchial and hypobranchial muscles are "developed from the anterior myotomes of the body" and thus "are intrusive elements of the head"; they "retain a spinal innervation" and "do not receive any branches from the Vth, VIIth, IXth and Xth nerves" (Edgeworth, 1935: 189). It is worth noting that apart from the mandibular, hyoid, hypobranchial, epibranchial and branchial arches Edgeworth (1935: 5) referred to a primitive

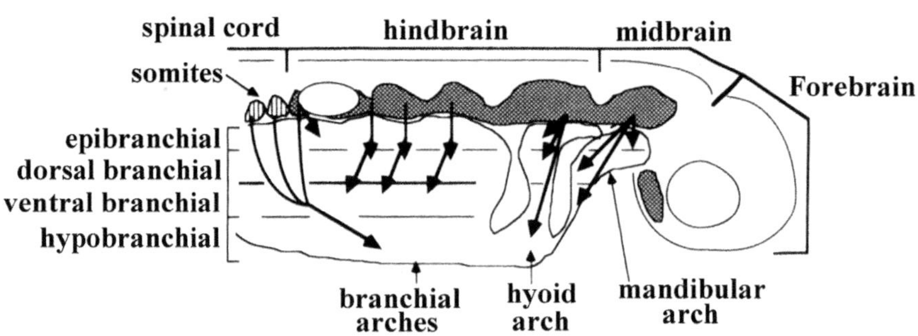

Figure 115 Schematic presentation of embryonic origin of cranial muscles in gnathostomes based on Edgeworth (1902, 1911, 1923, 1926abc, 1928, 1935); premyogenic cells originate from the paraxial mesoderm (hatched areas) and several somites (areas with vertical bars); large arrows indicate a contribution of cells in segments of the mesoderm to muscle formation of different cranial arches (modified from Miyake et al., 1992; the nomenclature of the structures illustrated basically follows that of these authors). For more details, see text.

"premandibular arch" in "which passed the IIIrd nerve". This IIIrd nerve, together with the IVth and VIth nerves—which according to Edgeworth (1935: 5) are "not segmental nerves; they innervate muscles of varied segmental origin and are, phylogenetically, of later development than are the other cranial nerves"—innervate the external ocular muscles of most extant osteichthyans.

The discussions provided in this chapter mainly concern the mandibular, hyoid and hypobranchial muscles, that is, the "superficial cranial muscles" sensu Diogo and Vandewalle (2003), although references will also be made to certain ocular and/or branchial muscles (note: the epibranchial muscles sensu Edgeworth, 1935, are absent in extant osteichthyans). Figures 116 to 121 provide an updated version of the diagrams of Miyake et al. (1992). These figures include representatives of some osteichthyan groups not considered in Miyake et al.'s paper, such as amphibians and amniotes; the chondrichthyans, not analyzed in the present work, are excluded from those diagrams. The information provided in these figures is complemented with that given in Tables 2 to 7. As can be seen in these tables, in those cases in which a single muscle is given a name X by researchers working with, for example, plesiomorphic bony fishes and a name Y by researchers working with, for example, tetrapods, I opted to use the name X. In my opinion, this option is methodologically more appropriate than to continue to designate that muscle by both names Y and X, as this is precisely one of the main reasons for the historical confusions concerning the homologies and evolution of osteichthyan muscles. I also consider that it makes more evolutionary sense than to use only the name Y, because tetrapods originated from bony fishes, and not the reverse.

Mandibular Muscles (Tables 2, 3; Figs. 116, 117)

According to Edgeworth (1935) the embryonic mandibular muscle gives rise to the premyogenic condensation constrictor dorsalis, dorsally, to the premyogenic condensation adductor mandibulae, medially, and to the intermandibularis, ventrally (no description of a ventral mandibular premyogenic condensation was given by Edgeworth) (see Figs. 115, 116 and 117). He stated that a constrictor dorsalis is absent in extant gnathostome taxa such as holocephalan, chondrichthyans, dipnoans and amphibians. Since he considered that the chondrichthyans, actinopterygians and tetrapods (sensu this work) were derived from an "early dipnoan stock" (see Chapter 1), he stated that the constrictor dorsalis was plesiomorphically absent in the gnathostomes and then independently acquired in some taxa within this group.

Table 2 Mandibular muscles of adults of representative actinopterygian taxa; the nomenclature of the muscles shown in bold follows that of the present work, "ad. mand." meaning adductor mandibulae (in order to facilitate comparisons, certain names used by other authors are also shown); data compiled from evidence provided by developmental biology, comparative anatomy, functional morphology, paleontology, experimental embryology and molecular biology, innervation and phylogeny (in studies by the author and other researchers). For more details, see text (see also Fig. 116).

Probable plesiomorphic osteichthyan condition	Cladistia: *Polypterus bichir* (Bichir)	Chondrostei: *Psephurus gladius* (Chinese swordfish)	Ginglymodi: *Lepisosteus osseus* (Longnose gar)	Halecomorphi: *Amia calva* (Bowfin)	Teleostei—basal: *Elops saurus* (Ladyfish)	Teleostei—clupeocephalan: *Danio rerio* (Zebrafish)
Intermandibularis anterior (*intermandibularis anterior and posterior plesiomorphically present in osteichthyans? See text)	**Intermandibularis**	**Intermandibularis**	**Intermandibularis**	**Intermandibularis anterior**	**Intermandibularis anterior**	**Intermandibularis anterior**
Intermandibularis posterior (*see cell above)	--------	--------	--------	**Intermandibularis posterior**	**Intermandibularis posterior** (*forming, together with interhyoideus, the protractor hyoidei)	**Intermandibularis posterior** (*see cell on the left)

--------	--------	--------	--------	--------	**Protractor hyoidei** (*including intermandibularis posterior and interhyoideus; it is thus derived from both the mandibular and hyoid muscle plates)	**Protractor hyoidei** (*see cell on the left)
Ad. mand. A3'	**Ad. mand. A3'** (ad. mand. of e.g. Lauder, 1980a)	--------	**Ad. mand. A3'** (preorbitalis superficialis of e.g. Lauder, 1980a)	**Ad. mand. A3'**	--------	--------
Ad. mand. A3"	**Ad. mand. A3"** (ad. mand. pterygoideus of e.g. Lauder, 1980a)	--------	**Ad. mand. A3"** (preorbitalis profundus of e.g. Lauder, 1980a)	**Ad. mand. A3"**	--------	--------
Ad. mand. A2	**Ad. mand. A2** (ad. mand. posterolateral of e.g. Lauder, 1980a)	**Ad. mand. A2** (ad. mand. of e.g. Carroll and Wainwright, 2003)	**Ad. mand. A2** (ad. mand. posterolateral of e.g. Lauder, 1980a)	**Ad. mand. A2**	**Ad. mand. A2**	**Ad. mand. A2**
--------	--------	--------	**Palatomandibularis minor and major**	--------	--------	--------

--------	--------	--------	--------	**Levator maxillae superioris 3 and 4**	--------	--------
--------	--------	--------	--------	--------	--------	**Ad. mand. A1-OST**
--------	--------	--------	--------	--------	--------	**Ad. mand. A0**
Ad. mand. Aω	**Ad. mand. Aω**	--------	--------	**Ad. mand. Aω**	**Ad. mand. Aω**	**Ad. mand. Aω**
Levator arcus palatini	**Levator arcus palatini**	**Protractor hyomandibulae** (*seemingly originated from the portion of the hyoid muscle plate from which the adductor arcus palatini and dilatator operculi of other actinopterygians originate)	**Levator arcus palatini**	**Levator arcus palatini**	**Levator arcus palatini**	**Levator arcus palatini**
--------	**Dilatator operculi**	-------- (*dilatator operculi absent as a separate element, but see cell above)	**Dilatator operculi**	**Dilatator operculi**	**Dilatator operculi**	**Dilatator operculi**

Table 3 Mandibular muscles of adults of representative sarcopterygian taxa; the nomenclature of the muscles shown in bold follows that of the present work, "ad. mand." meaning adductor mandibulae (in order to facilitate comparisons, certain names used by other authors are also shown); data compiled from evidence provided by developmental biology, comparative anatomy, experimental embryology and molecular biology, innervation and phylogeny (in studies by the author and other researchers). For more details, see text (see also Fig. 117).

Probable plesiomorphic osteichthyan condition	Actinistia: *Latimeria chalumnae* (Coelacanth)	Dipnoi: *Lepidosiren paradoxa* (South American lungfish)	Amphibia: *Ambystoma ordinarium* (Michoacan stream salamander)	Amniota: *Timon lepidus* (Ocellated lizard)
Intermandibularis anterior (*see Table 2)	**Intermandibularis anterior**	**Intermandibularis**	**Intermandibularis anterior**	**Intermandibularis anterior**
Intermandibularis posterior (*see Table 2)	**Intermandibularis posterior**	--------	**Intermandibularis posterior**	**Intermandibularis posterior**
Ad. mand. A3'	**Ad. mand. A3'** (ad. mand. 'moyen' of e.g. Millot and Anthony, 1958)	**Ad. mand. A3'** (ad. mand. anterior of e.g. Bemis and Lauder, 1986)	**Ad. mand. A3'** (pseudotemporalis posterior of e.g. Iordansky, 1992)	**Ad. mand. A3'** (pseudotemporalis superficialis of e.g. Abdala and Moro, 2003)
Ad. mand. A3"	**Ad. mand. A3"** (ad. mand. 'profond' of e.g. Millot and Anthony, 1958)	--------	**Ad. mand. A3"** (pseudotemporalis anterior of e.g. Iordansky, 1992)	**Ad. mand. A3"** (pseudotemporalis profundus of e.g. Abdala and Moro, 2003)
Ad. mand. A2	**Ad. mand. A2** (ad. mand. 'superficiel' of e.g. Millot and Anthony, 1958)	**Ad. mand. A2** (part of ad. mand. posterior of e.g. Bemis and Lauder, 1986)	**Ad. mand. A2** (ad. mand. externus of e.g. Iordansky, 1992)	**Ad. mand. A2** (ad. mand. externus of e.g. Abdala and Moro, 2003)

--------	--------	**Ad. mand. A2-PVM** (part of ad. mand. posterior of e.g. Bemis and Lauder, 1986)	**Ad. mand. A2-PVM** (ad. mand. posterior of e.g. Iordansky, 1992)	**Ad. mand. A2-PVM** (ad. mand. posterior of e.g. Abdala and Moro, 2003)
--------	--------	**Retractor anguli oris** (*seemingly derived from lateral portion of ad. mand.)	--------	--------
--------	--------	--------	--------	**Levator anguli oris** (*present, somewhat mixed with A2; seemingly derived from lateral portion of ad. mand, it may eventually be derived/modified from the retractor anguli oris, or at least from the portion of the mandibular muscle plate originating that muscle in other osteichthyan taxa)
Ad. mand. Aω	**Ad. mand. Aω** (intramandibular adductor of e.g. Lauder, 1980b)	--------	--------	**Ad. mand. Aω** (*in *Timon* the ad. mand. has an anteroventral section that is lodged in the 'adductor fossa' and that is very similar to the Aω of other osteichthyan taxa: is this section homologous to the Aω of those taxa? See text)
--------	--------	--------	--------	**Pterygomandibularis** (*seemingly derived from mesial portion of ad. mand.)
Levator arcus palatini	**Levator arcus palatini**	--------	--------	**Levator pterygoidei** (*it may well be derived/modified from the levator arcus palatini, or at least from the portion of the mandibular muscle plate originating that muscle in other osteichthyan taxa)
--------	--------	--------	--------	**Protractor pterygoidei** (*same as cell above)

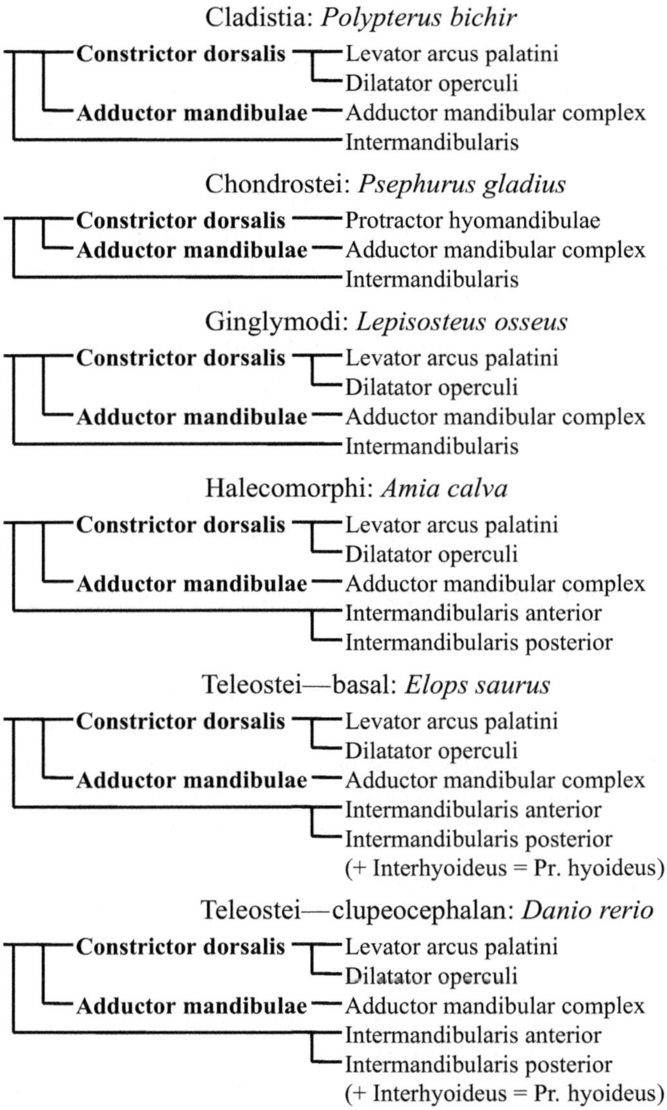

Figure 116 Developmental lineages of mandibular muscles in actinopterygians; Edgeworth's presumptive premyogenic condensations are in bold face (modified from Miyake et al., 1992); the nomenclature of the muscles listed on the right of the figure follows that of the present work, "Pr. hyoideus" meaning protractor hyoideus. Data compiled from evidence provided by developmental biology, comparative anatomy, experimental embryology and molecular biology, innervation and phylogeny (in studies by the author and other researchers). For more details, see text.

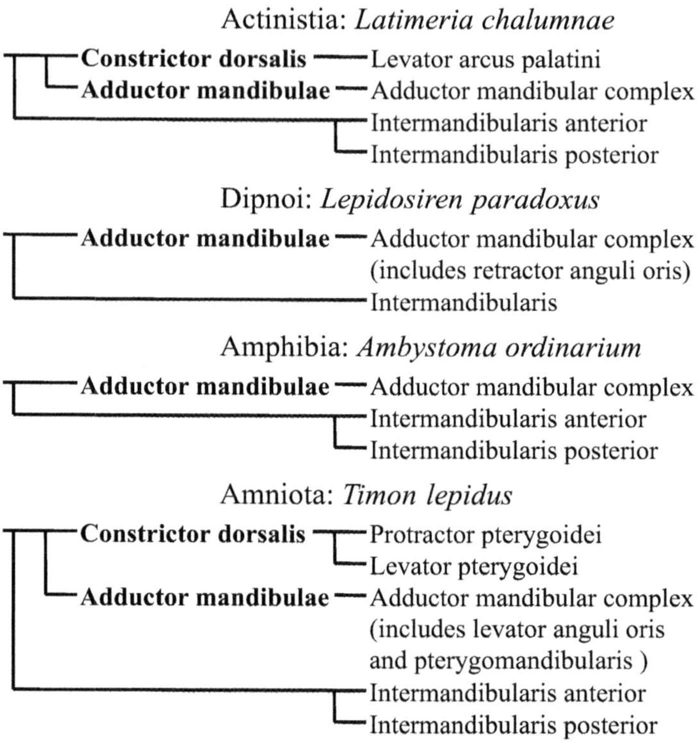

Figure 117 Developmental lineages of mandibular muscles in sarcopterygians, Edgeworth's presumptive premyogenic condensations are in bold face (modified from Miyake et al., 1992); the nomenclature of the muscles listed on the right of the figure follows that of the present work. Data compiled from evidence provided by developmental biology, comparative anatomy, experimental embryology and molecular biology, innervation and phylogeny (in studies by the author and other researchers). For more details, see text.

This is a case in which one of Edgeworth's conclusions is put in question by evidence now available. In fact, very few researchers would now accept that chondrichthyans, actinopterygians and tetrapods originated from early dipnoans (see Chapter 1). The phylogenetic results obtained in Chapter 3 indicate that the constrictor dorsalis was secondarily lost within osteichthyans: it was either independently lost within dipnoans and amphibians or lost in the node leading to non-actinistian sarcopterygians and then reacquired in amniotes (see Figs. 3 and 4). Although these two options appear equally parsimonious, I am inclined to favor the first one, that is that the constrictor dorsalis condensation was independently lost in amphibians and dipnoans. This is because this premyogenic condensation is remarkably similar in amniotes and in extant non-dipnoan sarcopterygian fishes (e.g., Brock, 1938). Thus, in my opinion it seems more likely that such a condensation was lost in dipnoans and in amphibians than to consider that it was lost in non-actinistian sarcopterygians and that a strikingly similar condensation was then independently acquired in amniotes. It is important to stress that the adult muscles derived from the constrictor dorsalis in amniotes such as lizards, that is, the levator and protractor pterygoideus, are in fact also strikingly similar to the adult muscle derived from the constrictor dorsalis in sarcopterygians such as *Latimeria*, that is, the levator arcus palatini (they essentially occupy the same position, they run from the neurocranium to the dorsal/dorsolateral margin of the palatoquadrate and thus are usually related to the elevation of this latter structure, they have the same innervation, etc.) (see Section 3.2). A detailed analysis of the presence/absence of dorsal mandibular muscles in plesiomorphic dipnoan and amphibian fossils as well as in other sarcopterygian fossils in which the presence/absence of myological structures could eventually be discerned is needed to clarify the taxonomic distribution of these dorsal mandibular muscles within the Sarcopterygii.

In most extant actinopterygians the constrictor dorsalis gives rise to the levator arcus palatini and also to the dilatator operculi (Table 2; Figs. 6, 15, 19, 27, 74, 116). The former muscle is usually related to the abduction of the suspensorium/palatoquadrate, while the latter is usually mainly associated with the abduction of the opercle. However, in extant acipenseriforms the constrictor dorsalis gives rise to a single, peculiar muscle that is mainly related to the protraction of the hyomandibula, the protractor hyomandibulae (see Fig. 12) (e.g., Danforth, 1913; Luther, 1913; Sewertzoff, 1928; Edgeworth, 1935; Kesteven, 1942-1945; Miyake et al., 1992; Carroll and Wainwright, 2003; this work). In view of the phylogenetic results obtained in Chapter 3, the presence of a separate dilatator operculi in adults could thus be seen as a feature acquired in the node leading to all extant actinopterygians and then reversed in a node leading to extant acipenseriforms or, instead, as a feature independently acquired in cladistians and in neopterygians. As explained in Section 3.1, although these two options appear equally parsimonious, I consider that there are reasons to suggest that a separate dilatator operculi was present in the ancestor of extant actinopterygians. In fact, the dilatator operculi of *Polypterus* and the dilatator operculi of neopterygians have the same developmental origin (the dorsal part of the mandibular muscle plate), the same innervation (the Vth nerve), the same function (essentially related to opercle abduction), and a very similar overall configuration (e.g., Pollard, 1892; Allis, 1897, 1922; Edgeworth, 1935; Winterbottom, 1974; Lauder,

1980a; Lauder and Liem, 1983; Miyake et al., 1992; this work). The absence of a distinct dilatator operculi in adult extant acipenseriforms may well be related to the fact that these fishes peculiarly lack an opercular bone and/or to the fact that extant acipenseriforms are often indicated as an example of paedomorphic taxa (e.g., Bemis et al., 1997; Findeis, 1997). This could help to explain why, unlike most other living actinopterygians, in which the constrictor dorsalis becomes ontogenetically differentiated into two muscles, adult acipenseriforms remain with a single, undivided dorsal mandibular muscle, the protractor hyomandibulae. Miyake et al. (1992: 221) stated that a "spiracularis was described in *Polypterus* by Edgeworth (1935), but confirmation of its actual existence in *Polypterus* is needed". No spiracularis was found in the *Polypterus* specimens or in any other osteichthyan specimens examined in the present work.

The ventral portion of the mandibular muscle plate gives rise to the intermandibularis in all major osteichthyan groups listed in Tables 2 and 3. In adult extant members of the Actinistia, Chondrostei, Ginglymodi and Dipnoi the intermandibularis is mainly undivided (see Figs. 6, 12, 15, 91), whereas in adult specimens of *Amia*, *Latimeria*, and numerous amphibian, amniote and teleostean genera this structure is divided into an intermandibularis anterior and an intermandibularis posterior (e.g., Figs. 19, 34, 86, 99, 110). As its name indicates, the intermandibularis is usually a transversal muscle connecting the two mandibles. In most teleosts the intermandibularis and the interhyoideus (see below) form a muscle protractor hyoidei (e.g., Figs. 29, 34), which is thus derived from both the mandibular and hyoid muscle plates. The protractor hyoidei is innervated by both the Vth and the VIIth nerves and functionally it is a complex muscle: Osse (1969), for example, demonstrated that its anterior and posterior sections may contract differently during different phases of respiration. Nonetheless, in general it can be said that this muscle plays a primary role in the elevation (protraction) of the hyoid bars, as well as in the depression of the mandible (e.g., Stiassny, 2000). Although a protractor hyoidei is absent in a few adult teleosts such as those of the genera *Albula* and *Mormyrus* (see Fig. 35; Greenwood, 1971; Winterbottom, 1974), the ancestors of extant teleosts did seemingly have a protractor hyoidei (see Section 3.1). Hunter and Prince (2002: 378), on the basis of their observations of morpholino-mediated Hox PG2 (*hoxa2b* and *hoxa2a*) knock-down zebrafish larvae (*Danio rerio*), suggested that at least in *Danio* "the basihyal may be important for the proper ontogenetic organization" of the intermandibularis posterior and the interhyoideus, and, thus, for the association of their fibers and the formation of the protractor hyoidei. Further studies are needed to check whether this is so and whether it may be a general feature within the Teleostei.

It is rather difficult to discern whether the intermandibularis was plesiomorphically divided in adult osteichthyans. I tentatively hypothesize that it was, because a divided intermandibularis is found in numerous chondrichthyans, actinopterygians and sarcopterygians (Tables 2, 3). However, a detailed analysis of the taxonomic distribution of this feature in the Chondrichthyes, and, if possible, in key osteichthyan and non-osteichthyan gnathostome fossils, is needed to clarify this issue.

The adductor mandibulae condensation is found in members of all major osteichthyan groups (Tables 2, 3; Figs. 116, 117). However, the number of structures originating from this condensation is highly variable within the Osteichthyes (Tables 2, 3). The adductor mandibulae A3' and A3'' sensu the present work (which correspond to the "mesial adductor mandibulae divisions" of Lauder, 1980a) are seemingly found plesiomorphically in osteichthyans (Tables 2, 3; Figs. 6, 7, 15, 16, 19, 21, 100). These two sections may, however, be missing in osteichthyan taxa such as extant acipenseriforms and various teleosts (Table 2; Figs. 12, 27, 74). The adductor mandibulae A2 and Aὼ were seemingly also present in basal osteichthyans (e.g., Lauder, 1980a,b; Tables 2, 3; Figs. 7, 21, 29, 75). The Aὼ may be missing in extant osteichthyans such as acipenseriforms, lepisosteiforms, various teleosts and most tetrapods (Table 2; Figs. 16, 62, 63, 68, 92, 100). As noted in Section 3.2, in the adult specimens of the lizard genus *Timon* examined the adductor mandibulae has a large anteroventral portion that is lodged in the "adductor fossa" (sensu, e.g., Lauder, 1980b) and that is very similar to the Aw of bony fishes. Such an anteroventral portion of the adductor mandibulae has also been described in other extant amniotes, such as *Crocodilus* (e.g., Edgeworth, 1935). It is difficult to discern, in view of the data available, whether the Aw of amniotes such as *Timon* and *Crocodilus* is in fact homologous to the Aw of bony fishes (see Table 3). I plan to undertake a detailed analysis of this issue in a future work.

In Lauder's (1980a) table II, it is suggested that the sections palatomandibularis minor and major of adult members of *Lepisosteus* are likely homologous to the levator maxillae superioris 3 and 4 of adult specimens of *Amia*, since these structures represent an "anterior division" of the adductor mandibulae. However, the overall configuration, position and attachments of the palatomandibularis minor and major of *Lepisosteus* are markedly different from those of the levator maxillae superioris 3 and 4 of *Amia* (see Section 3.2). The phylogenetic results of the present work contradict a sister-group relationship between the Ginglymodi and the Halecomorphi (Figs. 3, 4); this thus seems to indicate that the palatomandibularis minor and major of *Lepisosteus* and the levator maxillae 3 and 4 of *Amia* may be non-homologous.

Apart from the adductor mandibulae divisions mentioned in the above paragraphs, other divisions may eventually be found in adult osteichthyans, such as the A1-OST and A0 (e.g., Fig. 74), the A2-PVM (e.g., Figs. 92, 100), the retractor and levator anguli oris (e.g., Figs. 91 and 109), and the pterygomandibularis (e.g., Fig. 111) (Tables 2, 3). As explained in Chapter 3, I found, in the adult dipnoan specimens analyzed, a peculiar adductor mandibular section that has some fibers associated with those of the adductor mandibulae A2 but is well differentiated from it. As in these fishes this section is somewhat posterior, ventral, and medial to the main body of the A2, I have named it adductor mandibulae A2-PVM (the PVM meaning posteroventromesial: e.g., Fig. 92). Edgeworth (1935), Bemis (1986), Bemis and Lauder (1986), Miyake et al. (1992), and other authors did not mention the presence of such an adductor mandibulae section in extant dipnoans (the A2-PVM should not be confused with the retractor anguli oris of these authors, which corresponds to the retractor anguli oris of the present work and which is usually situated posteroventrolaterally to the A2, being, thus, mainly superficial, and not mesial, to the A2; both an A2-PVM and an A2 can actually be found in the same taxon, as, for example, in *Lepidosiren*: see Figs. 91, 92). The A2-PVM of dipnoans seemingly corresponds to the structure that is often named "adductor mandibulae posterior" by researchers working with amphibian and amniote tetrapods (e.g., Brock, 1938; Carroll and Holmes, 1980; Iordansky, 1992; Moro and Abdala, 2000; Montero et al., 2002; Abdala and Moro, 2003) (see Chapter 3 and Table 3). All extant non-dipnoan bony fishes I examined lack an A2-PVM. The textual descriptions and the illustrations of the adductor mandibulae of *Latimeria* provided by, for example, Millot and Anthony (1958) and Adamicka and Ahnelt (1992) indicate that such an A2-PVM is also seemingly absent in this taxon.

Although the structures originating from the adductor mandibulae condensation are usually related to the adduction of the mandible, this is not always the case. For instance, some of these divisions may attach to structures other than the mandible such as the maxilla (e.g., the adductor mandibulae A0 of cypriniforms: see Fig. 74) and thus not be directly related to mandible adduction, while others may inclusively be related to the opening, and not the closing, of the mouth (e.g., the "abductor mandibulae" of saccopharyngiforms: see Section 3.2).

Hyoid Muscles (Tables 4, 5; Figs. 118, 119)

According to Edgeworth (1935) the constrictor hyoideus condensation usually gives rise to dorso-medial and ventral derivatives. Two dorso-medial hyoid muscles were apparently present in plesiomorphic

Table 4 Hyoid muscles of adults of representative actinopterygian taxa; the nomenclature of the muscles shown in bold follows that of the present work (in order to facilitate comparisons, certain names used by other authors are also shown); data compiled from evidence provided by developmental biology, comparative anatomy, experimental embryology and molecular biology, innervation and phylogeny (in studies by the author and other researchers). For more details, see text (see also Fig. 118).

Probable plesiomorphic osteichthyan condition	Cladistia: *Polypterus bichir* (Bichir)	Chondrostei: *Psephurus gladius* (Chinese swordfish)	Ginglymodi: *Lepisosteus osseus* (Longnose gar)	Halecomorphi: *Amia calva* (Bowfin)	Teleostei—basal: *Elops saurus* (Ladyfish)	Teleostei—clupeocephalan: *Danio rerio* (Zebrafish)
Interhyoideus	**Interhyoideus**	**Interhyoideus**	**Interhyoideus**	**Interhyoideus**	**Interhyoideus** (*forming, together with intermandibularis posterior, the protractor hyoidei: see Table 2)	**Interhyoideus** (*see cell on the left)
--------	**Hyohyoideus**	**Hyohyoideus**	**Hyohyoideus**	**Hyohyoideus inferior**	**Hyohyoideus inferior**	**Hyohyoideus inferior**
--------	--------	--------	--------	**Hyohyoideus abductor** (*often considered as part of a hyohyoideus superior)	**Hyohyoideus abductor** (*see cell on the left)	**Hyohyoideus abductor** (*see cell on the left)
--------	--------	--------	--------	**Hyohyoidei adductores** (*often considered part of a hyohyoideus superior)	**Hyohyoidei adductores** (*see cell on the left)	**Hyohyoidei adductores** (*see cell on the left)

Adductor operculi	**Adductor operculi**	**Adductor operculi** (opercularis of e.g. Carroll and Wainwright, 2003)	**Adductor operculi**	**Adductor operculi**	**Adductor operculi**	**Adductor operculi**
Adductor arcus palatini	**Adductor arcus palatini**	**Retractor hyomandibulae** (*seemingly originated from the portion of the hyoid muscle plate from which the adductor arcus palatini of other actinopterygians originates)	**Adductor arcus palatini**	**Adductor arcus palatini**	**Adductor arcus palatini**	**Adductor arcus palatini**
--------	--------	--------	--------	--------	--------	**"Adductor hyomandibulae X"** (*seemingly not homologous with the "adductor hyomandibulae Y" of Table 5)
--------	--------	--------	--------	**Levator operculi** (*seemingly not homologous with the "levator operculi" of Table 5)	**Levator operculi** (*see cell on the left)	**Levator operculi** (*see cell on the left)

Table 5 Hyoid muscles of adults of representative sarcopterygian taxa; the nomenclature of the muscles shown in bold follows that of the present work (in order to facilitate comparisons, certain names used by other authors are also shown); data compiled from evidence provided by developmental biology, comparative anatomy, experimental embryology and molecular biology, innervation and phylogeny (in studies by the author and other researchers). For more details, see text (see also Fig. 119).

Probable plesiomorphic osteichthyan condition	Actinistia: *Latimeria chalumnae* (Coelacanth)	Dipnoi: *Lepidosiren paradoxa* (South American lungfish)	Amphibia: *Ambystoma ordinarium* (Michoacan stream salamander)	Amniota: *Timon lepidus* (Ocellated lizard)
Interhyoideus	**Interhyoideus** ("géniohyoïdien" plus "hyohyoïdien" of e.g. Millot and Anthony, 1958)	**Interhyoideus**	**Interhyoideus** (interhyoideus anterior plus interhyoideus posterior of e.g. Bauer, 1992, and Ericsson and Olsson, 2004)	**Interhyoideus** (constrictor colli of e.g. Herrel et al., 2005)
--------	--------	--------	**Branchiohyoideus** (branchiohyoideus externus of e.g. Edgeworth, 1935, and Ericsson and Olsson, 2004)	-------- (*as noted by e.g. Edgeworth, 1935, the "branchiohyoideus" of lepidosaurs seemingly corresponds to the branchial muscle subarcualis rectus 1 of amphibians, and not to the hyoid muscle branchiohyoideus of the present work)

Adductor arcus palatini	**Adductor arcus palatini**	-------- (*does the portion of the hyoid muscle plate that gives rise to the levator hyoideus/ depressor mandibulae eventually correspond to that giving rise to the adductor arcus palatini of other osteichthyans? See text)	-------- (*see cell on the left)	-------- (*see cell on the left)
--------	--------	**Levator hyoideus**	**Depressor mandibulae posterior** (*the fibers corresponding to those of the levator hyoideus of dipnoans become also attached on the mandible, forming the depressor mandibulae posterior; the depressor mandibulae anterior thus seemingly corresponds to the depressor mandibulae of dipnoans)	**Depressor mandibulae (part)** (*the fibers corresponding to those of the levator hyoideus of dipnoans become also attached on the mandible, forming part of the depressor mandibulae)
--------	--------	**Depressor mandibulae**	**Depressor mandibulae anterior** (*see cell above)	**Depressor mandibulae (part)** (*see cell above)
--------	--------	--------	--------	**Cervicomandibularis** (*seemingly derived from the levator hyoideus/ depressor mandibulae: e.g. Edgeworth, 1935)

--------	**"Adductor hyomandibulae ▾"** (*seemingly not homologous to the "adductor hyomandibulae X" of Table 4)	--------	--------	--------
Adductor operculi	**Adductor operculi**	-------- (*absent as a separate element in adults, but see text)	--------	--------
--------	**Latimeria's "levator operculi"** (*seemingly not homologous to the levator operculi of Table 4)	--------	--------	--------

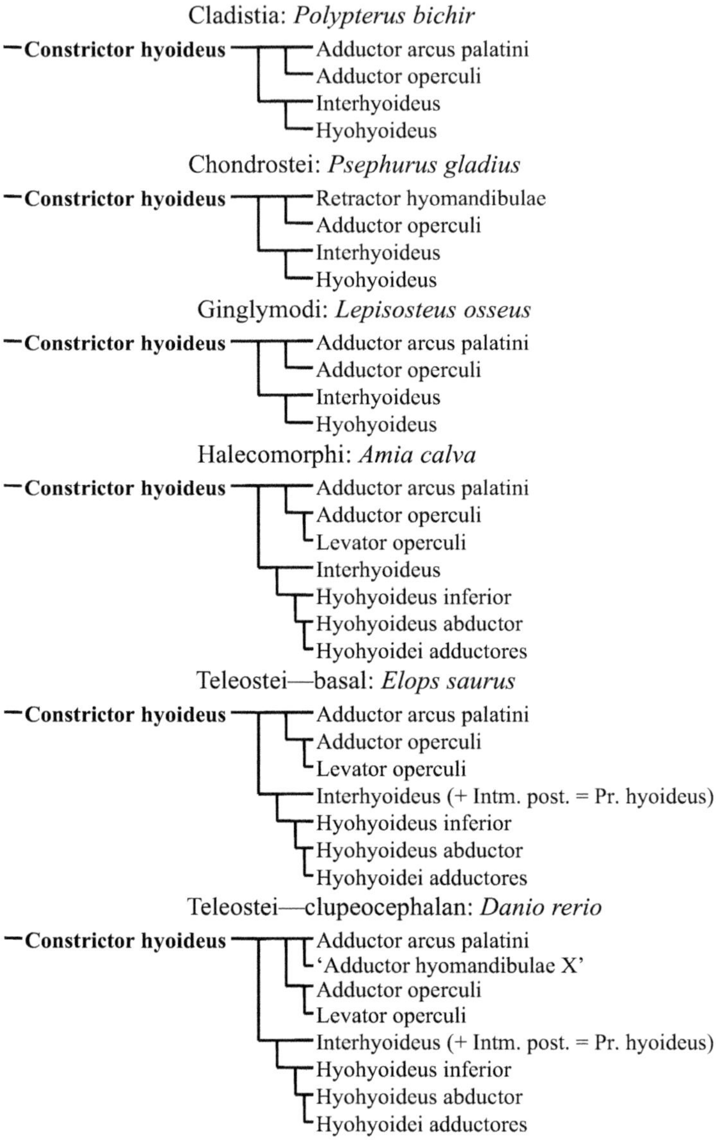

Figure 118 Developmental lineages of hyoid muscles in actinopterygians; Edgeworth's presumptive premyogenic condensations are in bold face (modified from Miyake et al., 1992); the nomenclature of the muscles listed on the right of the figure follows that of the present work, "Pr. hyoideus" meaning protractor hyoideus and "Intm. post." meaning intermandibularis posterior. Data compiled from evidence provided by developmental biology, comparative anatomy, experimental embryology and molecular biology, innervation and phylogeny (in studies by the author and other researchers). For more details, see text.

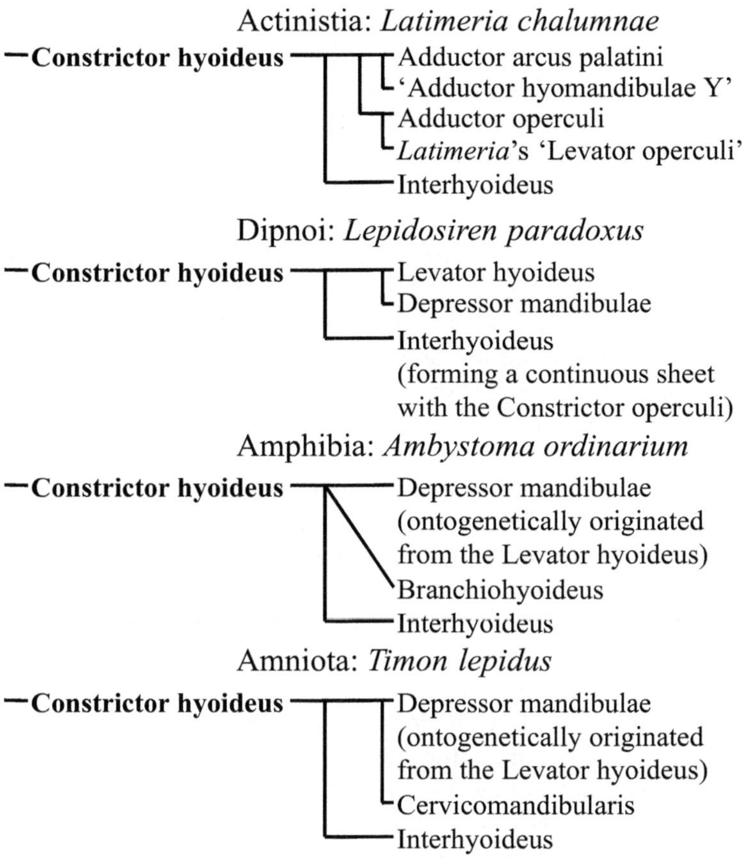

Figure 119 Developmental lineages of hyoid muscles in sarcopterygians, Edgeworth's presumptive premyogenic condensations are in bold face (modified from Miyake et al., 1992); the nomenclature of the muscles listed on the right of the figure follows that of the present work. Data compiled from evidence provided by developmental biology, comparative anatomy, experimental embryology and molecular biology, innervation and phylogeny (in studies by the author and other researchers). For more details, see text.

osteichthyans: the adductor arcus palatini and the adductor operculi (Tables 4, 5; Figs. 6, 19, 27, 74). These two muscles, which as their names indicate are usually related to the abduction of the suspensorium/ palatoquadrate and of the opercle, respectively, are found in *Latimeria* and in most living actinopterygians (Tables 4, 5; Figs. 118, 119). It should, however, be noted that some actinopterygians, such as the saccopharyngiform teleosts, may lack an adductor operculi (see Chapter 3, Section 3.2). It should also be noted that in living chondrosteans the dorso-medial portion of the hyoid muscle plate gives rise to a peculiar retractor hyomandibulae (see Fig. 12), and not to an adductor arcus palatini as found in most other actinopterygians (Table 4; Fig. 118) (e.g., Danforth, 1913; Luther, 1913; Edgeworth, 1935; Kesteven, 1942-1945; Miyake et al., 1992; Carroll and Wainwright, 2003; this work).

Apart from the adductor arcus palatini, other muscles connecting the neurocranium to the palatoquadrate/suspensorium and promoting the adduction of this latter structure may be found in osteichthyans. Examples of this are the muscles "adductor hyomandibulae" of Winterbottom (1974). As explained by Winterbottom (1974: 239), at least some of these muscles are seemingly non-homologous, since they may "separate 1) either from the posterior region of the adductor arcus palatini or 2) from the anterior fibers of the adductor operculi". I found separate, distinct muscles "adductor hyomandibulae" sensu Winterbottom in various teleosts (e.g., Fig. 53). An "adductor hyomandibulae" sensu Winterbottom is seemingly also found in *Latimeria* (e.g., the descriptions of Millot and Anthony, 1958). The phylogenetic results of the present work seem to support Winterbottom's idea that at least some of these "adductor hyomandibulae" were acquired independently. This is, for example, the case of the "adductor hyomandibulae" found in *Latimeria* and in teleosts such as *Danio*, which I have distinguished in Tables 4 and 5 by using the names "adductor hyomandibulae Y" and "adductor hyomandibulae X", respectively. Miyake et al. (1992) suggested that other key osteichthyan genera listed in Tables 4 and 5 such as *Amia* and *Lepisosteus* also have "adductor hyomandibulae" muscles sensu Winterbottom (note: the "adductor hyomandibulae" and "adductor arcus palatini" of Winterbottom and of the present work correspond respectively to the "adductor arcus palatini" and "adductor hyomandibulae" of Miyake et al.). However, in the *Amia* and *Lepisosteus* specimens examined there is no separate, well-differentiated "adductor hyomandibulae" sensu Winterbottom. This is also the case of the specimens of these genera analyzed by, for example, Lauder (1980a: 289), who has stated that "in *Lepisosteus* the adductor arcus palatini (= "adductor hyomandibulae" of Winterbottom) and the adductor hyomandibulae (= "adductor arcus palatini" of Winterbottom) form a continuous sheet of parallel-fibered muscle" and that "the adductor arcus palatini (= "adductor hyomandibulae" of Winterbottom) is absent in *Amia*".

Apart from the muscles adductor arcus palatini, adductor operculi and "adductor hyomandibulae" (sensu this work) other dorso-medial hyoid muscles may be found in living osteichthyans (see Tables 4, 5; Figs. 118, 119). For instance, *Amia* and most extant teleosts have a levator operculi that is usually related to a peculiar mechanism mediating lower jaw depression via the so-called "four-bar linkage system" in which the force of contraction of this muscle is transmitted through the opercular series and the interopercular ligament to the lower jaw (e.g., Stiassny, 2000). A levator operculi may be missing in certain teleosts, such as saccopharyngiforms (e.g., Tchernavin, 1947a,b, 1953; this work). Millot and Anthony (1958) stated that *Latimeria* has a "levator operculi". As can be seen in the descriptions and the figures provided by these authors (e.g., their plate VII), and as explicitly recognized in their page 61, the fibers of their "levator operculi" are deeply mixed with those of the adductor operculi. As explained in Chapter 3, Section 3.2, this has led Lauder (1980c) and other authors to be skeptical about the presence, in *Latimeria,* of a distinct levator operculi such as that found in *Amia* and teleosts. However, Adamicka and Ahnelt (1992: 108) have reaffirmed, seemingly on the basis of their own observations of *Latimeria,* that this taxon "does have a levator operculi muscle differentiated out of the adductor (operculi)". According to the phylogenetic results of Chapter 3 the "levator operculi" of *Latimeria* is not homologous with the levator operculi of *Amia* and teleosts (Tables 4, 5). It is also important to stress that the function of the "levator operculi" of *Latimeria* is not similar to that of the levator operculi of *Amia* and teleosts, because *Latimeria* does not have an interopercular ligament and, thus, does not have an opercular mechanism mediating mandible depression like that found in *Amia* and teleosts (e.g., Millot and Anthony, 1958; Alexander, 1973; Anthony, 1980; Lauder, 1980c). In order to distinguish the "levator operculi" of *Latimeria* and the levator operculi of *Amia* and teleosts, I named the former muscle *Latimeria*'s "levator operculi" in Table 5.

In an illustration of the dipnoan *Neoceratodus,* Kardong (2002: fig. 10.39B) shows a muscle that he designated as "levator operculi". However, as explained in Chapter 3, in the dipnoan specimens examined in the present work, as well as in those described by, for example, Bischoff (1840), Owen (1841), Luther (1914), Edgeworth (1935), Kesteven (1942-1945), Bemis (1986), Bemis and Lauder (1986) and Bartsch (1994), there is no structure resembling the "levator operculi" of *Latimeria* or the levator operculi of *Amia* and teleosts. The "levator operculi" of Kardong's (2002) fig. 10.39B seems to correspond, in fact, to the constrictor operculi of, for example, Bemis and Lauder (1986), which may eventually correspond to the adductor operculi of other bony fishes but forms, in living adult dipnoans, a continuous sheet together with other cranial muscles (see Fig. 91) (there is thus no separate adductor operculi in adult dipnoans, as shown in Table 5).

Examples of other dorso-medial hyoid muscles found in osteichthyans are the levator hyoideus and the depressor mandibulae (Table 5; Fig. 119). The levator hyoideus is usually related to the elevation of the posterodorsal portion of the ceratohyal (e.g., Fig. 93), while the depressor mandibulae is usually related to the opening of the mouth (e.g., Figs. 91, 99, 109). These muscles are found in at least some developmental stages of extant dipnoans and of numerous extant tetrapods (see Chapter 3). Edgeworth (1935) considered that, contrary to what is suggested in Table 5, the depressor mandibulae of adult dipnoans such as *Protopterus* and *Lepidosiren* is not homologous with part of the depressor mandibulae of adult tetrapods. This is because, in the developmental series of these dipnoans that he observed, the levator hyoideus and the depressor mandibulae seemingly appear at the same time, while in those of tetrapods the depressor mandibulae seemingly appears as a modification of part, or all, of the levator hyoideus. However, Forey (1986) and others suggested that the depressor mandibulae/levator hyoideus of adult members of *Protopterus* and *Lepidosiren* are homologous with the depressor mandibulae/levator hyoideus of adult tetrapods, and that this actually provides support for a close relationship between dipnoans and tetrapods. The observations, comparisons and phylogenetic results of the present work strongly support Forey's hypothesis: the innervation, position, relations with other structures, and function of the depressor mandibulae of adult dipnoans are strikingly similar to those of the posterior part of the depressor mandibulae of adult tetrapods, and in all the most parsimonious trees obtained in the cladistic analysis of Chapter 3 the dipnoans do appear as the closest living relatives of tetrapods (Figs. 3, 4). In fact, it is important to stress that even authors who have adhered to Edgeworth's (1935) view admit that the depressor mandibulae of adult dipnoans is "functionally and topographically comparable" to the anterior part of the depressor mandibulae of adult tetrapods such as salamanders (e.g., Bauer, 1997: 79).

Given the above, the following hypothesis on the evolution of the levator hyoideus and depressor mandibulae may be tentatively proposed. The first evolutionary step was seemingly the differentiation of the levator hyoideus (possibly from the portion of the hyoid muscle plate giving the adductor arcus palatini in other osteichthyans, but this is far from clear: Table 5; see below). A configuration such as this, that is, in which there is no depressor mandibulae and in which the levator hyoideus consists in a single mass of fibers attaching on the hyoid arch, is found, for example, in early development stages of the dipnoan *Neoceratodus* and of various tetrapods (e.g., Edgeworth, 1935: fig. 313; Bartsch, 1994: fig. 2B). The second evolutionary step may have been the attachment of some fibers of the levator hyoideus on the mandible and the ultimate differentiation of these fibers in a depressor mandibulae. A configuration such as this, that is, in which there is both a levator hyoideus inserting on the hyoid arch and a depressor mandibulae inserting on the mandible, is found in early development stages of various tetrapods and of dipnoans such as *Lepidosiren* and *Protopterus*, as well as in juveniles and adults of these two dipnoan genera and of amphibians such as *Siren* (e.g., Figs. 91, 93; Edgeworth, 1935: fig. 327). These two evolutionary steps probably occurred before the splitting between dipnoans and tetrapods, since at least some members of both these groups have depressor mandibulae fibers attaching on the mandible (see Table 5). In other words, the ancestors of dipnoans and tetrapods probably had, in at least some stages of their development, both a levator hyoideus and a depressor mandibulae. From this point, the evolution of these muscles was seemingly rather diverse and complex. For instance, in dipnoans such as *Neoceratodus* the levator hyoideus becomes ontogenetically mixed with other hyoid muscles, being absent as a separate element in adults; a separate, distinct depressor mandibulae is missing in early and late developmental stages of this taxon. In dipnoans such as *Lepidosiren* and *Protopterus* and in tetrapods such as *Siren* these two muscles remain as separate elements until the adult stage (e.g., Figs. 91, 93). As in *Neoceratodus*, the levator hyoideus is also absent as a separate element in numerous adult tetrapods, but for a different reason: because, as shown by Edgeworth (1935) and others, in these tetrapods all the fibers of this muscle become ontogenetically attached to the mandible and, thus, integrated in a depressor mandibulae (e.g., Figs. 99, 109, 110; Table 5).

The levator hyoideus of *Protopterus* and *Lepidosiren* (Fig. 93) as well as of adult tetrapods such as *Siren* thus seemingly corresponds to the depressor mandibulae posterior of adult amphibians such as *Ambystoma* (Fig. 99) and to part of the depressor mandibulae of adult amniotes such as *Timon* (e.g., Fig. 109) (Table 5). The depressor mandibulae of adult *Protopterus* and *Lepidosiren* (Fig. 91) and of adult tetrapods such as *Siren* seemingly corresponds to the depressor mandibulae anterior of adult amphibians such as *Ambystoma* (Fig. 99) and to part of the depressor mandibulae of adult amniotes such as *Timon* (Fig. 109). The statements of Edgeworth concerning the different development of the depressor mandibulae of dipnoans and of the anterior part of the depressor mandibulae of tetrapods such as *Ambystoma* (see above) may thus be due to two main factors: (1) the mode of appearance of these structures may be somewhat different in dipnoans and in tetrapods (which, in view of what will be said below as well as of the recent discoveries made in the field of evolutionary developmental biology, does not completely invalidate the hypothesis that they are homologous: e.g., Gould, 2002; West-Eberhard, 2003; Carroll et al., 2005; Kirschner and Gerhart, 2005) and (2) the mode of appearance of the depressor mandibulae of dipnoans is, in reality, similar to that of the anterior part of the depressor mandibulae of tetrapods (i.e., it appears ontogenetically after the levator hyoideus, resulting from the differentiation of part of its fibers), but the youngest dipnoan specimens observed by Edgeworth (1935) were too old (i.e., the differentiation had already occurred and thus both the levator hyoideus and the depressor mandibulae were already present, giving the idea that these structures originated ontogenetically at the same time). Further detailed comparative analyses on the development of the hyoid muscles of dipnoans and of other osteichthyans are needed to clarify this issue.

Such detailed analyses are also needed to clarify whether the portions of the constrictor hyoideus from which the levator hyoideus/depressor mandibulae and the constrictor operculi of extant dipnoans originate may eventually correspond to the portions from which the adductor arcus palatini and the adductor operculi of other bony fishes originate, as suggested by, for example, Edgeworth and in Chapter 3. In fact, Edgeworth (1935: 102) stated that the levator hyoideus of dipnoans originates from the portion of the constrictor hyoideus that gives rise to the retractor hyomandibulae in extant acipenseriforms, which seemingly corresponds to the adductor arcus palatini of other bony fishes (Table 4; see above). It is important to note that in living dipnoans the hyomandibula is very reduced or even missing and the palatoquadrate is fused to the neurocranium, being thus much less mobile than that of most other bony fishes. Therefore, the portion of the hyoid muscle plate that gives rise to the adductor arcus palatini of other bony fishes may have lost its usual attachment to the hyomandibula and/or palatoquadrate and become attached on the ceratohyal; the dorsal surface of this bone lies in fact more dorsal in dipnoans than in most other bony fishes, occupying a position somewhat similar to that of the hyomandibula of these latter fishes (e.g., Fig. 93; Rosen et al., 1981; Forey, 1986; Bauer, 1997). As explained above, my observations of juvenile and adult specimens of *Lepidosiren* pointed out that in these specimens the levator hyoideus actually attaches on the dorsal surface of the ceratohyal but also on part of its dorsomesial margin (e.g., Fig. 93). Thus, as the adductor arcus palatini of other bony fishes usually attaches on the dorsomesial margin of the hyomandibula/palatoquadrate in order to adduct these structures, it seems that the levator hyoideus might not only elevate but also adduct the dorsal surface of the ceratohyal. Regarding the constrictor operculi of dipnoans (= "constrictor hyoideus dorsalis" of Miyake et al., 1992), this structure could originate from the same portion of the constrictor hyoideus from which the adductor operculi of other bony fishes originates (e.g., Fig. 91). In fact, according to, for example, Campbell and Barwick (1986), some basal fossil dipnoans do seemingly exhibit well-defined scars on the mesial margins of the opercular bones for the attachment of muscles that seem to have been somewhat similar to the adductor operculae of other fishes. Be that as it may, in adult extant dipnoans the constrictor operculae forms a continuous sheet together with other muscles such as the interhyoideus. Therefore, even if these structures eventually originate from the portion of the constrictor hyoideus that gives rise to the adductor operculae in other bony fishes, in adult extant dipnoans there are no separate, distinct adductor operculae such as those found in adult members of *Latimeria*, actinopterygians and other taxa (see Table 5 and Chapter 3).

The plesiomorphic condition for osteichthyans is seemingly that in which the ventral portion of the hyoid muscle gives rise to a single division, which is designated here as interhyoideus (Tables 4, 5). In most extant actinopterygians part of the interhyoideus becomes ontogenetically differentiated into a well-differentiated, separate muscle, the hyohyoideus (e.g., Table 4; Figs. 12, 15, 19, 29, 34, 77, 118). In *Amia* and most teleosts the hyohyoideus becomes differentiated into three distinct divisions: the hyohyoideus inferior, the hyohyoideus abductor and the hyohyoidei adductores (e.g., Fig. 77), the two latter divisions being often considered parts of a hyohyoideus superior (Table 4, Fig. 118). As stated by Stiassny (2000: 122) "there is little commentary in the literature regarding the function of HhI (hyohyoideus inferior) but adduction of the hyoid bar is suggested by its position and presumed line of action" (e.g., Fig. 77). Regarding the hyohyoideus abductor and the hyohyoidei adductores, they are usually related to the expansion and constriction of the branchiostegal membranes, respectively (e.g., Fig. 77). As stated above, the interhyoideus, which as its name indicates usually connects the two hyoid bars, has become associated with the intermandibularis posterior in teleosts, forming the peculiar protractor hyoideus of these fishes (see Tables 4, 5).

In Miyake et al.'s (1992) table 1, it is suggested that in extant cladistians, chondrosteans and ginglymodians the hyohyoideus is differentiated into two well-differentiated divisions, a hyohyoideus inferior and a hyohyoideus superior. However, in the adult specimens of these three groups examined in the present work the hyohyoideus is not divided into well-differentiated, separate parts as in *Amia* and teleosts, consisting instead of a continuous sheet of fibers. These observations are supported by those of other authors. For instance, Lauder (1980a: 293) wrote that "in *Lepisosteus* the hyohyoideus superioris ... also (as in *Polypterus*) is continuous with the fibers of the hyohyoideus inferioris" (i.e., the fibers of the hyohyoideus form a continuous sheet, as is the case in the *Lepisosteus* and *Polypterus* specimens examined in the present work). The hyohyoideus of the chondrosteans described by, for example, Danforth (1913: *Polyodon*) and Carroll and Wainwright (2003: *Scaphirhynchus*) is also constituted by a continuous sheet of fibers, as is the case with the *Psephurus* and *Acipenser* specimens analyzed in the present work (note: the interhyoideus of this work corresponds to the "geniohyoideus posterior" and the "constrictor ventralis posterior" of Danforth, 1913, and of Carroll and Wainwright, 2003, respectively).

There is seemingly no well-differentiated, separate hyohyoideus in extant Sarcopterygii (Table 5, Figs. 86, 91, 99, 119). That is, there are some sarcopterygians in which the portion of the hyoid muscle plate giving rise to the interhyoideus and hyohyoideus in actinopterygians might eventually be somewhat differentiated into bundles that resemble, in certain aspects, those two muscles, but these bundles remain deeply mixed throughout the development of these sarcopterygians. This is, for example, the case of the interhyoideus anterior and the interhyoideus posterior of various salamanders (Table 5; Fig. 99; Lubosch, 1914; Luther, 1914; Edgeworth, 1935; Jarvik, 1963; Larsen and Guthrie, 1975; Carroll and Holmes, 1980; Bauer, 1992; Haas, 2001). The same applies to the "géniohyoïdien" and the "hyohyoïdien" described by Millot and Anthony (1958) in *Latimeria*, which, according to these authors, are deeply mixed in the adult members of this genus and which thus seem to correspond to the interhyoideus of the present work (Table 5).

Apart from the muscles mentioned in the paragraphs above, other hyoid muscles are found in osteichthyans (Tables 4, 5). For instance, numerous urodeles have a peculiar muscle branchiohyoideus connecting the hyoid and branchial arches in at least some developmental stages (e.g., Fig. 99). As stressed by, for example, Ericsson and Olsson (2004: 136), ontogenetically the branchiohyoideus appears situated between the interhyoideus and the levator hyoideus/depressor mandibulae, a position that "makes it difficult to determine if it belongs to the dorsal or ventral (hyoid) muscles" (see Fig. 119). According to, for example, Lauder and Shaffer (1985: 308) the function of the branchiohyoideus is to "mediate hyoid retraction and possibly also produce ceratobranchial abduction in the absence of antagonistic activity". Edgeworth (1935) and Lauder and Shaffer (1988) stated that the branchiohyoideus becomes lost after metamorphosis in most urodeles, but the metamorphosed specimens of *Ambystoma ordinarium* examined in the present work do have a distinct muscle branchiohyoideus (see Fig. 99). As explained by Edgeworth (1935), the structure that is often named "branchiohyoideus" in amniotes is seemingly homologous to the branchial muscle subarcualis rectus 1 of amphibians, and not to the hyoid muscle branchiohyoideus of the present work (which he designated as "branchiohyoideus externus") (see Figs. 99, 110). Jarvik (1963) suggested that at least some of the sarcopterygian fossil fishes placed in his "porolepiform-stock" might eventually have had a branchiohyoideus such as that of urodeles. However, as stressed by, for example, Rosen et al. (1981), some of Jarvik's interpretations of fossils, principally those concerning soft structures, should be viewed with much caution, as they were profoundly influenced by the author's strong, and rather heterodox, beliefs. For instance, in this specific case concerning the supposed development in his "porolepiforms" of a branchiohyoideus similar to that of urodeles, this has much to do with his profound conviction that these two taxa were sister-groups and, thus, that amphibians are not monophyletic, a view to which almost no author would adhere nowadays ("these great resemblances ... cannot possibly be due to parallel evolution and they prove definitively that the urodeles are closely related to and descendants of porolepiforms": Jarvik, 1963: 61). A detailed, updated comparative analysis is thus needed to investigate whether or not some sarcopterygian fish fossils might eventually have had a muscle branchiohyoideus similar to that of urodeles.

Hypobranchial Muscles (Tables 6, 7; Figs. 120, 121)

According to Edgeworth (1935), there are two major lineages of muscles originating from the hypobranchial muscle plate: his genio-hyoideus and his rectus cervicus (Figs. 118, 119). It is not clear whether Edgeworth's genio-hyoideus and rectus cervicus represent separate premyogenic condensations or later states of muscle development. The plesiomorphic condition for osteichthyans is seemingly that found in adult members of the Actinistia and the Dipnoi, in which there is a coracomandibularis and a sternohyoideus (sensu this work: Table 7; Figs. 86, 91). According to, for example, Edgeworth (1935), Kesteven (1942-1945), Wiley (1979a,b), Jollie (1982), Mallat (1997), Wilga et al. (2000), and Johanson (2003), these muscles were originally mainly related to the opening of the mouth.

Extant tetrapods exhibit various hypobranchial muscles that are absent in other living osteichthyans, such as the omohyoideus and the specialized glossal muscles related to the movements of the tongue (Table 7; Figs. 99, 110, 121). Edgeworth (1935) stated that the omohyoideus and the genioglossus are seemingly derived from the sternohyoideus and the coracomandibularis, respectively (e.g., Fig. 99). However, his statements regarding the origin of the hyoglossus are somewhat confusing: in his page 196 it is stated that in amphibians such as salamanders this muscle originates from the sternohyoideus, but in his page 211 it is claimed that "the hypobranchial muscles of Amphibia, Sauropsida and Mammalia are

Table 6 Hypobranchial muscles of adults of representative actinopterygian taxa; the nomenclature of the muscles shown in bold follows that of the present work; data compiled from evidence provided by developmental biology, comparative anatomy, experimental embryology and molecular biology, innervation and phylogeny (in studies by the author and other researchers). For more details, see text (see also Fig. 120).

Probable plesiomorphic osteichthyan condition	Cladistia: *Polypterus bichir* (Bichir)	Chondrostei: *Psephurus gladius* (Chinese swordfish)	Ginglymodi: *Lepisosteus osseus* (Longnose gar)	Halecomorphi: *Amia calva* (Bowfin)	Teleostei—basal: *Elops saurus* (Ladyfish)	Teleostei—clupeocephalan: *Danio rerio* (Zebrafish)
Coracomandi bularis	**Branchiomandi bularis** (*modified from coracomandibularis)	**Branchiomandi bularis** (*see cell on the left)	--------	**Branchiomandi bularis** (*see cell on the left)	--------	--------
Sternohyoideus	**Sternohyoideus**	**Sternohyoideus**	**Sternohyoideus**	**Sternohyoideus**	**Sternohyoideus**	**Sternohyoideus**

Table 7 Hypobranchial muscles of adults of representative sarcopterygian taxa; the nomenclature of the muscles shown in bold follows that of the present work (in order to facilitate comparisons, certain names used by other authors are also shown); data compiled from evidence provided by developmental biology, comparative anatomy, experimental embryology and molecular biology, innervation and phylogeny (in studies by the author and other researchers). For more details, see text (see also Fig. 121).

Probable plesiomorphic osteichthyan condition	Actinistia: *Latimeria chalumnae* (Coelacanth)	Dipnoi: *Lepidosiren paradoxa* (South American lungfish)	Amphibia: *Ambystoma ordinarium* (Michoacan stream salamander)	Amniota: *Timon lepidus* (Ocellated lizard)
Coracomandibularis	**Coracomandibularis**	**Coracomandibularis** (geniothoracicus of e.g. Miyake et al., 1992)	**Coracomandibularis** (geniohyoideus of e.g. Bemis et al., 1983, and Lauder and Shaffer, 1988)	**Coracomandibularis** (geniohyoideus and/or at least part of mandibulohyoideus of e.g. Edgeworth, 1935, and Herrel et al., 2005)
--------	--------	--------	**Genioglossus** (*according to e.g. Edgeworth, 1935, the genioglossus of salamanders such as *Ambystoma* is derived from the coracomandibularis)	**Genioglossus** (*according to e.g. Edgeworth, 1935, the genioglossus of lizards such as *Timon* is derived from the coracomandibularis)

--------	--------	--------	**Hyoglossus** (*the statements of Edgeworth, 1935, concerning the origin of this muscle in salamanders such as *Ambystoma* are somewhat confused: in his page 196 he states that it originates from the sternohyoideus but in his page 211 he seems to indicate that, as in other amphibians as well as in amniotes, it derives from the coracomandibularis)	**Hyoglossus** (*according to e.g. Edgeworth, 1935, the hyoglossus of lizards such as *Timon* is derived from the coracomandibularis)
Sternohyoideus	**Sternohyoideus**	**Sternohyoideus** (rectus cervicis of e.g. Bemis and Lauder, 1986)	**Sternohyoideus** (rectus cervicis of e.g. Lauder and Shaffer, 1988)	**Sternohyoideus** (rectus cervicis of e.g. Kardong, 2002)
--------	--------	--------	**Omohyoideus** (*seemingly derived from the sternohyoideus)	**Omohyoideus** (*see cell on the left)

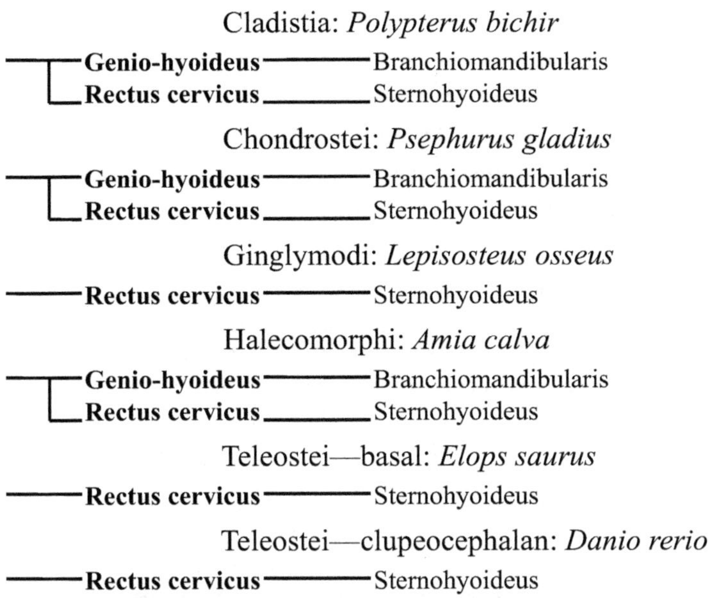

Figure 120 Developmental lineages of hypobranchial muscles in actinopterygians, Edgeworth's presumptive premyogenic condensations are in bold face (modified from Miyake et al., 1992); the nomenclature of the muscles listed on the right of the figure follows that of the present work. Data compiled from evidence provided by developmental biology, comparative anatomy, experimental embryology and molecular biology, innervation and phylogeny (in studies by the author and other researchers). For more details, see text.

essentially similar … a genioglossus and a hyoglossus are developed from the genio-hyoideus (= coracomandibularis of the present work)". Jarvik (1963: 41) reanalyzed this issue and argued that the hyoglossus of salamanders "seems to be to be an anterior portion of the rectus cervicis (= sternohyoideus of this work)". Larsen and Guthrie (1975) suggested that the hyoglossus of salamanders originated from "part of the genioglossus complex" (and, thus, from the coracomandibularis of this work: see above) but stated that it was not possible to confirm this hypothesis "without examining early ontogenetic stages". The information provided in more recent publications referring to the cranial musculature of salamanders (e.g., Carroll and Holmes, 1980; Lauder and Shaffer, 1985, 1988; Reilly and Lauder, 1989, 1991; Iordansky, 1992; Bauer, 1992, 1997; Kardond and Zalisko, 1998; Haas, 2001; Kardong, 2002; Ericsson and Olsson, 2004; Ericsson et al., 2004) does not fully clarify whether the hyoglossus of salamanders is derived from the sternohyoideus or from the coracomandibularis (or eventually from both). The analysis of the

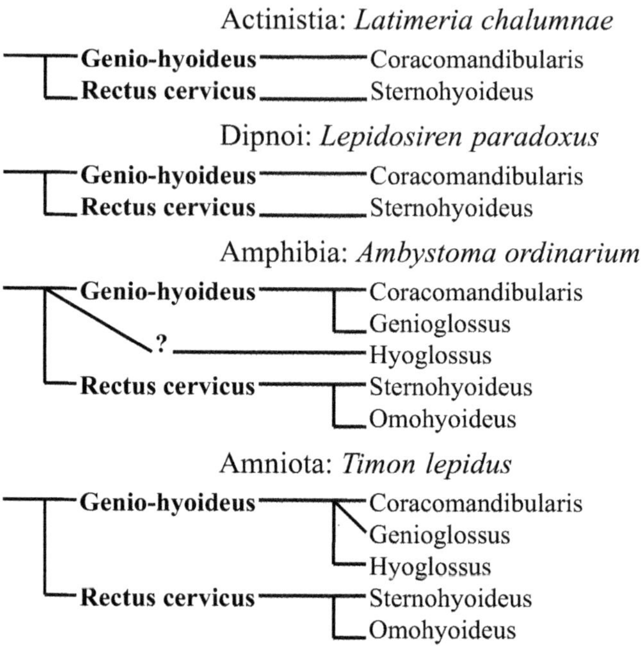

Figure 121 Developmental lineages of hypobranchial muscles in sarcopterygians, Edgeworth's presumptive premyogenic condensations are in bold face (modified from Miyake et al., 1992); the nomenclature of the muscles listed on the right of the figure follows that of the present work. Data compiled from evidence provided by developmental biology, comparative anatomy, experimental embryology and molecular biology, innervation and phylogeny (in studies of the author and of other researchers). For more details, see text.

salamander specimens examined in the present work unfortunately did not help to clarify this issue (see Fig. 121); detailed comparative analyses of the development of the hyoglossus in salamanders and in other tetrapods are needed. I plan to undertake such analyses in a future work.

In extant cladistians, chondrosteans and halecomorphs the coracomandibularis is modified into a peculiar muscle branchiomandibularis connecting the branchial arches to the mandible (e.g., 6, 19); in living ginglymodians and teleosts this muscle is absent (e.g., Figs. 15, 29, 77). Wiley (1979a,b), Lauder and Liem (1983), and other authors proposed that the ancestors of extant actinopterygians probably had a branchiomandibularis, and that the absence of this muscle in living ginglymodians and teleosts is due to a secondary loss. The present works supports this view (see Chapter 3).

General Remarks

In view of the information summarized in Tables 2 to 7 and in Figs. 116 to 121 and of the discussion provided above, it thus seems that plesiomorphically in osteichthyans the mandibular muscle plate gave rise to the levator arcus palatini, dorsally, to the adductor mandibulae A2, A3′, A3″ and Aω, medially, and to a divided intermandibularis, ventrally. Within osteichthyan evolution, the posterior part of the intermandibularis became associated with the interhyoideus in teleosts, forming the peculiar muscle protractor hyoidei (Table 2). The number of divisions of the adductor mandibulae complex has changed during the evolutionary history of osteichthyans, some peculiar divisions being exclusively found in extant members of certain groups, such as the A1-OST (ostariophysans), A0 (cypriniforms), palatomandibularis minor and major (ginglymodians), levator maxillae superioris 3 and 4 (halecomorphs), A2-PVM (non-actinistian sarcopterygians) and pterygomandibularis (amniotes) (Tables 2, 3). The number of dorsal mandibular muscles is also variable within adult osteichthyans: some taxa have a single muscle (e.g., *Latimeria*, in which there is only a levator arcus palatini and which seems to represent the plesiomorphic condition for osteichthyans: see above); some have more than one muscle (e.g., the numerous actinopterygians exhibiting a levator arcus palatini and a dilatator operculi or the numerous amniotes exhibiting a levator pterygoidei and a protractor pterygoidei); and some have none (e.g., living dipnoans and amphibians) (Tables 2, 3).

Regarding the hyoid muscle plate, the plesiomorphic condition for osteichthyans is seemingly that in which this plate gave rise to a single ventral division, which is designated here as interhyoideus, and to two dorsomedial divisions, the adductor arcus palatini and the adductor operculi (Tables 4, 5). In actinopterygians a portion of the interhyoideus became differentiated into a separate muscle, the hyohyoideus, which then became divided into three separate, clearly recognizable divisions in halecostomes, the hyohyoideus inferior, the hyohyoideus abductor and the hyohyoidei adductores (Table 4). Other hyoid muscles may be found in osteichthyans, for example, the branchiohyoideus of urodeles (Table 5) and the levator operculi of halecostomes (Table 4). *Latimeria* has a muscle "levator operculi" that in certain aspects resembles, but is seemingly not homologous to, that of halecostomes (Table 5). *Latimeria* also has an "adductor hyomandibulae" that somewhat resembles, but is seemingly also not homologous to, the muscles "adductor hyomandibulae" found in certain other osteichthyans such as actinopterygians (Tables 4, 5). As explained above, there are many uncertainties concerning the homologies and evolution of the levator hyoideus/depressor mandibulae of dipnoans and tetrapods. According to the hypothesis proposed in the present work both these muscles seem to have been found in at least some ontogenetic stages of the ancestors that later gave rise to dipnoans and tetrapods (see above) (Table 5). Interestingly, works such as Köntges and Lumsden (1996) have shown that in tetrapods such as birds the posterior region of the mandible in which the depressor mandibulae attaches is constituted by neural crest derivatives of the hyoid arch, and not of the mandibular arch. This is, in fact, one of the various examples given by these authors to illustrate the highly constrained pattern of cranial skeletomuscular connectivity that they found in these tetrapods, in which each rhombomeric neural crest population remains coherent throughout ontogeny, forming both the connective tissues of specific muscles and their respective attachment sites on to the neuro- and viscerocranium. It would thus be interesting to investigate whether the depressor mandibulae of dipnoans such as *Protopterus* and *Lepidosiren* also attaches in a region of the mandible constituted by neural crest derivatives of the hyoid arch. If that is the case, and if further investigations eventually support the position that the mandible of extant non-dipnoan bony fishes is exclusively formed by mandibular neural crest derivatives, this would indicate that the presence of a depressor mandibulae in tetrapods and dipnoans might be related to an evolutionary change in which hyoid neural crest derivatives have become incorporated in the formation of the lower jaw.

Concerning the hypobranchial muscles, the plesiomorphic condition for osteichthyans seemingly corresponds to that found in extant actinistians and dipnoans, in which there is a coracomandibularis and a sternohyoideus (Table 7). Changes to this plesiomorphic condition occurred within osteichthyan evolution. For instance, in actinopterygians the coracomandibularis became modified into a peculiar branchiomandibularis (Table 6), while in sarcopterygians such as tetrapods the hypobranchial muscle plate became differentiated into various muscles that are absent in other extant osteichthyans, such as the omohyoideus and the glossal muscles of the tongue (Table 7). In mammals, the hypobranchial muscle plate has become divided into an even greater number of muscles; this is also the case of the mandibular and hyoid muscle plates, which in mammals usually give rise to more muscles than in other tetrapods (e.g., Edgeworth, 1935; Brock, 1938; Jarvik, 1963, 1980; Gorniak, 1985; Pough et al., 1996; Kardong and Zalisko, 1998; Gibbs et al., 2000, 2002; Kardong, 2002; Kisia and Onyango, 2005).

It should be emphasized that, as mentioned above, the discussions and the hypotheses advanced in this Section are based on data compiled from evidence provided by developmental biology, comparative anatomy, functional morphology, paleontology, experimental embryology and molecular biology, innervation, and phylogeny. In fact, as stressed by, for example, Edgeworth (1935: 222), in order to provide a well-grounded analysis on the homologies and evolution of a certain muscle within different taxa it is imperative to take into consideration all the available lines of evidence, since "no one criterion is sufficient, not even two". This is because, as Edgeworth noted, none of these lines of evidence is infallible. For instance, although the innervation of a muscle generally remains constant and corresponds to its segment of origin (e.g., Luther, 1913, 1914; Edgeworth, 1935; Kesteven, 1942-1945; Köntges and Lumsden, 1996), there are cases in which a single muscle may have different innervations in different taxa. One of the examples provided by Edgeworth (1935: 221) to illustrate this concerns the intermandibularis of extant dipnoans, which "is innervated by the Vth and VII nerves, though wholly of mandibular origin".

Also, there are eventually cases in which a single muscle may originate from different regions and/or segments of the body in different taxa. An example provided by Edgeworth (1935: 221) concerns the branchial muscle protractor pectoralis (his "cucullaris"), which "has diverse origins in *Ornithorhynchus*, *Talusia* and *Sus*; in the first-named it is developed from the 3rd, in the second from the 2nd and in the last from the 1st branchial muscle-plate; these changes are secondary to the non-development of the branchial muscle-plates, from behind forwards; the muscles are homologous and have a constant primary innervation from the Xth nerve". This is one of the reasons I mentioned above that even if the development of the depressor mandibulae of dipnoans is not completely similar to the development of the anterior part of the depressor mandibulae of tetrapods such as salamanders, this does not necessarily mean that these structures are not homologous. This is because, in this specific case, there are actually several different lines of evidence supporting their homology: (1) innervation (e.g., they are innervated by the VIIth nerve); (2) adult anatomy (e.g., they occupy a similar position and have similar relations to other structures); (3) functional morphology (e.g., they have a similar function, being mainly related to the depression of the lower jaw); (4) ontogeny (even if their development is eventually not completely similar, as suggested by Edgeworth, most aspects concerning this development are actually similar, for example, they originate from the dorso-medial portion of the hyoid muscle plate); and (5) phylogeny (e.g., the phylogenetic results of the present work support the idea that dipnoans are the closest living relatives of tetrapods).

As stressed by Edgeworth (1935: 224), there are also cases in which "an old structure may be lost" (e.g., the branchiomandibularis is lost in extant ginglymodians and teleosts), in which "new muscles may be developed" (e.g., the glossal muscles of tetrapods), and in which "an old structure or group of structures may be transformed" (e.g., the levator hyoideus "may be transformed, either partially or wholly, into a depressor mandibulae"). The occurrence of such phenomena thus raises further difficulties for comparative analyses within different clades. And there are also cases in which "similar secondary developments occur in separate genera or phyla", that is, cases of convergence and parallelism (see, e.g., Diogo, 2005, for a discussion of these two concepts). This emphasizes the importance of turning to the evidence provided by phylogenetic analyses in order to recognize the occurrence of such cases. For instance, the "levator operculi" of *Latimeria* has some similarities (e.g., innervation, position, relations to other structures) with the levator operculi of halecostomes, but the phylogenetic results of this work and of other recent studies, together with other lines of evidence (e.g., functional morphology), indicate that these similarities are probably due to homoplasy (see above).

The examples provided in the paragraphs above thus illustrate the risks of discussing the homologies of structures such as muscles on the basis of a single line of evidence, even if it concerns innervation or development. But they also illustrate that when various lines of evidence are available (e.g., developmental biology, comparative anatomy, functional morphology, paleontology, experimental embryology, innervation, and/or phylogeny) and when all these are taken into consideration, it is possible to establish well-grounded hypotheses of homology. In other words, the hard work, in this case, does seem to be rewarding. In fact, a better understanding of the muscles of a certain taxon allows a much more detailed, integrative analysis of the comparative anatomy, functional morphology and evolution of that taxon. In the specific case of the present work, it is hoped that the information provided here may help to contribute to a better understanding of the muscles of osteichthyans and, thus, to a better knowledge of the evolution of these gnathostomes and of vertebrates in general (see also Sections 4.3 and 4.4).

4.3 CRANIAL MUSCLES, ZEBRAFISH, AND EVOLUTIONARY DEVELOPMENTAL BIOLOGY

As explained in the work of Carroll (2005), the intimate connection between development and evolution has long been appreciated in biology. For instance, Darwin and Huxley "leaned heavily on the facts of embryology (as they were in the mid-nineteenth century) to connect man to animal kingdom and for indisputable evidence of evolution" (Carroll, 2005: 5). But "while

Darwin and Huxley were right about development as a key to evolution, for more than one hundred years after their chief works, virtually no progress was made in understanding the mysteries of development" (Carroll, 2005: 6). Thus, during that time different kinds of biologists were approaching evolution at dramatically different scales. Embryologists "were preoccupied with phenomena that could be studied by manipulating the eggs and the embryos of a few species, and the evolutionary framework faded from embryology's view"; evolutionary biologists were "studying genetic variation in populations, ignorant of the relationship between genes and form" (Carroll, 2005: 7). However, this began to change in the 1970s, when "voices for the reunion of embryology and evolutionary biology made themselves heard" (Carroll, 2005: 7). As stressed by Carroll (2005), one example of this is Gould, whose book *Ontogeny and Phylogeny* (1977) revived discussion of the ways in which the modification of development may influence evolution. From this reunion between developmental biology and evolutionary biology resulted the so-called evolutionary developmental biology, or evo-devo. Since then, we have contributed to a real "evo-devo revolution". Among the numerous important contributions of this revolution, one of the most remarkable was to reveal that despite their great differences in appearance and physiology, all complex animals share a common "tool kit" or "master" genes that govern the formation and patterning of their bodies and body parts (e.g., Gould, 2002; West-Eberhard, 2003; Carroll, 2005; Carroll et al., 2005; Kirschner and Gerhart, 2005).

The discovery of this "tool kit", like most of the first discoveries associated with the evo-devo revolution, resulted from experiments on fruit flies. However, many other organisms are now the subject of study by evolutionary developmental biologists. Of these organisms, one of the most commonly studied is the zebrafish (*Danio rerio*), a freshwater teleostean fish of the order Cypriniformes (see Fig. 4). A combination of advantages makes this fish an ideal organism for researchers interested in embryonic development, comparative anatomy, physiology, and disease, such as its small size, its large number of offspring, its short generation time, or its very rapid and synchronous embryonic development (e.g., Cubbage and Mabee, 1996; Schilling and Kimmel, 1997; Blagden et al., 1997; Nüsslein-Volhard et al., 2002; Schilling, 2002). As stated by Cubbage and Mabee (1996), comparisons between zebrafish and other vertebrates are often made in developmental studies, the zebrafish being often taken as a "good representative" of teleosts, of actinopterygians and/or even of bony fishes. Several of those developmental studies deal with zebrafish myology, but only a few of them focus on cranial muscles (e.g., Easter and Nicola, 1996; Schilling and Kimmel, 1997; Hatta et al., 1990; Schilling; 2002; Hernandez et al., 2002, 2005; Hunter and Prince, 2002). And these mainly concern larval stages. In fact, as stressed by Schilling (2002: 73), "no study has carefully described the anatomy of the musculature of the adult zebrafish." This is surprising given that the musculature of other adult members of the order Cypriniformes has been described in detail in the literature (e.g., Takahasi, 1925; Edgeworth, 1935; Winterbottom, 1974; Vandewalle, 1975, 1977; Howes, 1978, 1979; Gosline, 1986, 1989). Schilling (2002) provided a short summary of the myology of the adult zebrafish. But, as he recognized, this was mainly based on an extrapolation from his "own observations of larval cranial muscles" and from "studies in other teleosts", and not from direct dissection of adult specimens of *Danio rerio* (Schilling, 2002: 73).

Apart from the poor knowledge of the late stages of development of zebrafish cranial muscles there are also problems with the homologation between some of these muscles and those of other vertebrates. To give just an example, in a recent paper it is stated that "while the rat geniohyoideus is composed predominantly of fast fibers, a trait shared by the homologous intermandibularis posterior of the zebrafish, the pattern of fiber type distribution is quite different; slow fibers in the rat geniohyoideus are not regionalized but rather mixed in among fast fibers; such a mosaic distribution characterizes mammalian muscles" (Hernandez et al., 2005: 332). However, the muscle that is often called "geniohyoideus" in mammals and, thus, in the rat, does not correspond to the intermandibularis posterior of the zebrafish, but with the coracomandibularis, a muscle that is missing in the zebrafish (see Section 4.2). This example illustrates how an insufficient knowledge of zebrafish cranial muscles and of their homologies may weaken the discussions and evolutionary hypotheses advanced in evo-devo studies.

One of the main aims of evo-devo is to compare the development of different organisms and to discuss the implications of the data obtained in an evolutionary context (e.g., Gould, 2002; West-Eberhard, 2003; Carroll, 2005; Carroll et al., 2005; Kirschner and Gerhart, 2005). Without a good knowledge of the structures being examined and without well-grounded hypotheses of homology between these structures and those of other organisms, it is thus difficult to attain this aim. As stressed by, for example, Hunter and Prince (2002), one of the main problems in interpreting and comparing data obtained in the zebrafish is related to insufficient knowledge of its cranial muscles and their homologies. For instance, these authors recognized that it was rather difficult to compare the effect of Hox PG2 loss of function on the development of the individual cranial muscles of zebrafish and of tetrapods, as "the precise homologies between individual pharyngeal arch muscles are difficult to determine, because of the differing organization and number of muscles" (Hunter and Prince, 2002: 383).

The main aim of this Section 4.3 is therefore to provide a solid basis for future molecular, developmental and evo-devo works concerning zebrafish cranial muscles, by addressing four main questions: (1) How are the mandibular, hyoid and hypobranchial muscles of zebrafish developed until they reach their adult form? (2) To which muscles of other osteichthyans do these muscles correspond? (3) Is there a correspondence between the ontogeny of these muscles in the zebrafish and their evolutionary history within the Osteichthyes? (4) Regarding these muscles, is it appropriate to consider the zebrafish a "good representative" of teleosts, of actinopterygians and/or of bony fishes? **It is important to emphasize that this Section is based on the results of an original research done by me and Simon Hughes. Therefore, it can be said that the data provided in this Section is, in a certain manner, co-authored by the two of us.** It should also be noted that the cranial muscles that will be discussed here correspond to the "superficial cranial muscles" sensu Diogo and Vandewalle (2003), that is, to the mandibular, hyoid and hypobranchial muscles. The branchial and ocular muscles sensu Edgeworth will thus not be considered in this Section (note: these muscles have been described/illustrated in zebrafish larvae by, e.g., Easter and Nicola, 1996; Schilling and Kimmel, 1997; Hatta et al., 1990; Schilling, 2002; Hernandez et al., 2002, 2005; Hunter and Prince (2002); the configuration of these muscles in adults is essentially similar to that found in the larvae and, thus, to that described/illustrated by these authors).

Development of Zebrafish—Mandibular, Hyoid, and Hypobranchial Muscles

The mandibular, hyoid, and hypobranchial muscles of zebrafish larvae have been described with some detail by, for example, Schilling and Kimmel (1997), Hatta et al. (1990), Schilling (2002), Hernandez et al. (2002, 2005), and Hunter and Prince (2002). Those descriptions concern the first five (e.g., Schilling and Kimmel, 1997; Schilling, 2002; Hernandez et al., 2002; Hunter and Prince, 2002) or six (e.g., Hernandez et al., 2005) days of development. Schilling and Kimmel (1997) provided detailed information on the time of appearance of these muscles and their subsequent development until a stage of 120 hours. It was thus decided to analyze six selected stages that cover the development of the main features of these muscles between the stages described by Schilling and Kimmel (1997) and the adult configuration: 4-day-old larvae (96 hours, 10 larvae, mean total length 3.2 mm), 9-day-old larvae (216 hours, 10 larvae, mean total length 4.0 mm), 14-day-old larvae (336 hours, 10 larvae, mean total length 4.5 mm), 24-day-old larvae (576 hours, 10 larvae, mean total length of 6.5 mm), 35 day-old juveniles (986 hours, 10 juveniles, mean total length 7.4 mm) and adults (10 adults from the collection of the Museo Nacional de Ciencias Naturales de Madrid, about 1-year-old, mean total length 45.2 mm). The observations of the muscular muscles in the 4-day-old larvae examined in the present work essentially corroborate those of Schilling and Kimmel (1997), unless otherwise stated. The larval and juvenile specimens examined are wild-type fishes from the MRC Centre for Developmental Neurobiology. They were bleached in 1% H_2O_2 5% formamide solution to remove pigment, processed for immunohistochemistry with anti-myosin antibody (A4.1025) as previously described (Schilling and Kimmel, 1997), and viewed and photographed on a Zeiss Axiophot.

Mandibular Musculature

According to Schilling and Kimmel (1997) five bilateral mandibular muscles are formed in the first three days of zebrafish development: the intermandibularis anterior, the intermandibularis posterior, the adductor mandibulae, the levator arcus palatini, and the dilatator operculi. In their study, the adductor mandibulae was formed at 53 hours; the other four mandibular were formed at 62 hours. These five muscles are found in the 4-day-old (96 hours) larvae examined in the present work (Plates 1, 2). These mandibular muscles are innervated by the 5th nerve (see Section 4.2).

Throughout development, the intermandibularis posterior becomes deeply associated with the hyoid muscle interhyoideus, forming the protractor hyoideus (Fig. 77, Plates 1, 2, 3, 5, 6; See Section 4.2). In contrast with the 4-, 9-, 14- and 24-day-old larvae examined (Plates 1, 2, 3), in the 35-day-old juvenile and in the adult specimens analyzed two protractor hyoideus sections—dorsal and ventral—are recognized (Fig. 77, Plates 5, 6). In adults the ventral section connects the anterior ceratohyal and ventral hypohyal to the ventromesial surface of the dentary bone of the mandible, while the dorsal one runs from the anterior ceratohyal to the ventromesial margin of this dentary bone. As explained in Section 4.2, the protractor hyoideus is a complex muscle innervated by both the 5th and 7th nerves; its anterior and posterior portions (which correspond respectively to the intermandibularis anterior and to the interhyoideus: see, e.g., Plates 1, 2) may contract differently during different phases of respiration. Nonetheless, as a broad generality the protractor hyoideus can be said to play a primary role in the elevation (protraction) of the hyoid bars, as well as in the depression of the mandible (see Section 4.2). The overall configuration of the intermandibularis anterior remains rather constant throughout the development stages examined in the present work; in adults this structure runs from the dentary bone to the dentary bone of the opposite side, thus joining the two mandibles (Fig. 77, Plates 1, 2, 3).

As in the 5- and 6-day-old larvae described by, e.g., Schilling and Kimmel (1997), Hunter and Prince (2002), and Hernandez et al. (2002, 2005), in the 4- and 9-day-old larvae analyzed in the present work the adductor mandibulae comprises a single mass of fibers (Plates 1, 2). In the 14- and 24-day-old larvae analyzed three different sections of the adductor mandibulae are recognized (Plates 3, 4); these seem to correspond to the A2, the Aw, and the A1-OST plus A0 of adults (Figs. 74, 75, Plates 3, 4). In the 35-day-old juveniles examined there is seemingly some differentiation between the A1-OST and the A0 (Plate 5); in the adults observed these sections are well differentiated (Fig. 74). It can thus be said that the overall configuration of the adult adductor mandibulae, divided into four distinct sections, is rather different from the undivided adductor mandibulae found in early larvae. The adult A0 inserts posteriorly on the preopercle and quadrate and anteriorly, by means of a thick tendon, on the maxilla (Fig. 74). The adult A1-OST, mesial to the A0, runs from the preopercle and quadrate to the posterodorsal margin of the mandible, namely to the angulo-articular and dentary bones (Figs. 74, 75). The adult A2 (Figs. 74, 75) lies mesially to the A1-OST and connects the preopercle, hyomandibula and metapterygoid to the small coronomeckelian bone lodged on the mesial surface of the mandible. The adult Aw attaches anteriorly on the mesial surface of both the angulo-articular and dentary bones and posteriorly on the tendon of the A2 (Fig. 75). As its names indicates, the adductor mandibulae is mainly associated with the adduction of the mandible. However, because the adult A0 is attached on the maxilla and not on the mandible, it is not directly associated with mandibular adduction: it is instead directly associated with the adduction of the maxilla and, thus, of the upper jaw, participating in the peculiar mechanisms of mouth protraction/retraction found in the zebrafish and in other extant cypriniforms (e.g., Takahasi, 1925; Gosline, 1973a,b, 1989; Vandewalle, 1975, 1977, 1978; Diogo and Chardon, 2000). Hernandez et al. (2005: 327) stated that the adult zebrafish has an Aw, an A1, an A2 and an A3. The A1, A2 and A3 of these authors may thus correspond to the A0, A1-OST and A2 of the present work, respectively. However, the statements of these authors concerning these adductor mandibulae sections are somewhat confused. For instance, they affirm that their A1 is situated between the A2 and the A3. But, according to Vetter's (1878) original definition, and according to definitions followed by most authors, the term A1 is used to designate a section that is lateral, and not mesial, to the A2 (in some cases the position of these sections has changed during evolution, but this is not the case in the zebrafish: see above). Therefore, it is difficult to assess whether or not the A1, A2 and A3 of Hernandez et al. (2005) correspond to the A0, A1-OST and A2 of the present work.

Each of the two dorsal mandibular muscles of the zebrafish, the levator arcus palatini and dilatator operculi, remains undivided throughout development (Fig. 74, Plates 1, 2, 3, 4, 5, 7). The adult levator arcus palatini connects the sphenotic to the metapterygoid and hyomandibula and is mainly associated with the elevation/abduction of the suspensorium (Fig. 74). The adult dilatator operculi lies laterally to the levator arcus palatini and connects the lateral surfaces of the frontal, pterotic and hyomandibula to the anterodorsal surface of the opercle (Fig. 74). It is mainly associated with the abduction of the opercle.

Hyoid Musculature

Five paired hyoid muscles are formed in the first four days of development: the interhyoideus, hyohyoideus, adductor hyomandibulae, adductor operculi, and levator operculi (Schilling and Kimmel, 1997). According to Schilling and Kimmel (1997), the interhyoideus and hyohyoideus form at 58 hours, the adductor hyomandibulae and adductor operculi at 68 hours, and the levator operculi at 85 hours. These five muscles, innervated by the VII nerve, are found in the 4-day-old (96 hours) larvae examined in the present work (Plates 1, 2, 7).

However, another hyoid muscle can be recognized in these 4-day-old larvae, as well as in the 9-, 14-, 24- and 35-day-old and in the adult specimens analyzed: the adductor arcus palatini (Fig. 74, Plates 2, 3, 4, 5, 7). As explained in Section 4.2, there is much confusion in the literature concerning the muscles adductor arcus palatini and adductor hyomandibulae. Most teleosts have a single muscle connecting the neurocranium to the mesial surface of the suspensorium and thus promoting the adduction of this latter structure. Winterbottom (1974) opted to designate this muscle "adductor arcus palatini" and not "adductor hyomandibulae" because using the name "adductor hyomandibulae" becomes inappropriate in the numerous taxa in which this muscle is expanded anteriorly along the floor of the orbit and attaches on elements of the suspensorium other than the hyomandibula such as the metapterygoid and/or entopterygoid (as is the case in adult zebrafishes: see below). Winterbottom (1974), therefore, used the name "adductor hyomandibulae" to designate a muscle that is only found in a few osteichthyans (one of them being the zebrafish) and that usually situates posteriorly to his adductor arcus palatini, connecting the neurocranium to the mesial surface of the hyomandibula; this nomenclature is followed in the present work (see Section 4.2). Thus, the adductor arcus palatini and "adductor hyomandibulae" of the present work correspond respectively to the anterior and posterior suspensorial adductors found in the zebrafish (Fig. 74, Plates

2, 3, 4, 5, 7; see Section 4.2). In adults, the "adductor hyomandibulae" connects the neurocranium to the mesial margin of the hyomandibula; the adductor arcus palatini is broader than the adductor hyomandibulae, extending from the neurocranium to the mesial sides of the hyomandibula, metapterygoid and entopterygoid (Fig. 74). Although in the larval and juvenile zebrafish specimens examined in this work these adductors lie close to each other, they do seem to constitute distinct muscles, as is the case in zebrafish adults and in a few other adult teleostean and non-teleostean osteichthyans analyzed (see Section 4.2). This observation is supported by the examination of 5-day-old alpha-actin GFP (green fluorescent protein) transgenic zebrafish larvae, in which the adductor arcus palatini and the adductor hyomandibulae also seem to constitute separate muscles (Hinits, Diogo and Hughes, work in progress).

As in most other adult teleosts, in the adult zebrafish specimens examined the hyohyoideus is divided into three paired structures: the hyohyoideus inferior, running from the anterior ceratohyals to a mesial aponeurosis in which it meets its counterpart; the hyohyoideus abductor, running from the first branchiostegal ray to a mesial aponeurosis that is attached by means of two thin tendons to the ventral hypohyals and in which it meets its counterpart; and the hyohyoidei adductores, connecting the branchiostegal rays, the opercle and the subopercle of one side of the fish (Fig. 77; see Section 4.2). As explained above, there is little commentary in the literature regarding the function of the hyohyoideus inferior but adduction of the hyoid bar is suggested by its position and presumed line of action. Regarding the hyohyoideus abductor and the hyohyoidei adductores, which are often considered parts of a hyohyoideus superior, they are usually related to the expansion and constriction of the branchiostegal membranes, respectively (see Fig. 77). A reference point that is thus often used in the literature to distinguish the hyohyoideus abductor and the hyohyoidei adductores is the presence of at least some branchiostegal rays: the hyohyoideus abductor situates mesially to the most mesial ray; the hyohyoidei adductores situates laterally to it (see Fig. 77). The hyohyoideus of the 9-, 14-, 24- and 35-day-old zebrafish specimens analyzed seems to be divided into hyohyoideus inferior, hyohyoideus abductor and hyohyoidei adductores, as in adults (Fig. 77, Plates 3, 5, 6). The presence of these two latter divisions in those developmental stages is supported by the fact that in those stages the branchiostegal rays are usually already ossified (e.g., Cubbage and Mabee, 1996). In the 4-day-old larvae examined the hyohyoideus inferior is clearly divided into an anterior part, the hyohyoideus inferior, and a posterior part, named here hyohyoideus superior (Plates 1, 2; see Section 4.2). However, it is difficult to discern whether or not this latter part is differentiated into hyohyoideus abductor and hyohyoidei adductores. As pointed out by Cubbage and Mabee (1996), some 4-day-old zebrafish larvae may have ossified branchiostegal rays, while others may not. Therefore, it is conceivable that the division between the hyohyoideus abductor and the hyohyoidei adductores is found in some 4-day-old zebrafish larvae but not in others.

Both the adductor operculi and the levator operculi remain undivided throughout development (Fig. 74, Plates 1, 2, 3, 4, 5, 7). The adult adductor operculi lies mesial to the levator operculi and connects the pterotic to the posterodorsal surface of the opercle (Fig. 74). As its name indicates, it is mainly associated with the adduction of the opercle. The adult levator operculi runs from the ventrolateral margin of the pterotic to the dorsomesial edge of the opercle (Fig. 74). The teleostean levator operculi is usually related to a peculiar mechanism mediating lower jaw depression via the so-called "four-bar linkage system" in which the force of contraction of this muscle is transmitted through the opercular series and the interopercular ligament to the lower jaw (see Section 4.2). As mentioned above, in the zebrafish, as well as in most other teleosts, the hyoid muscle interhyoideus becomes associated with the mandibular muscle intermandibularis posterior, forming the protractor hyoideus.

Hypobranchial Musculature

There is a single hypobranchial muscle sensu Edgeworth in the zebrafish: the sternohyoideus (Fig. 77, Plates 1, 2). In the study of Schilling and Kimmel (1997) this muscle, innervated by the anterior branches of the occipito-spinal nerves, appeared at 53 hours. In early stages the sternohyoideus is markedly divided longitudinally, its right and left parts only meeting anteriorly, near the region of the hyohyoideus inferior (Plates 1, 2). As described by Schilling and Kimmel (1997), each of these parts consists of three myomeres separated by two myocommata (sensu, e.g., Stiassny et al., 2000). In older stages of development the right and left parts become closer to each other; in adults they are connected mesially through all their lengths, forming a large cone-shaped structure originating from the anterior region of the cleithrum and passing dorsally to the hyohyoideus inferior and hyohyoideus abductor in order to attach on the urohyal (Fig. 77). As in numerous other teleosts, some fibers of the sternohyoideus are associated posteriorly with fibers of the hypaxialis. The sternohyoideus plays a major role in hyoid depression and, through a series of mechanical linkages, in mouth opening and suspensorial abduction (e.g., Stiassny, 2000; Hernandez et al., 2002; see Section 4.2).

Ontogeny/Evolution

It is now possible to address the following question: is there a correspondence between the development of the mandibular, hyoid, and hypobranchial muscles in the zebrafish and the evolution of these muscles within the Osteichthyes?

As explained above, within osteichthyan evolutionary history the mandibular muscles intermandibularis anterior, intermandibularis posterior, adductor mandibulae, and levator arcus palatini were seemingly found in basal osteichthyans; the dilatator operculi was acquired later in evolution, being exclusively found in actinopterygians (Tables 2, 3). However, according to the times of appearance provided by Schilling and Kimmel (1997: table 3), ontogenetically the dilatator operculi, the levator arcus palatini, the intermandibularis anterior, and the intermandibularis posterior appear at about the same time in the zebrafish, that is, at 62 hours (the adductor mandibulae appears at 53 hours). In contrast, the development of the zebrafish adductor mandibulae divisions does seem to follow the order in which these divisions were acquired in evolution. In fact, the A2 and Aw were acquired earlier in evolution, being plesiomorphically found in osteichthyans; the A1-OST and A0 were acquired later, namely in the nodes leading to ostariophysans and to cypriniforms, respectively (Tables 2, 3). During zebrafish development, the A2 and Aw also form earlier, being already separated in the 9-day-old larvae examined; the A1-OST and A0 can only be recognized in the 35-day-old juveniles and the adults observed (see above).

The order in which the hyoid muscles were acquired in evolution is as follows: first, the interhyoideus, adductor operculi, and adductor arcus palatini (plesiomorphically found in osteichthyans); then, the hyohyoideus (found only in extant actinopterygians); then, the levator operculi (found only in extant halecomorphs and teleosts); and, last, the "adductor hyomandibulae" (found in some teleosts, seemingly not homologous with the "adductor hyomandibulae" of osteichthyans such as *Latimeria*) (see Tables 4, 5). According to Schilling and Kimmel (1997: table 3), in the zebrafish the interhyoideus and the hyohyoideus appear at 58 hours, the adductor operculi and adductor hyomandibulae at 68 hours, and the levator operculi at 85 hours (as explained above, Schilling and Kimmel listed only one adductor of the suspensorium in their table 3, which they named adductor hyomandibulae, and thus did not refer to an adductor arcus palatini in that table). Thus, as in evolution, in zebrafish ontogeny the levator operculi forms later than the interhyoideus, the hyohyoideus and the adductor operculi. However, contrary to evolution, the zebrafish hyohyoideus seemingly develops earlier than the adductor operculi.

As explained above, there is a single hypobranchial muscle in the zebrafish, the sternohyoideus, which appeared at 53 hours in the study of Schilling and Kimmel (1997). Interestingly, during zebrafish development the overall configuration of this muscle is changed in a manner that resembles the changes that occurred in actinopterygian evolution. In fact, in basal adult actinopterygians the sternohyoideus is longitudinally divided into left and right parts that are easily distinguished from each other (e.g., Figs. 6, 15, 19). This plesiomorphic overall configuration was, however, changed in the node leading to the Teleostei: in adult teleosts, including the zebrafish, the sternohyoideus is a cone-shaped structure in which the left and right parts are hardly distinguished from each other. In early zebrafish larvae the sternohyoideus is longitudinally divided into left and right parts that are easily distinguished from each other, thus resembling the configuration found in basal adult actinopterygians (see above). Thus, regarding its overall configuration, there is some correspondence between the development of the sternohyoideus in the zebrafish and the evolution of this muscle within the actinopterygian osteichthyans.

The examples provided above, therefore, illustrate that although in certain cases there is a correspondence between the ontogeny of the mandibular, hyoid, and hypobranchial muscles in the zebrafish and the evolution of these muscles within Osteichthyes, this is clearly not always the case. This seems also to apply to other zebrafish cranial muscles, as well as to other cranial structures such as cartilage and bones. For instance, as shown in Plates 1 and 2, in 4-day-old zebrafish larvae the muscle levator arcus branchialis 5 is already much broader than the other branchial muscles. Thus, the broadening of this muscle occurs earlier in development than, for example, the differentiation of the adductor mandibulae in different sections (see above). However, in evolution the hypertrophy of the levator arcus branchialis 5 occurred only in the node leading to cypriniforms, thus much later than the division of the adductor mandibulae in different sections (see Tables 2, 3). The modification of the muscle levator arcus branchialis 5, as well as of the skeletal structure that is moved by this muscle, the ceratobranchial 5, is related with the specialized peculiar feeding mechanisms of cypriniforms (e.g., Takahasi, 1925; Edgeworth, 1935; Winterbottom, 1974; Vandewalle, 1975, 1977; Howes, 1978, 1979; Schilling and Kimmel, 1997). In fact, in cypriniforms the ceratobranchial 5 bears teeth and ossifies earlier than other ceratobranchials (e.g., Cubbage and Mabee, 1996). Schilling and Kimmel (1997: 2958-2959) considered the early ossification of the ceratobranchial 5 in cypriniforms a case of "acceleration" of development. According to these authors, "the control of size of a particular element might be accomplished by acceleration or retarding when differentiation begins; the same hold for muscles since cartilages and their muscles develop together, and larger cartilages tend to be associated with larger muscles". Therefore, coordinated ontogenetic timing changes might ensure proper size relationships between skeletal and myological structures.

Zebrafish as a Case Study

As explained above, because the zebrafish is the most commonly studied model organism among osteichthyan fishes, it is often taken as a "good representative" of teleosts, of actinopterygians, and even of bony fishes in developmental and molecular studies. However, regarding its mandibular, hyoid, and hypobranchial muscles, to what extent is it appropriate to consider the zebrafish a "good representative" of teleosts, actinopterygians and/or bony fishes?

As can be seen in Tables 2 to 7, of the 13 mandibular, hyoid, and hypobranchial muscles found in the adult zebrafish (intermandibularis anterior, protractor hyoideus, adductor mandibulae, levator arcus palatini, dilatator operculi, hyohyoideus inferior, hyohyoideus abductor, hyohyoidei adductores, adductor arcus palatini, adductor hyomandibulae, adductor operculi, levator operculi, and sternohyoideus), 13 are found in at least some other living teleosts, and 12 are found in at least some other extant actinopterygians (the protractor hyoideus being the exception). Therefore, although the zebrafish occupies a rather derived phylogenetic position within the Actinopterygii and even within the Teleostei (see Fig. 4), with respect to these muscles, it seems justified to consider it a potential representative of these two groups. However, of these 13 muscles only 6 are found in at least some extant sarcopterygian fishes (intermandibularis anterior, adductor mandibulae, levator arcus palatini, adductor arcus palatini, adductor operculi, and sternohyoideus). Therefore, with respect to these muscles, caution is required if this fish is taken as a representative of bony fishes as a whole. Lastly, it should be stressed that of these 13 muscles only 3 are found in at least some extant adult tetrapods (intermandibularis anterior, adductor mandibulae, and sternohyoideus). Therefore, among the cranial muscles discussed in this paper, these three latter muscles are the most appropriate ones for a proper comparison between the results obtained in molecular and developmental studies of the zebrafish and the data obtained from model tetrapod organisms from clades such as Amphibia and/or Amniota. It is hoped that the information provided here may form a solid basis for future analyses on zebrafish cranial muscles and for a proper comparison between these muscles and those found in other osteichthyans.

4.4 COMPARATIVE ANATOMY, HOMOLOGIES AND EVOLUTION OF OSTEICHTHYAN PECTORAL MUSCLES

As is the case with the cranial muscles (see Section 4.2), the most comprehensive comparative analyses of osteichthyan pectoral muscles that were actually based on a direct observation of actinopterygian and sarcopterygian taxa as varied as Teleostei, Halecomorphi, Ginglymodi, Chondrostei, Cladistia, Dipnoi, Amphibia, and Amniotes (see Fig. 4), and not mainly on a recompilation from the literature, were provided long ago, in works such as Romer (1922-1944) and Howell (1933-1936). Thus, despite the quality of these works, their authors could not access information now available, for example, information on the pectoral musculature of *Latimeria chalumnae* and the essential role of neural crest cells in the development and patterning of the cranial muscles (see Sections 4.2 and 4.3) and seemingly also pectoral muscles (e.g., McGonnell, 2001). Also, some of hypotheses proposed in those works regarding the homologies and evolution of osteichthyan pectoral muscles were based on phylogenetic scenarios that have been contradicted by numerous studies. For instance, Romer (1944) held that the cladistian *Polypterus* is more closely related to tetrapods than are the extant dipnoans, a view to which very few authors would adhere nowadays (see Chapter 3).

This Section provides an updated discussion on the homologies and evolution of osteichthyan pectoral muscles. **It is important to emphasize that this discussion is based on the results of an original work done by both me and Virginia Abdala. Therefore, it can be said that the data given in this Section is, in a certain manner, co-authored by the two of us.** It should also be noted that the pectoral muscles discussed in this Section basically correspond to the "appendicular" muscles of Kardong and Zalisko (1998), Kardong (2002), and other authors and not to all the muscles attaching to the pectoral girdle and/or to the pectoral fins/limbs. Therefore, hypobranchial muscles such as the sternohyoideus and branchial muscles such as the protractor pectoralis, as well as muscles such as the hypaxialis and the epaxialis, which usually attach to one or more pectoral skeletal structures, are not discussed in the present Section (see Sections 4.2 and 4.3).

Actinopterygii (Table 8)

As can be seen in Table 8, the plesiomorphic condition for actinopterygians seems to be that in which there are only two pectoral muscles related to the movements of the pectoral fins, the abductor and the adductor, which may be subdivided into different bundles. Such a condition is found, for example, in extant chondrosteans (Fig. 13) and in extant cladistians (Fig. 8). In fact, this condition seems to be plesiomorphic for the osteichthyans as a whole, since it is also found in extant actinistian and dipnoan sarcopterygians (Table 9; Figs. 87, 94, 95; see below) as well as in non-osteichthyan gnathostomes such as extant chondrichthyans (e.g., Romer, 1924; Jarvik, 1965, 1980; Kardong and Zalisko, 1998; Kardong, 2002). In both actinopterygian (e.g., Fig. 78) and sarcopterygian (e.g., Fig. 87) fishes the adductor and abductor of the pectoral fin are often subdivided into adductor superficialis and adductor profundus and into abductor superficialis and abductor profundus, respectively. In the present Section the adductor superficialis and adductor profundus are considered as bundles of the adductor of the pectoral fin, and not as separate muscles; the abductor superficialis and abductor profundus are considered as bundles of the abductor of the fin, and not as individual muscles (Tables 8, 9). As their names indicate, in these fishes the adductor and the abductor are mainly related to the adduction and abduction of the pectoral fin, respectively (e.g., Bischoff, 1840; Owen, 1841; Pollard, 1892; Romer, 1924; Howell, 1933b; Millot and Anthony, 1958; Greenwood and Thomson, 1960; Jessen, 1972; Winterbottom, 1974; Kaseda and Nomura, 1975; Brosseau, 1978a,b; Lauder and Liem, 1983; Adriaens et al., 1993; Pough et al., 1996; Kardong and Zalisko, 1998; Diogo et al., 2001; Kardong, 2002; Westneat et al., 2004; Kisia and Onyango, 2005; Thorsen and Hale, 2005; Thorsen and Westneat, 2005).

In contrast to extant cladistians and chondrosteans, as well as to extant sarcopterygian fishes, the living ginglymodians, the living halecomorphs, and the vast majority of the living teleosts exhibit a separate, well-distinguished muscle arrector dorsalis, which may be subdivided into different bundles (Table 8; e.g., Figs. 17, 23, 30B, 41, 44, 41, 57, 58, 59, 78). This arrector dorsalis usually originates on the mesial surface of the pectoral girdle, laterally to the adductor of the fin and to the mesocoracoid arch (when this structure is present) and inserts on the proximal head of the first and eventually of the second pectoral fin rays (e.g., Figs. 17, 23, 30B, 78). The arrector dorsalis is ontogenetically derived from the adductor of the pectoral fin (e.g., Jessen, 1972; Winterbottom, 1974; Thorsen and Hale, 2005). The arrector dorsalis was seemingly phylogenetically acquired in the node leading to the Neopterygii (see Fig. 4; Table 8).

Apart from the arrector dorsalis, other separate, well-distinguished arrector muscles may be present in neopterygians (Table 8). One of these muscles is the arrector ventralis, which is found in the great majority of extant teleosts, and which was seemingly phylogenetically acquired in the node leading to the Teleostei (Fig. 4; Table 8; see Chapter 3). The arrector ventralis usually originates lateral to the abductor and inserts on the first

Table 8 Pectoral muscles of adults of representative actinopterygian taxa; the nomenclature of the muscles shown in bold follows that of the present work (in order to facilitate comparisons, in some cases certain names used by other authors are also shown). For more details, see text.

Probable plesiomorphic osteichthyan condition	Cladistia: *Polypterus bichir* (Bichir)	Chondrostei: *Psephurus gladius* (Chinese swordfish)	Ginglymodi: *Lepisosteus osseus* (Longnose gar)	Halecomorphi: *Amia calva* (Bowfin)	Teleostei—basal: *Elops saurus* (Ladyfish)	Teleostei—clupeocephalan: *Danio rerio* (Zebrafish)
Abductor (of pectoral fin)	**Abductor (of pectoral fin)**	**Abductor (of pectoral fin)**	**Abductor (of pectoral fin)**	**Abductor (of pectoral fin)**	**Abductor (of pectoral fin)**	**Abductor (of pectoral fin)**
Adductor (of pectoral fin)	**Adductor (of pectoral fin)**	**Adductor (of pectoral fin)**	**Adductor (of pectoral fin)**	**Adductor (of pectoral fin)**	**Adductor (of pectoral fin)**	**Adductor (of pectoral fin)**
--------	--------	--------	**Arrector dorsalis**	**Arrector dorsalis**	**Arrector dorsalis**	**Arrector dorsalis**
--------	--------	--------	--------	--------	**Arrector ventralis**	**Arrector ventralis**
--------	--------	--------	--------	--------	--------	**Arrector 3** (= large external bundle of superficial abductor of e.g. Brosseau, 1978a,b)

pectoral ray (e.g., Figs. 30A, 41, 46, 56, 65, 74). It is ontogenetically derived from the abductor of the pectoral fin (e.g., Jessen, 1972; Winterbottom, 1974; Thorsen and Hale, 2005). The other muscle is the small muscle arrector 3, which usually connects the pectoral girdle to the ventrolateral surface of the first pectoral ray (e.g., Fig. 74). This small muscle has been frequently neglected in the literature (Diogo, in press). That is, it has frequently been overlooked or considered as a bundle of the arrector ventralis or of the abductor of the fin (e.g., Brosseau, 1978a,b; see Table 8). However, as explained in Chapter 3, the arrector 3 is found in numerous otocephalans, and in at least some euteleosts (Figs. 46, 56, 65, 74), thus constituting a potential synapomorphy of the Clupeocephala (see Fig. 4; Table 8).

As explained in Chapter 3, in previous works by the author and colleagues (e.g., Diogo et al., 2001a; Diogo, 2004a) the arrector 3 (sensu the present work) was named "arrector ventralis". One of the main reasons for this confusion was precisely the fact that the small muscle arrector 3 was not described by, for example, Winterbottom (1974). Owing to this confusion, the names attributed to the teleostean pectoral muscles by Diogo et al. (2001a) and Diogo (2004a) were substantially different from those proposed by Winterbottom (1974). In order to solve this problem, in the present work I opted to designate this small muscle as arrector 3, and to use the nomenclature proposed by Winterbottom (1974) to designate the other teleostean pectoral muscles. Thus, in order to facilitate comparisons with previous works such as Diogo et al. (2001a) and Diogo (2004a), it is worth noting that the "arrector ventralis", "arrector dorsalis", "abductor superficialis 1", "abductor superficialis 2", "adductor superficialis 1", "adductor superficialis 2", and "abductor profundus" of those works correspond respectively to the arrector 3, arrector ventralis, abductor superficialis, abductor profundus, adductor superficialis, adductor profundus, and arrector dorsalis of the present volume.

As explained in a paper by Thorsen and Hale (2005: 149) the arrectors of the pectoral fin "initiate the movement of the fin at the leading edge", while the adductor and the abductor "power the upstroke and downstroke." Besides the abductor, the adductor, and the arrector muscles, some derived teleosts (e.g., certain neoteleosts) may eventually exhibit other pectoral muscles, such as the coracoradialis, adductor radialis, interradialis pectoralis, and adductor medialis (e.g., Fig. 33; Winterbottom, 1974).

Sarcopterygii (Table 9)

The plesiomorphic condition for sarcopterygians seems also to be that in which there are only two distinct pectoral muscles associated with the movements of the pectoral fins, the abductor and the adductor; this condition is found in extant dipnoans (Figs. 94, 95) and seemingly also in extant actinistians (Fig. 87; see below). However, in contrast with the configuration found in actinopterygians and in non-osteichthyan gnathostomes such as living chondrichthyans, in these two sarcopterygian groups the adductor and the abductor extend far into the pectoral fin, thus giving to this fin its characteristic "lobed" or "fleshy" appearance (e.g., Figs. 87, 94, 95; see also, e.g., Bischoff, 1840; Owen, 1841; Romer, 1924; Howell, 1933b; Millot and Anthony, 1958; Jessen, 1972; Pough et al., 1996; Kardong and Zalisko, 1998; Kardong, 2002; Kisia and Onyango, 2005).

Millot and Anthony (1958) suggested that apart from an adductor and an abductor of the pectoral fin, *Latimeria* exhibits various "pronator" and "supinator" muscles (see Fig. 87). As explained in Chapter 2, it was not possible to undertake a detailed dissection of *Latimeria* because of the difficulty of finding specimens of this taxon available for muscular examination. However, from the textual descriptions and the illustrations provided by Millot and Anthony (1958), it seems that their "pronator" and "supinator" muscles are not well-separated, functionally independent muscles but are, instead, bundles of the adductor and of the abductor of the fins (Table 8; see, e.g., in Fig. 87 the seemingly poor differentiation between Millot and Anthony's "supinators" 1 and 2). As in extant dipnoans (see, e.g., Figs. 94, 95), these adductor and abductor bundles of *Latimeria* may eventually resemble certain tetrapod pectoral muscles but are not as distinct and as functionally independent as are these latter muscles (e.g., Romer, 1924; this work). This view is indirectly supported by the results of the cladistic analysis of Chapter 3, which strongly support the position that tetrapods are more closely related to dipnoans than to actinistians (Fig. 4). Thus, according to these results it would seem rather unsound that *Latimeria* may exhibit several distinct pectoral muscles such as those found in extant tetrapods, since this would imply: (1) that such muscles were independently acquired twice in evolution or (2) that such muscles were acquired only once and were present in the last common ancestor of actinistians, dipnoans, and tetrapods but were secondarily lost within the Dipnoi (see Fig. 4). An apparently more plausible scenario would be to consider that in this last common ancestor of actinistians, dipnoans, and tetrapods the abductor and the adductor of the fins were eventually already differentiated in certain subdivisions, but that these subdivisions, as well as other subdivisions acquired later in evolution, became well-separated, independent muscles only during the subsequent evolutionary transitions leading to the origin of tetrapods. However, only a detailed, updated analysis of the pectoral muscles of *Latimeria* can clarify whether the members of this taxon exhibit (as

Table 9 Pectoral muscles of adults of representative sarcopterygian taxa; the nomenclature of the muscles shown in bold follows that of the present work (in order to facilitate comparisons, in some cases certain names used by other authors are also shown). Note: the listing of the numerous pectoral muscles of the tetrapods *Ambystoma ordinarium* and *Timon lepidus* mainly follows the order used in Walthall and Ashley-Ross' (2006) table I. For more details, see text.

Probable plesiomorphic osteichthyan condition	Actinistia: *Latimeria chalumnae* (Coelacanth)	Dipnoi: *Lepidosiren paradoxa* (South American lungfish)	Amphibia: *Ambystoma ordinarium* (Michoacan stream salamander)	Amniota: *Timon lepidus* (Ocellated lizard)
Abductor (of pectoral fin)	**Abductor (of pectoral fin)**	**Abductor (of pectoral fin)**	-------- (*there is no separate muscle abductor)	-------- (*see cell on the left)
Adductor (of pectoral fin)	**Adductor (of pectoral fin)**	**Adductor (of pectoral fin)**	-------- (*there is no separate muscle abductor)	-------- (*see cell on the left)
--------	--------	--------	**Pectoralis**	**Pectoralis**
--------	--------	--------	**Supracoracoideus**	**Supracoracoideus**
--------	--------	--------	**Procoracohumeralis**	**Procoracohumeralis** (deltoides claviculars of e.g. Romer, 1944)
--------	--------	--------	**Coracobrachialis longus**	**Coracobrachialis longus**
--------	--------	--------	**Anconaeus coracoideus**	**Anconaeus coracoideus** (*seemingly present, making part of the triceps group of e.g. Romer, 1944)
--------	--------	--------	**Anconaeus scapularis medialis**	**Anconaeus scapularis medialis** (*seemingly present, making part of the triceps group of e.g. Romer, 1944)
--------	--------	--------	**Anconaeus humeralis lateralis**	**Anconaeus humeralis lateralis** (*seemingly present, making part of the triceps group of e.g. Romer, 1944)

--------	--------	--------	**Anconaeus humeralis medialis**	**Anconaeus humeralis medialis** (*seemingly present, making part of the triceps group of e.g. Romer, 1944)
--------	--------	--------	**Latissimus dorsi**	**Latissimus dorsi**
--------	--------	--------	**Dorsalis scapulae**	**Dorsalis scapulae** (deltoides scapularis of e.g. Romer, 1944)
--------	--------	--------	**Humeroantebrachialis**	**Humeroantebrachialis** (part, or all, of biceps of e.g. Romer, 1944)
--------	--------	--------	-------- (*seemingly absent in *Ambystoma*, but may be found in other salamanders)	**Brachialis inferior**
--------	--------	--------	**Flexor digitorum communis**	**Flexor digitorum communis** (part, or all, of flexor digitorum longus of e.g. Abdala and Moro, 2006)
--------	--------	--------	**Flexor antebrachii et carpi radialis**	**Flexor antebrachii et carpi radialis** (flexor carpi radialis of e.g. Abdala and Moro, 2006)
--------	--------	--------	--------	**Pronator teres** (*seemingly differentiated from part of the flexor antebrachii et carpi radialis)
--------	--------	--------	**Flexor antebrachii et carpi ulnaris**	**Flexor antebrachii et carpi ulnaris** (flexor carpi ulnaris of e.g. Abdala and Moro, 2006)
--------	--------	--------	**Extensor digitorum communis**	**Extensor digitorum communis** (extensor digitorum longus of e.g. Abdala and Moro, 2006)

--------	--------	--------	**Extensor antebrachii et carpi radialis**	**Extensor antebrachii et carpi radialis** (extensor carpi radialis of e.g. Abdala and Moro, 2006)
--------	--------	--------	**Extensor antebrachii et carpi ulnaris**	**Extensor antebrachii et carpi ulnaris** (extensor carpi ulnaris of e.g. Abdala and Moro, 2006)
--------	--------	--------	**Contrahentium caput longum**	-------- (*seemingly absent as an independent muscle in *Timon*, but may eventually be found in some other lizards)
--------	--------	--------	**Flexor accessorius lateralis**	Present? It may eventually have given part, or all, of the pronator accessorius of e.g. Abdala and Moro, 2006
--------	--------	--------	**Flexor accessorius medialis**	Present? It may eventually have given part, or all, of the pronator accessorius of e.g. Abdala and Moro, 2006
--------	--------	--------	**Pronator profundus**	**Pronator profundus**
--------	--------	--------	**Epitrochleoanconeus** (*our dissections indicate that it is seemingly present in at least some members of the genus *Ambystoma*)	**Epitrochleoanconeus**
--------	--------	--------	**Flexores breves superficiales**	**Flexores breves superficiales** (flexores digiti brevis superficialis of e.g. Abdala and Moro, 2006)
--------	--------	--------	**Intermetacarpales**	**Intermetacarpales I** (intermetacarpalis I of e.g. Abdala and Moro, 2006)

--------	--------	--------	--------	**Intermetacarpales II** (intermetacarpalis II of e.g. Abdala and Moro, 2006)
--------	--------	--------	**Extensores digitorum breves**	**Extensores digitorum breves** (extensores digiti brevis of e.g. Abdala and Moro, 2006)
--------	--------	--------	**Abductor et extensor digiti I**	**Abductor et extensor digiti I** (abductor longus pollici of e.g. Abdala and Moro, 2006)
--------	--------	--------	**Extensor lateralis digiti IV**	**Abductor digitorum V** (*see text)
--------	--------	--------	**Contrahentes digitorum**	**Contrahentes digitorum**
--------	--------	--------	--------	**Lumbricalis**
--------	--------	--------	**Flexores breves profundi**	**Flexores breves profundi** (flexores digiti brevis profundus of e.g. Abdala and Moro, 2006)
--------	--------	--------	--------	**Abductor brevis pollicis**
--------	--------	--------	--------	**Dorsometacarpalis**
--------	--------	--------	--------	**Flexor digitorum V transversus I**
--------	--------	--------	--------	**Flexor digitorum V transversus II**

suggested by Millot and Anthony, 1958), or not (as suggested here), well-differentiated pectoral muscles other than the abductor and the adductor of the fin.

As explained above, I will not provide in this Section a detailed account on the configuration and function of each of the numerous tetrapod pectoral muscles (see, e.g., Table 9; Figs. 103-106, 112-114). Detailed, updated accounts on the pectoral muscles of members of the representative amphibian and amniote groups shown in Table 9, that is of urodeles and lizards, are, for instance, given in the studies of Walthall and Ashley-Ross (2006) and of Abdala and Moro (2006), respectively. For the purpose of the present Section, I prefer instead to focus on the major differences between the pectoral musculature of these representative amphibian and amniote groups. In fact, contrary to what the rather different names generally used in amphibian and amniote literature to designate the pectoral muscles of the members of these groups might indicate, the observations and comparisons of the present work pointed out that the overall configuration of these muscles in amphibians such as urodeles and in amniotes such as lizards is in reality very similar (e.g., Table 9). In order to simplify the comparisons between these two groups, I will use the works of Walthall and Ashley-Ross (2006) and of Abdala and Moro (2006) as a main example of how different nomenclatures are applied to designate homologous muscles in the amphibian and the amniote literature (e.g., Table 9).

Some major differences between the pectoral musculature of adult members of a representative urodele species, *Ambystoma ordinarium*, and of adult members of a representative lizard species, *Timon lepidus*, are summarized in Table 9. As can be seen in that table, a few muscles found in the lizard are absent in the urodele, for example, the intermetacarpales II, the flexor digitorum V transversus I, the flexor digitorum V transversus II, the branchialis inferior, the pronator teres, the abductor brevis pollicis, and the dorsometacarpal and lumbrical muscles. In turn, the contrahentium caput longum, present in the urodele (e.g., Fig. 105), is seemingly absent in the lizard *Timon*. However, as noted in Table 9, this muscle does seem to be present in some other lizards examined by Virginia Abdala, connecting the flexor plate to the carpus (e.g., *Liolaemus*, *Teius*). Other muscles found in the urodele that may eventually be missing in the lizard *Timon* are the flexor accessorius lateralis and/or the flexor accessorius medialis (e.g., Fig. 105). But the observations and comparisons made by Virginia Abdala and me indicate that at least part of these muscles may well have given part, or all, of the muscle that is often named pronator accessorius in amniote literature; this latter muscle is present in the lizard *Timon* (e.g., Fig. 114; Table 9).

It should be noted that, contrary to what seems to be the case in members of certain other urodele genera (e.g., *Taricha*: Walthall and Ashley-Ross, 2006), our dissections revealed that a muscle epitrochleoanconeus such as that found in lizards is also found in at least some members of the genus *Ambystoma* (e.g., *Ambystoma texanum*). They also indicated that the muscle extensor lateralis digiti IV of the urodele seemingly corresponds to the lizard muscle that is frequently named abductor digitorum V (e.g., Abdala and Moro, 2006). However, in contrast to what is done for other muscles in Table 9, in this case I prefer not to attribute the name frequently used in the literature for amphibians to designate the corresponding lizard muscle. This is because the lizard muscle is attached to digit V, and not to digit IV, as is the case in the urodele (which has four digits, and not five like the lizard: see Fig. 105). It thus seems inappropriate to designate a muscle that in the lizard attaches on digit V as "extensor lateralis digiti IV". Consequently, in Table 9, I keep the name that is often used for amniotes, that is abductor digitorum V, to designate this lizard muscle.

Apart from the differences mentioned above, there are other differences concerning the configuration of certain homologous muscles in amphibians such as urodeles and in amniotes such as lizards. To give just one illustration, in urodeles the pronator profundus usually extends distally in order to attach on digit I (e.g., Fig. 105), while in lizards the proximal insertion of this muscle is usually on the radius, that is, it does not reach the digits (e.g., Fig. 114). However, as stated above, in a general way it can be said that the overall configuration of the pectoral musculature of these taxa is rather similar. In fact, as shown in Table 9, the vast majority of the individual pectoral muscles found in urodeles are present in lizards (e.g., at least 26 of the 29 *Ambystoma* pectoral muscles listed in that table are seemingly present in *Timon*). It should be noted that apart from the muscles listed in that table, other pectoral muscles may be eventually found in extant tetrapods: for instance, Shellswell and Wolpert (1977) mentioned that some birds may have about 50 distinct muscles in the wing, a number that is significantly greater than the number of pectoral muscles found in tetrapods such as urodeles and lizards (see Table 9; e.g., Romer, 1922-1944; Howell, 1933-1936; Sullivan, 1962; George and Berger, 1966; Jarvik, 1980; Pough et al., 1989; Kardong and Zalisko, 1998; Gibbs et al., 2000, 2002; Kardong, 2002; Abdala and Moro, 2003, 2006; Moro and Abdala, 2004; Kisia and Onyango, 2005; Walthall and Ashley-Ross, 2006). Together with Virginia Abdala, I plan to provide, in a future work, a detailed account of the pectoral musculature in all the major tetrapod subgroups.

General Remarks

In view of the discussion above and of the information summarized in Tables 8 and 9, it can thus be said that from a similar plesiomorphic overall configuration (i.e., the presence of only two pectoral muscles, the abductor and the adductor of the fin), the evolution of the pectoral musculature has been rather different within the actinopterygian and the sarcopterygian clades. Within the Actinopterygii, different arrector muscles were acquired at different evolutionary stages: an arrector dorsalis is present only in extant neopterygians, an arrector ventralis is present only in extant teleosts, and an arrector 3 is present only in extant clupeocephalans. These three arrector muscles, together with the adductor and the abductor, for example, are present in one of the model organisms that is most studied among actinopterygians, as well as among osteichthyan fishes in general (see Section 4.3), the zebrafish *Danio rerio* (Table 8). As explained above, apart from these five muscles, other pectoral muscles may eventually be found in some derived actinopterygians, for example, in certain neoteleosts. Within sarcopterygians, the most significant evolutionary transformations concerning the pectoral musculature occurred in the transitions that had led to the origin of the tetrapods: the configuration of the pectoral musculature of all living tetrapods is markedly different from that found in living sarcopterygian fishes (Table 9). As mentioned above, certain living tetrapods may have up to 50 individual pectoral muscles.

As a result of the different evolutionary routes followed within the actinopterygian and the sarcopterygian clades, none of the individual muscles found, for example, in derived actinopterygians such as teleosts is found in derived sarcopterygians such as tetrapods (Tables 8 and 9). Thus, concerning the pectoral musculature, much caution should be observed when, for instance, one takes a model actinopterygian organism such as the teleostean zebrafish to be representative of osteichthyan fishes as a whole or to be representative of the plesiomorphic condition found in these fishes, as is unfortunately done in some recent developmental and molecular studies (see Section 4.3). Also, much caution should be exercised when the results obtained in developmental and molecular studies concerning the pectoral muscles of a model actinopterygian such as the zebrafish are compared with those concerning the pectoral muscles of model sarcopterygians from clades such as the Amphibia and/or the Amniota (see, e.g., the work of Thorsen and Hale, 2005). A proper knowledge of the osteichthyan pectoral musculature, thus, is not only important to increase our general understanding of the comparative anatomy, functional morphology, and evolution of this group, but also to provide a solid basis for the comparisons and extrapolations made in such developmental and molecular studies. It is hoped that the information provided in this Section will contribute to a better knowledge of the pectoral muscles of osteichthyans as well as help pave the way for future comparative, functional, evolutionary, molecular, and/or developmental works concerning this group.

4.5 ORIGIN, HOMOLOGIES AND EVOLUTION OF THE WEBERIAN APPARATUS

The Weberian apparatus is one of the most remarkable and enigmatic structural complexes of osteichthyans (Weber, 1820). It is found in extant teleostean otophysans, that is, Cypriniformes, Characiformes, Siluriformes, and Gymnotiformes. However, as will be explained below, it was seemingly present in a "rudimentary" form in basal, now extinct, otophysan fishes such as †*Chanoides macropoma*, †*Clupavus maroccanus*, †*Santanichthys diasii*, and †*Lusitanichthys characiformis* (see Figs. 3, 4) (e.g., Gayet, 1981, 1985, 1986a,b; Patterson, 1984; Taverne, 1995, 2005; Cavin, 1999; Filleul and Maisey, 2004). The apparatus (Figs. 69, 70, 71, 72, 73) is essentially a mechanical device improving audition, consisting of a double chain of ossicles joining the swimbladder to the inner ear, associated with a modification of the pars inferior of the labyrinths of the inner ear and of the anterior portion of the swimbladder (camera aerea Weberiana) (e.g., Weber, 1820; Sagemehl, 1885; Bridge and Haddon, 1893, 1894; Schreiber, 1935; Franz, 1937; Poggendorf, 1952; Kleerekoper and Roggenkamp, 1959; Alexander, 1961a,b, 1964a,b, 1965; Chardon, 1968; Weiss et al., 1969; Vandewalle, 1975; Chardon and Vandewalle, 1997).

Chardon et al. (2003) provided an overview on the comparative anatomy, functional morphology and evolution of the Weberian apparatus. Some parts of the review provided in the present Section are based on Chardon et al.'s (2003) work. However, there are significant differences between this Section and that work. One main difference is that this review also takes into account several studies that have provided relevant information on the subject and that were published after the writing of Chardon et al.'s (2003) overview (e.g., Grande and Shardo, 2002; Grande and Braun, 2002; Coburn and Chai, 2003; De Pinna and Grande, 2003; Filleul and Maisey, 2004; Grande and Young, 2004; Grande and De Pinna, 2004; Taverne, 2005). But perhaps the most original contribution of the present Section is that it provides a discussion on the homologies and evolution of the Weberian apparatus that is directly based on the phylogenetic results of the cladistic analysis of Chapter 3, which included not only representatives of the four extant otophysan orders, of the Gonorynchiformes and of the Clupeomorpha, but also ostariophysan fossils such as †*Chanoides macropoma*, †*Clupavus maroccanus*, †*Santanichthys diasii*, †*Lusitanichthys characiformis*, and †*Sorbininardus apuliensis* (Figs. 3, 4). This thus allows a broader, more comprehensive discussion on the phylogeny and evolution of the otophysans and their closely related groups, allowing us, for instance, to cladistically trace the homologies and evolution of some key Weberian structures. It should be noted that as a basis for the discussion provided in this Section, I did a preliminary cladistic analysis that corroborated that clupeomorph fossils such as *Diplomystus* appear as the sister-group of the extant clupeiforms examined (see Chapter 2), as suggested by, for example, Grande (1985a), and that the fossil *Tischlingerichthys* appears as the sister-group of the clade including all the fossil and extant Ostariophysi examined (see Chapter 2), as proposed by, for example, Arratia (1997).

Homologies of the Weberian Ossicles

With the exception of the shape of the os suspensorium (see below), the overall configuration of the Weberian apparatus is somewhat constant in extant otophysans (Figs. 69, 70D, 71B, C, D, 83B) (e.g., Weber, 1820; Sagemehl, 1885; Bridge and Haddon, 1893; Chranilov, 1927, 1929; Alexander, 1961b, 1964a,b; Popper, 1971; De la Hoz and Chardon, 1984; Chardon and Vandewalle, 1997; Grande and Shardo, 2002; Chardon et al., 2003). The curved posterior portion of the tripus, that is, the transformator tripodis, is, on the one hand, in contact with the swimbladder and, on the other hand, connected by the suspensor ligament to the os suspensorium (Figs. 69, 70D, 73B). The anterior portion of the tripus is usually connected to the intercalarium and this latter connected to the scaphium and, thus, to the concha scaphii, by the interossicular ligament (e.g., Figs. 69, 70D, 71B, C, D, 73B) (note: although, unlike the os suspensorium, the intercalarium and the scaphium exhibit a characteristic, rather similar configuration in plesiomorphic extant Otophysi, the configuration of these elements can eventually vary in some derived otophysans, being, for example, very small or even absent in some catfishes: e.g., Chardon, 1968; Chardon et al., 2003). Such a configuration of the Weberian ossicles and associated ligaments means that, when the volume of the swimbladder is changed, the tripus and, consequently, the anterior Weberian ossicles, are displaced; these displacements are therefore transmitted to the sinus impar and to the labyrinths of the inner ear (Fig. 69, 70D). It is important to note that above the scaphium usually lies a small ossicle, the claustrum (Fig. 71B, C, D). This ossicle does not really make part of the vibration-transmitting chain between the swimbladder and the inner ear. Grande and De Pinna (2004) and others suggested that it might instead form a dorsolateral bony wall around the anterior portion of the neural canal, in the space between the back of the skull and the beginning of the neural arch series (see below).

Developmental studies have described somewhat different ontogenetic origins of the Weberian ossicles of extant otophysans, thus suggesting different homologies between these ossicles and the postcranial structures of other teleosts (e.g., Weber, 1820; Watson, 1939; Bamford, 1948; Butler, 1960; Hoedemann, 1960a,b; Rosen and Greenwood, 1970; Kulshrestha, 1977; Fink and Fink, 1981, 1996; Radermaker et al., 1989; Vandewalle et al., 1989, 1990; Bogutskaia, 1991; Fukushima et al., 1992; Ichiyanagi et al., 1993, 1996, 1997; Coburn and Futey, 1996; Chardon and Vandewalle, 1997; Coburn and Grubach, 1998; Heyd and Pfeiffer, 2000; Grande and Shardo, 2002; Chardon et al., 2003; Coburn and Chai, 2003; De Pinna and Grande, 2003; Grande and De Pinna, 2004; Grande and Young, 2004).

Three main general hypotheses on the homologies of the Weberian ossicles have been proposed in the literature: that these ossicles are homologous with the mammalian ear bones (e.g., Weber, 1820); that they represent modified portions of the first free vertebrae (e.g., Bamford, 1948; Rosen and Greenwood, 1970); and that they originate from modified portions of the first free vertebrae as well as from ossification of other structures such as mesenchyme, ligaments, and/or the swimbladder (e.g., Watson, 1939; Kulshrestha, 1977; Chardon and Vandewalle, 1997; Chardon et al., 2003). The first of these hypotheses (Weber, 1820) has been discarded by the accumulation of data on vertebrate phylogeny, ontogeny, and evolution. However, there is still controversy on whether the Weberian ossicles are exclusively, or just mainly, derived from modifications of certain bony vertebral structures found in other teleosts, and even about which of these bony structures might have been modified in order to be included in the Weberian apparatus of extant otophysans (e.g., neural arches, supraneurals, ribs, and/or parapophyses of the first, second, and/or third free vertebra) (e.g., Chardon et al., 2003; De Pinna and Grande, 2003; Grande and De Pinna, 2004; Grande and Young, 2004).

Scaphium

Some authors stated that the scaphium is exclusively formed in ontogeny from the neural arch 1 (e.g., Fig. 73; Bamford, 1948; Butler, 1960; Rosen and Greenwood, 1970). Others maintained that this ossicle is formed from the neural arch 1 plus ossification of mesenchyme (e.g., Watson, 1939; Kulshrestha, 1977; Radermaker et al., 1989; Vandewalle et al., 1990; Grande and Young, 2004). According to some of these latter authors (e.g., Watson, 1939), the mesenchyme contributes to the formation of the concha scaphii. In a paper by Grande and De Pinna (2004) it is suggested that the scaphium of plesiomorphic extant otophysans (e.g., Cypriniformes) is derived mainly from the neural arch 1 but also from a mesenchymous contribution, which may be a phylogenetic remnant of the cartilages positioned between the exoccipitals and the first neural arch in fishes such as gonorynchiforms.

The trees shown in Figs. 3 and 4, including basal otophysan fossils such as †*Chanoides macropoma*, †*Lusitanichthys characiformis*, †*Santanichthys diasii*, and †*Clupavus marocannus*, support a single, unique origin of the characteristic scaphium of extant otophysans (see Chapter 3). And, although the information provided by these fossils cannot clarify whether or not the mesenchyme contributes to a small portion of the scaphium, it does provide support for a major contribution of the first neural arch to the scaphium of extant otophysans, showing a somewhat transitional stage between non-otophysans and living otophysans (e.g., Patterson, 1984; Fink and Fink, 1996; Filleul and Maisey, 2004). In fact, the "rudimentary" scaphia of species such as †*Chanoides macropoma* and †*Santanichthys diasii* show typical features of the scaphia of extant otophysans, such as their overall shape and the well-defined, thin ventral articulation with the first centrum. But they also exhibit typical features of the first neural arches of extant non-otophysans, such as the presence of foramina for dorsal and ventral nerve roots (Figs. 71A, 72D) (see below).

Intercalarium

Some authors argued that the intercalarium develops exclusively from the second neural arch (e.g., Bamford, 1948; Vandewalle et al., 1990; Bogutskaia, 1991; Fukushima et al., 1992; Ichiyanagi et al., 1993, 1996; Grande and De Pinna, 2004). Others suggested that it develops from the second neural arch plus an ossification of the interossicular ligament (see, e.g., Fig. 70C, D; Watson, 1939; Butler, 1960; Kulshrestha, 1977; Chardon et al., 2003).

In my opinion, understanding the origin of the intercalarium is crucial for the understanding of the origin and evolution of the Weberian apparatus as a whole. In fact, the two main functional evolutionary hypotheses proposed so far to explain the origin and evolution of the chain of Weberian ossicles and ligaments connecting the swimbladder and the inner ear are based on two rather different interpretations of the intercalarium. One of these evolutionary hypotheses, which I call "*indirect* hypothesis", was formulated by Rosen and Greenwood (1970) and other authors (see Fig. 73). It states that initially there was an indirect ligamentous connection between the tripus and the scaphium: the tripus was connected by a ligament to the neural arch 3 (Fig. 73A: lig. E); this neural arch was connected by a second ligament to the modified neural arch 2 (Fig. 73A: lig. C); this latter neural arch was connected by a third ligament to the modified neural arch 1 (Fig. 73A: lig. B). According to this hypothesis, the connection between the tripus and the third neural arch was lost only later in evolution, giving thus the characteristic configuration seen in most extant otophysans: a ligament between the tripus and the intercalarium, and another between the intercalarium and the scaphium (see Fig. 73B).

The other hypothesis, which I will call the "*direct* hypothesis", was formulated by Watson (1939), Chardon et al. (2003) and others. It states that the connection between the tripus and the scaphium was initially realized by a single ligament (see Fig. 70C). Part of this ligament then became ossified, giving the manobrium of the intercalarium, which attached only later in evolution to the modified second neural arch, thus forming the complete intercalarium (see Fig. 70D). Therefore, an ontogenetic origin of the intercalarium exclusively from the second neural arch, with a connection between this bone and the interossicular arriving only later in development, might be used by some researchers as an argument in favor of the direct hypothesis. Alternatively, an initial ontogenetic origin of the intercalarium from two different parts, one from the second neural arch and the other from an ossification of the ligament interossicular, with the connection between these two parts occurring only later in development, might be used by some authors as an argument in favor of the indirect hypothesis. In my opinion, developmental studies on the Weberian apparatus should thus pay special attention to the formation of the interossicular ligament and its connections to the surrounding Weberian structures.

Although the interossicular ligament has not been conserved in fossils such as †*Chanoides macropoma*, †*Chanoides chardoni*, †*Lusitanichthys characiformis*, †*Lusitanichthys africanus*, †*Santanichthys diasii*, and †*Clupavus maroccanus*, these fossils do shed some light on the homologies of the intercalarium of extant otophysans (note: an interossicular ligament is shown in Patterson's 1984 reconstruction of †*Chanoides macropoma* illustrated in Fig. 71A, but this was due to a functional interpretation of Patterson, and not to a real preservation of this ligament in the specimens examined by him). First, these fossils do support the hypothesis that the second neural arch contributes to the formation of the intercalarium of extant otophysans. This is because, as will be seen below, although the "rudimentary" intercalaria of these fossils clearly show some features in common with the intercalaria of extant otophysans, they also exhibit typical features of the second neural arch of other fishes. For instance, they are large structures that cover almost all the dorsal surface of the second centrum, and they exhibit foramina for ventral and dorsal nerve roots in †*Chanoides macropoma* (Figs. 71A, 72) (see below). Second, the overall configuration of the Weberian ossicles of these fossils seems to support the direct evolutionary hypothesis mentioned above. For example, the manobrium of the intercalarium of †*Chanoides macropoma* drawn by Patterson (1984) seems to be somewhat separated from the main bone of the intercalarium, thus being seemingly a sesamoid ossification, as proposed by the defenders of that direct hypothesis (see Fig. 71A). Moreover, the intercalarium of †*Lusitanichthys characiformis*, †*Santanichthys diasii*, and †*Clupavus maroccanus* either lacked a manobrium or had a small manobrium that, by being somewhat separated from the main body of the intercalarium, as is the case in †*Chanoides macropoma*, was possibly lost (e.g., Fig. 72; Gayet, 1985; Gayet and Chardon, 1987; Chardon and Vandewalle, 1997; Chardon et al., 2003). Any of these two latter options would support the direct hypothesis, because they suggest that in those basal otophysan fossils with an already "rudimentary", and perhaps somewhat functional, Weberian apparatus the manobrium was still missing or was, as in †*Chanoides macropoma*, possibly present but still loosely attached to the main body of the intercalarium (suggesting that the interossicular ligament would already be partly ossified but the resulting sesamoid ossification was still not firmly attached to the main body of the intercalarium: see Fig. 70C, D).

As stressed by Chardon et al. (2003), from a purely morphofunctional point of view, if one applies the indirect hypothesis suggested by, for example, Rosen and Greenwood (1970) to basal otophysan fossils such as †*Lusitanichthys characiformis* or †*Santanichthys diasii*, one obtains, in fact, a rather non-functional, unlikely Weberian apparatus. As can be seen in Fig. 72B and 72D, these fossils seemingly had large and rather immobile neural arches 2 (= main body of intercalarium) and 3. Moreover, these structures essentially lay dorsally to the tripus. Thus, with such a configuration, an eventual ligament joining the tripus to these neural arches (Fig. 73A: lig. E), as proposed in the indirect hypothesis, would be essentially oriented in a dorsoventral, and not in a rather oblique, direction. Therefore, with such a ligament, the movements of the tripus would hardly provoke an anteroposterior displacement of the third neural arch and/or of the main body of the intercalarium, and thus are still less likely to provoke, by means of still another ligament between these latter structures and the scaphium (Fig. 73A: lig. B), any significant displacement of the latter.

Since the intercalarium of basal otophysan fossils such as †*Lusitanichthys characiformis*, †*Santanichthys diasii*, and †*Clupavus maroccanus* does not lie within the line of action between the tripus and the scaphium (see Fig. 72), there are apparently only two ways in which the Weberian apparatus of these fossils might have been functionally efficient. One is by having a direct ligament between the tripus and the scaphium (see Fig. 70C); the other is by having a sesamoid ossification of this ligament eventually associated with the main body of the intercalarium, as seems to have been the case in †*Chanoides macropoma* (Figs. 70D, 71A). Therefore, I would say that in light of the morphofunctional and palaeontological data available, I am inclined to favor the direct hypothesis mentioned above for the origin and evolution of the Weberian interossicular ligament. I am also inclined to agree with an evolutionary origin of the intercalarium from both the second neural arch and a sesamoid ossification of the interossicular ligament. My own comparisons between extant otophysans and extant representatives of closely related groups such as the gonorynchiforms and clupeomorphs (see Figs. 3, 4) also raise objections for the indirect hypothesis of, for example, Rosen and Greenwood (1970). This is because, among all the extant members I have examined from these two latter groups, I never found a well-defined ligament between the parapophysis and/or rib of the third free vertebra and the third and/or second arch as proposed (see Fig. 73A) by these authors.

Developmental data is, of course, very important for discussing homologies. As explained above, however, the developmental data available so far does not help to clarify the ontogenetic origin of the intercalarium: some authors state that this ossicle develops from the second neural arch plus an ossification of the interossicular ligament, while authors state that it develops exclusively from the second neural arch. It is, however, important to stress here that developmental data is not the only, and in certain specific cases not even the most reliable, type of data to be considered in discussions of homology. Development does not always recapitulate evolution; evolutionary innovations can also be related, and often are, to non-terminal additions (e.g., Mabee, 1989a,b, 1993; Gould, 2002). This is well known in theory, but continues unfortunately to be too often neglected in practical discussions of homology, as pointed out by, for example, Gould (2002).

Claustrum

As stated by De Pinna and Grande (2003) and Grande and De Pinna (2004), the homology of the claustrum has also been, and continues to be, controversial. The data available from the known specimens of †*Chanoides macropoma*, †*Lusitanichthys characiformis*, †*Santanichthys diasii*, and †*Clupavus maroccanus* does not help clarify the homologies of this ossicle, since it is apparently missing in all these specimens (Figs. 71A, 72). According to Patterson (1984), Gayet (1981, 1985), Taverne (1995). and Filleul and Maisey (2004), the lack of this ossicle in †*Chanoides macropoma*, †*Lusitanichthys characiformis*, †*Santanichthys diasii*, and †*Clupavus maroccanus* might be due to a real absence of the ossicle or, perhaps more likely, to its poor preservation in the fossil record. The apparent presence, according to Taverne's (2005: fig. 12) interpretation, of a well-developed claustrum in †*Chanoides chardoni* seems to support this latter hypothesis.

Concerning the developmental data available, different authors have different interpretations. Fink and Fink (1981), for example, argued that the claustrum develops from a dissociated portion of the first neural arch. In turn, Watson (1939) and Vandewalle et al. (1990) stated that this ossicle develops from the ossification of mesenchyme. Coburn and Futey (1996) maintained that it develops from the first supraneural. De Pinna and Grande (2003) and Grande and De Pinna (2004) reviewed the arguments used to support each of these three hypotheses. They concluded that the developmental data available strongly support a fourth hypothesis: that the claustrum was originated from a modified accessory neural arch, a structure usually present in teleosts such as clupeomorphs (see, e.g., Fig. 70A). Apart from evidence from developmental data, De Pinna and Grande (2003) and Grande and De Pinna (2004) also provided topological and functional arguments in favor of their hypothesis. They argued that, as the accessory neural arch, the claustrum is topologically positioned as the first arch like a structure of the vertebral column, lying between the back of the skull and the neural arch of the first free vertebra and not associated with the centrum of this vertebra (see, e.g., Fig. 70A). Thus, as explained above, according to these authors the function of the claustrum is similar to that of the accessory neural arch: it forms a protective bony wall around the anterior portion of the neural canal, in the space between the back of the skull and the start of the neural arch series. The homology between the claustrum of extant otophysans and the accessory neural arch of teleosts such as clupeomorphs has also been considered to be plausible by Chardon et al. (2003). More detailed developmental, as well as palaeontological, data is needed to test this hypothesis of homology.

Tripus

Some authors affirmed that the tripus develops exclusively from the parapophysis of the third free vertebra (Radermaker et al., 1989; Vandewalle et al., 1990; Fukushima et al., 1992; Ichiyanagi et al., 1993, 1996). Others stated that it develops ontogenetically from both the parapophysis and the rib of this vertebra (Bamford, 1948; Rosen and Greenwood, 1970; Grande and Young, 2004; see, e.g., Fig. 73). Other researchers argued that part of this ossicle arises ontogenetically from the partial ossification of soft structures such as the swimbladder and/or of the interossicular ligament (e.g., Watson, 1939; Kulshrestha, 1977; Bogutskaia, 1991). This latter hypothesis was supported by Chardon et al. (2003), who, on the basis of a review of the literature available as well as a functional and structural analysis, argued that at least the characteristic posterior portion of the tripus of extant otophysans, that is, the transformator tripodis, originated by partial ossification of the swimbladder (see Fig. 71B, C, D).

The palaeontological data available for basal otophysan fossils does not help to clarify whether soft structures such as the interossicular ligament and/or the swimbladder might indeed have contributed to the formation of the tripus of extant otophysans. But Filleul and Maisey's (2004) observations on †*Santanichthys diasii* strongly support the view that both the parapophysis and the rib of the third free vertebra contributed to the formation of the tripus of extant otophysans. This is because both these structures are clearly incorporated in the "rudimentary" tripus of this species (see Fig. 72D).

Os Suspensorium

Regarding the homologies of the os suspensorium, some authors affirmed that this structure develops ontogenetically from the parapophysis of the fourth free vertebra (e.g., Vandewalle et al., 1990; Fukushima et al., 1992; Ichiyanagi et al., 1993, 1996), others argued that it develops from the rib, and not from the parapophysis, of the fourth free vertebra (e.g., Sagemehl, 1885; Bamford, 1948), and still others suggested that it develops from both the parapophysis and the rib of this vertebra (e.g., Rosen and Greenwood, 1970; see, e.g., Fig. 73).

The os suspensorium is the Weberian structure exhibiting a higher morphological diversity in extant otophysans (see, e.g., Fig. 71B, C, D). This structure also seemingly displays a considerable morphological diversity in basal otophysan fossils such as †*Chanoides macropoma*, †*Chanoides chardoni*, †*Lusitanichthys characiformis*, †*Lusitanichthys africanus*, and †*Clupavus maroccanus* (the os suspensorium was not found in †*Santanichthys diasii*) (see, e.g., Figs. 71A, 72A, B, C). Some comparisons can be made between these fossils and extant otophysans. For example, in †*Chanoides macropoma* the compound structure formed by the rib and the parapophysis of the fourth free vertebra appears to be somewhat similar to the compound structure formed by these two elements in the extant cypriniform *Opsariichthys uncirostris* (see Fig. 71A, B). However, no information is available on the presence and eventual configuration of a suspensor ligament or on the swimbladder and its connections in the preserved specimens of †*Chanoides macropoma* (Patterson, 1984). Consequently, it is difficult to discern whether in that fossil species either of the two ventral arms of this compound structure formed by the rib and the parapophysis of the fourth free vertebra had such connections to the tripus and/or to the swimbladder, as has the os suspensorium of *Opsariichthys uncirostris*.

Otocephalan Phylogeny and the Evolution of the Weberian Apparatus

Some authors suggested that the Weberian apparatus of extant members of the orders Cypriniformes, Characiformes, Gymnotiformes, and Siluriformes may have originated more than once (e.g., Gayet, 1981, 1985, 1986a,b; Gayet and Chardon, 1987). This is because, according to these authors, certain extant otophysans appeared to be more closely related to fossils with a "rudimentary" Weberian apparatus such as †*Lusitanichthys characiformis*, †*Chanoides macropoma*, or †*Clupavus maroccanus* than to other living otophysans. This hypothesis was, however, contradicted by Patterson (1984), Fink et al. (1984), Fink and Fink (1996), Chardon et al. (2003) and other researchers, who held that such fossils were very likely phylogenetically basal to the four orders of extant otophysans. One of the main causes for this controversy was the fact that, until then, the discussions on the phylogenetic position of these fossils were focused on disarticulated analyses of a few features that according to different authors were specially important for defending their positions. That is, these discussion were not based on the results of an explicit cladistic analysis including these fossils and representative extant otophysan taxa.

The cladistic analysis of Chapter 3, including various extant otophysan taxa, the fossils †*Chanoides macropoma*, †*Clupavus maroccanus*, †*Santanichthys diasii*, and †*Lusitanichthys characiformis*, and also members of closely related groups such as gonorynchiforms and clupeomorphs, can, therefore, shed light on this controversy. According to the results of that cladistic analysis, †*Lusitanichthys characiformis*, †*Chanoides macropoma*, and †*Clupavus maroccanus* form, together with another fossil exhibiting a "rudimentary" Weberian apparatus, †*Santanichthys diasii* (e.g., Filleul and Maisey, 2004), a monophyletic unit that is the sister-group of the clade including extant otophysans (Figs. 3, 4). Thus, according to these results, a "rudimentary" Weberian apparatus was acquired in the clade including fossils such as †*Chanoides macropoma*, †*Clupavus maroccanus*, †*Santanichthys diasii*, and †*Lusitanichthys characiformis* plus the four extant otophysan orders, which was then further modified in the characteristic, functional Weberian apparatus (e.g., Chardon et al. 2003) of extant otophysans in the node leading to these latter fishes (see Figs. 3, 4). In other words, the characteristic Weberian apparatus found in extant otophysans was acquired only once in evolution.

As noted above, the scaphia of fossils such as †*Chanoides macropoma* and †*Santanichthys diasii* seem more "rudimentary" than that of extant otophysans, because it bears foramina for dorsal and ventral nerve roots

(Fig. 71A). The scaphia of †*Lusitanichthys characiformis* and †*Clupavus maroccanus* are not well preserved, and it is thus not clear if these ossicles presented, or not, such foramina, but their overall configuration seems also less derived than that found in extant otophysans (e.g., Patterson, 1984; Fink et al., 1984; Fink and Fink, 1996; Chardon et al., 2003). The primary homology hypothesis (e.g., De Pinna, 1991; Diogo, 2004a) that the scaphia of fossils such as †*Lusitanichthys characiformis*, †*Clupavus maroccanus*, †*Chanoides macropoma*, and †*Santanichthys diasii* are less derived than those of extant otophysans is supported by the phylogenetic results of the cladistic analysis of Chapter 3.

Patterson (1984), Fink et al. (1984), Fink and Fink (1996), and Chardon et al. (2003) stated that the intercalaria of these fossils are apparently less derived than those of extant otophysans. For instance, unlike in extant otophysans, in †*Chanoides macropoma* the intercalarium has also foramina for dorsal and ventral nerve roots (see Fig. 71A, compare with Fig. 71B, C, D). Such foramina are apparently not found in the intercalaria of †*Lusitanichthys characiformis*, †*Clupavus maroccanus*, and †*Santanichthys diasii* (e.g., Fig. 72). However, the intercalaria of these three latter species are much larger than the intercalaria of extant otophysans, and apparently lack the ascendens processus typical of these latter fishes (Fig. 72, compare with Fig. 71B, C, D). The primary homology hypothesis that the intercalaria of †*Lusitanichthys characiformis*, †*Clupavus maroccanus*, †*Chanoides macropoma*, and †*Santanichthys diasii* are less derived than those of extant otophysans is also supported by the cladistic analysis of Chapter 3.

Patterson (1984), Fink et al. (1984), Fink and Fink (1996), Taverne (1999), Chardon et al. (2003), and Taverne (2005) called attention to significant morphological differences between the tripus of †*Lusitanichthys characiformis*, †*Clupavus maroccanus*, †*Chanoides macropoma*, †*Chanoides chardoni* and †*Santanichthys diasii* and that of extant otophysans. For instance, in †*Chanoides macropoma* the tripus is seemingly much smaller than the characteristic tripus of extant otophysans (the tripus of †*Chanoides chardoni* is also relatively small, although it is not as small as that of †*Chanoides macropoma*; Taverne, 2005: fig. 12). Moreover, the tripus of †*Chanoides macropoma* and †*Chanoides chardoni* seemingly lacks a transformator tripodis; in †*Chanoides macropoma* most of the lateral portion of the tripus is situated anteriorly to its mesial portion contacting the third centrum (see Fig. 71A, compare with Fig. 71B, C, D). It should, however, be noted that Fink and Fink (1996) mentioned that Patterson (unpublished results) found a structure similar to a transformator tripodis in some specimens of †*Chanoides macropoma*; therefore, there is some confusion about whether or not †*Chanoides macropoma* has a transformator tripodis.

The tripus of †*Santanichthys diasii* is apparently complete in the reconstruction of Filleul and Maisey (2004) (Fig. 72D). According to that reconstruction, the lateral portion of this bone mainly lies anterior to its mesial portion, and the transformator tripodis is seemingly lacking; this, again, is different to the characteristic condition found in extant otophysans (Fig. 72D, compare with Fig. 71B, C, D). Moreover, according to that reconstruction the parapophysis and the rib of the third free vertebra, that is, the two structures forming the tripus, are still completely separated (Fig. 72D, compare with Fig. 71B, C, D).

The tripus of the †*Lusitanichthys characiformis* specimen illustrated in Gayet's (1985) fig. 23 apparently does have a distal incurvation that somewhat resembles the characteristic transformator tripodis of extant otophysans (see Fig. 72B). However, this tripus is rather different from the tripus of other †*Lusitanichthys characiformis* specimens illustrated in Gayet's (1985) figs. 24 and 25, the latter being seemingly much less transformed from the plesiomorphic condition and lacking such a distal incurvation. The tripus of †*Lusitanichthys africanus* is unknown (Cavin, 1999).

The tripus of the †*Clupavus maroccanus* specimens examined by Taverne (1995) is poorly preserved (see, e.g., Fig. 72C). However, the analysis of the figures provided by Taverne (1995) seems to indicate that the tripus of this fossil species is also different from the characteristic tripus of extant otophysans, being apparently somewhat similar to a modified parapophysis of the third free vertebra like that found in derived clupeomorphs such as pristigasteroids (e.g., Grande and De Pinna, 2004). It should be noted that the phylogenetic results of Chapter 3 strongly suggest that the condition found in these derived clupeomorphs and in †*Clupavus maroccanus* is not homologous (Figs. 3, 4). These phylogenetic results support the view of, for example, Patterson (1984), Fink et al. (1984), Fink and Fink (1996), Taverne (1999, 2005), and Chardon et al. (2003), according to which †*Clupavus maroccanus* as well as †*Lusitanichthys characiformis*, †*Chanoides macropoma*, †*Chanoides chardoni*, and †*Santanichthys diasii* have a somewhat "rudimentary" tripus that is less derived than that of extant otophysans (see Chapter 3).

The claustrum is seemingly lacking in the specimens known of †*Chanoides macropoma*, †*Lusitanichthys characiformis*, †*Clupavus maroccanus*, and †*Santanichthys diasii* (see Figs. 71A, 72A, B, C, D). However, it is important to stress that in extant Otophysi the claustrum is often a very small bone (see, e.g., Fig. 71B, C), being inclusively missing in numerous fishes of this group, such as gymnotiforms and many siluriforms. Therefore, the apparent absence of a claustrum in the specimens of these four fossil species does not necessarily mean that this ossicle was missing in all members of those species. In fact, it should be noted that in Taverne (2005) documented that he found a well-developed claustrum in †*Chanoides chardoni*.

Concerning the os suspensorium, the phylogenetic results of Chapter 3 corroborate the opinion of, for example, Patterson (1984), Fink et al. (1984), Fink and Fink (1996), and Chardon et al. (2003), according to which the condition found in fossils such as †*Chanoides macropoma*, †*Lusitanichthys characiformis*, and †*Clupavus maroccanus* is less derived than that found in extant otophysans. It is interesting to note that, according to the drawings of Patterson (1984), Gayet (1985), and Taverne (1995), the os suspensorium of †*Chanoides macropoma*, †*Lusitanichthys characiformis*, and †*Clupavus maroccanus* does not form, together with the other ventral structures of the four free vertebra, a tunnel/bifurcation for the enclosing of part of the tripus, as is the case in extant otophysans (see Figs. 71A, 72A, B, C, compare with Fig. 71B, C, D). However, as explained above, Patterson apparently (unpublished results) found a transformator tripodis in †*Chanoides macropoma*. Thus, this latter structure might have eventually passed in the pronounced ventrolateral bifurcation of the compound formed by the parapophysis and the rib of the fourth free vertebra of this species (see Fig. 71A). One should also keep in mind that the shape of the os suspensorium of extant otophysans is rather variable and that, thus, unlike the other Weberian ossicles, it is somewhat difficult to define, in fact, what is its overall "characteristic" configuration for these fishes (see Fig. 71B, C, D; see above).

At this point one important clarification needs to be made: when it is said, as was done above, that the "functional" Weberian apparatus found in extant otophysans was acquired only once, this does not mean that all the numerous individual features integrated in the functioning of this apparatus appeared at the same time. In fact, as stressed by Rosen and Greenwood (1970), Fink and Fink (1996), Chardon and Vandewalle (1997), Chardon et al. (2003), Grande and De Pinna (2004), and others, it is now clear that many of the individual features that constitute the Weberian apparatus and many of the requirements necessary to allow the functionality of the whole apparatus in extant otophysans did not appear exclusively on the node leading to these fishes. For instance, the connection between the parapophyses and/or ribs of the first free vertebrae and the swimbladder, the constriction of this swimbladder into two chambers, and the anteriormost neural arch abutting the back of the skull are features also present in gonorynchiforms (e.g., Fig. 73A) and even in some clupeiforms such as pristigasteroids (e.g., Rosen and Greenwood, 1970; Grande and Poyato-Ariza, 1999; Grande and De Pinna, 2004). But the presence of such features in these non-otophysan taxa does not provide them a functional Weberian apparatus like that found in extant otophysans. Instead, the presence of these features in those taxa should probably be seen as a case of exaptation (e.g., Gould, 2002; Diogo, 2004a). That is, these features have likely been initially acquired in non-otophysan taxa for another reason, and only later in evolution were integrated in the functioning of the Weberian apparatus. Thus, as pointed out by, for example, Grande and Braun (2002), Chardon et al. (2003), De Pinna and Grande (2003), and Grande and De Pinna (2004), in order to provide a more comprehensive and contextual analysis on the origin and evolution of the various features evolved in the functioning of the Weberian apparatus, one should take into account not only the situation found in basal and extant otophysans, but also that found in other ostariophysans as well as in other teleosts such as clupeomorphs (see Figs. 3, 4). In view of the phylogenetic results of Chapter 3, the author's observations and comparisons, and a review of the literature available, it seems possible to provide here an updated, phylogenetically based discussion on how these features may have evolved, in major lines. This discussion is given below in three parts, concerning respectively the otocephalan clade, the ostariophysan clade, and the otophysan clade.

Otocephalan Clade

Some key features related to the peculiar configuration and the peculiar functioning of the Weberian apparatus of extant otophysans have seemingly been acquired in the node leading to the otocephalans as a whole (see Fig. 4). Examples of such features are the posterior and medial position of the saccular and lagenar otoliths and the presence of a silvery peritoneal tunic of the swimbladder covering at least the anterior portion of this bladder (e.g., Fink and Fink, 1981; Chardon et al., 2003; Grande and De Pinna, 2004; see Chapter 3).

One should, however, keep in mind that the second feature listed above refers to soft structures and that, although these soft structures seem to be plesiomorphically present in extant clupeomorphs and in extant ostariophysans (Grande and De Pinna 2004), one cannot be completely sure whether they were also present in basal fossils of these groups such as †*Diplomystus dentatus* or †*Tischlingerichthys viohli* (see above). Consequently, unless the eventual discovery of well-preserved basal otocephalan, clupeomorph and/or ostariophysan fossils allows us to discern whether such features were present in such fossils, one should be particularly careful about considering them synapomorphies for the Otocephala as a whole.

A feature that is also important for the subsequent formation of the Weberian apparatus and that might eventually constitute a synapomorphy of the Otocephala is the attachment of the anterior pleural ribs to the swimbladder. Rosen and Greenwood (1970) and Fink and Fink (1981, 1996) have suggested that this feature is a synapomorphy of ostariophysans. However, as pointed out by Grande and De Pinna (2004), in extant clupeiforms such as pristigasteroids and some engrauloids there is an attachment of the anterior ribs to the swimbladder that is somewhat similar to that found in extant gonorynchiforms and otophysans. Since it is difficult to discern whether such a configuration was eventually found in basal clupeomorphs such as †*Diplomystus dentatus* and in ostariophysan fossils such as †*Tischlingerichthys viohli* and †*Sorbininardus apuliensis*, it is difficult to assess whether it might have been plesiomorphically present in the Clupeomorpha (its absence in extant clupeiforms such as *Denticeps* being thus due to a secondary loss) and in the Otocephala as a whole.

Chardon and Vandewalle (1997) and Chardon et al. (2003) suggested that the first otocephalans might also have had another peculiar feature that was ultimately important for the formation of the Weberian apparatus: the presence of an anterior otophysic diverticula of the swimbladder penetrating the back of the skull and contacting the inner ear, as seen in, for example, many extant clupeiforms (see Fig. 70). As explained above, according to these authors the interossicular ligament of otophysans may well have originated from such swimbladder diverticula (Fig. 70). As an argument for supporting their hypothesis, they pointed out that a comparison between the fibers of the interossicular ligament and those of the tunica externa of the swimbladder reveal the same histological composition: both contain elastin and ichthyocoll, a special type I collagen that is usually absent in other ligaments. Therefore, they hypothesized that clupeiform-like diverticula transformed into the interossicular ligaments by losing their internal endoderm-splanchnopleura sheet (Fig. 70). According to their hypothesis, this sheet was probably inefficient in transporting gas (and pressure) instantaneously because the lumen of the diverticula was too narrow, while their external fibrous coating could transmit vibrations. Consequently, the "functional part" of the Weberian ossicles would be the result of intraligamentous ossifications: the neural arches 1 and 2 and the parapophysis of the third free vertebra "fused with the intraligamentous primordia later to form the complete ossicles in the same way as in ontogeny; their role seemingly being to sustain the ligament during lateral flexions of the anterior most backbone" (Chardon et al., 2003: 106).

According to, for example, Grande (1985a) and Grande and De Pinna (2004), the areas of thin inflated bone on the posteroventral region of the neurocranium of clupeomorph fossils such as †*Diplomystus* and †*Armigatus* seem to suggest that these fossils had prootic and pterotic bullae, which in turn suggests that these basal clupeomorphs had otophysic diverticula connecting the swimbladder to the inner ear. Thus, according to these authors, such a feature might be plesiomorphic for clupeomorphs. If this is so, and if an origin of the interossicular ligament of the otophysan Weberian apparatus from the swimbladder diverticula were to be accepted (see above), one could hypothesize that such diverticula were plesiomorphically present in otocephalans and secondarily lost in gonorynchiforms. Such a secondary loss in gonorynchiforms could eventually be explained by the fact that, from the plesiomorphic condition found in ostariophysans, otophysans and gonorynchiforms seem to have followed two different evolutionary pathways regarding their connection between the swimbladder and the posterior region of the skull (see below).

However, if one accepts that the Weberian interossicular ligament was derived from anterior diverticula of the swimbladder, as proposed by Chardon et al. (2003), one could as well postulate an alternative scenario, which is perhaps even more likely than that mentioned above: that the swimbladder diverticula that originated this ligament are not homologous with those found in clupeomorphs. That is, it might well be that both otophysans and clupeomorphs have independently acquired such otophysic diverticula connecting the swimbladder to the inner ear. Such a scenario is not unlikely because, as stressed by, for example, Grande and Braun (2002), anterior diverticula of the swimbladder have been independently acquired several times within teleosts. They are, for example, present in numerous non-otocephalan teleost families such as the Notopteridae, Mormyridae, Hiodontidae, Megalopidae, Gadidae, Holocentridae, and Cichlidae. Moreover, it is important to note that such anterior extensions of the swimbladder were also independently acquired in members of certain ostariophysan families. For example, unlike other all the other pimelodid and non pimelodid catfishes examined by Chardon (1968), the pimelodid *Calophysus macropterus* exhibits a modified swimbladder with two well-defined, anterior diverticula extending anteriorly under the prootic, near the level of the utriculus. These diverticula lie ventral to the Weberian ossicles and are almost parallel to the chain formed by these ossicles (Chardon, 1968: fig 71). Since swimbladder diverticula have been acquired various times in Teleostei, one cannot thus exclude the hypothesis that anterior diverticula such as those found in *Calophysus macropterus* or those found in clupeomorphs might have been acquired in the node leading to otophysans and then gave rise to the Weberian ossicular ligament. In fact, as explained above, the first otocephalans seemingly already had some peculiar features that may eventually have facilitated the evolution of an otophysic connection between the swimbladder and the region of the skull surrounding the inner ear in at least some of their descendants (see above). Thus, starting from ancestors that already had these peculiar features, it is conceivable that different otocephalan groups such as clupeomorphs and otophysans might have acquired, by parallel evolution (e.g., Gould, 2002; Diogo, 2004a), anterior swimbladder diverticula allowing the establishment of such an otophysic connection.

Chardon et al.'s (2003) hypothesis of an origin of the interossicular ligament from otophysic diverticula, illustrated in Fig. 70, is in opposition to the hypothesis proposed by Rosen and Greenwood (1970), which is illustrated in Fig. 73. According to this latter hypothesis, the Weberian interossicular ligament was derived from a combination of the ligaments joining the parapophysis/rib of the third free vertebra and the third neural arch, joining this neural arch to the second neural arch, and joining this latter neural arch to the first neural arch (Fig. 73). However, there are three main difficulties with this hypothesis. First, as mentioned in Chardon et al.'s (2003) paper and explained above, both the interossicular ligament and the tunica externa of the swimbladder contain elastin and ichthyocoll, a special type I collagen which is usually absent in other ligaments. Second, in all the numerous otocephalan fishes I have dissected so far, I never found a well-defined ligament joining the parapophysis and/or rib of the third free vertebra and the third neural arch, as postulated in Rosen and Greenwood's hypothesis (see Fig. 73A). Third, if this latter ligament was indeed present in the first ostariophysans, as postulated by these authors, both this ligament and the ligament joining the third neural arch to the second neural arch should have lost, during evolution, their connections to the third neural arch in order to give the interossicular ligament of otophysans (see Fig. 73). This is not theoretically impossible, but it is important to note that the recent advances in evolutionary developmental biology and vertebrate phylogeny have pointed out that such changes in the attachment points of ligaments and/or muscle tendons are clearly not as usual in evolution as previously assumed (e.g., Köntges and Lumsden, 1996; Shoshani et al., 1996; Collard and Wood, 2000; Gibbs et al., 2000, 2002; Diogo, 2004a,b; Graham, 2005).

Apart from the two hypotheses formulated by Chardon et al. (2003) and by Rosen and Greenwood (1970) to explain the origin of the interossicular ligament, mentioned in the paragraph above, another possibility might be pondered: that this ligament eventually originated from a muscular tendon and/or an unossified intermuscular bone. As stressed by Patterson and Johnson (1995), the importance of intermuscular bones has often been neglected in discussions on the evolution and homologies of teleosts. And the study of muscles and ligaments has, unfortunately, received much less attention in teleostean literature than that given to the analysis of bones (see above). As noted by, for example, Patterson and Johnson (1995), Chardon et al. (2003), and Diogo (2004a), among the studies carried out so far on the comparative anatomy and development of the postcranial region of otocephalans, very few have focused on the configuration of the muscles and ligaments of this region. Two arguments can be given to support an eventual origin of the Weberian interossicular ligament of otophysans from a tendon and/or an unossified intermuscular bone. The first is that intermuscular bones are commonly found in the postcranial region of clupeomorph and gonorynchiform fishes (e.g., Patterson and Johnson, 1995). Such structures might thus have been present in the first ostariophysans. And they are, in fact, present in some basal otophysan fossils such as †*Chanoides macropoma* and †*Lusitanichthys characiformis* (e.g., Patterson, 1984; Gayet 1981, 1985). The second argument is that in some adult otocephalans I have dissected, such as the clupeiform *Ilisha* or the gonorynchiform *Gonorynchus*, I did find some long, thin muscular tendons running from the parapophyses and/or ribs of the first free vertebrae to the posterior region of the skull.

Ostariophysan Clade

The discussion given in the paragraphs above has stressed that some key features that were subsequently integrated in the functioning of the Weberian apparatus of otophysans, such as the posteromedian position of the saccular and lagenar otoliths and the presence of a silvery peritoneal tunic of the swimbladder, as well as the eventual presence of postcranial intermuscular bones and the attachment of the anterior ribs to the swimbladder, might in fact have already been present in the first otocephalans. However, as might be expected, the configuration of the anterior free vertebrae and related structures of the first Otocephala was very likely not as similar to the configuration found in Otophysi as that found in the first Ostariophysi (sensu lato, that is, the clade including otophysans, gonorynchiforms, and fossils such as †*Tischlingerichthys viohli* and †*Sorbininardus apuliensis*: see above).

According to Rosen and Greenwood (1970), Fink and Fink (1981, 1996), Chardon and Vandewalle (1997), and Grande and De Pinna (2004), there was an acquisition, in the node leading to the Ostariophysi, of a large number of peculiar, derived features that were subsequently important for the formation and functioning of the Weberian apparatus. An example of this is the marked enlargement of the anterior neural arches (e.g., Figs. 71, 72, 73). According to, for example, Rosen and Greenwood (1970) and Fink and Fink (1981, 1996), the attachment of the first pleural ribs to the swimbladder might also be another example of these features, since such an attachment is present in extant gonorynchiforms and extant otophysans. However, as explained above, it is not possible to discern this feature in fossils and, thus, to discern whether it actually constitutes a synapomorphy of an eventual clade otophysans + †*Sorbininardus apuliensis*, of gonorynchiforms + otophysans + †*Sorbininardus apuliensis*, of these three latter taxa plus †*Tischlingerichthys viohli*, or even of the Otocephala as a whole (see Figs. 3, 4). Another feature that was perhaps acquired in the node leading to the Ostariophysi is the enlargement of the anterior supraneurals. Within ostariophysans, these supraneurals are usually enlarged in the Gonorynchoidei and in the Otophysi, as well as in the fossil †*Sorbininardus apuliensis*. Thus, the first members of the clade including all ostariophysans except †*Tischlingerichthys viohli* had very likely enlarged anterior supraneurals. Since in the †*Tischlingerichthys viohli* examined by Arratia (1997) the anterior supraneurals could not be observed, it is not clear whether the enlargement of the anterior supraneurals might be a synapomorphy of the whole Ostariophysi or, instead, of ostariophysans excepting this fossil species.

Some authors suggested that the shortening of the centra of the first three free vertebrae might be a synapomorphy of ostariophysans (e.g., Grande and De Pinna, 2004). However, if we take into account fossils such as †*Tischlingerichthys viohli* and †*Sorbininardus apuliensis*, this is not so clear. In †*Tischlingerichthys viohli*, which is seemingly the most basal ostariophysan taxon discovered so far (see above), the total length of the centra of the first three free vertebrae is not less than the total length of the centra of the free vertebrae 4, 5 and 6 (e.g., Arratia, 1997: fig. 65). In †*Sorbininardus apuliensis*, the length of the centra of the first three free vertebra is unknown (see Taverne, 1999). Therefore, one cannot say that the shortening of the centra of the first three free vertebrae constitutes an unambiguous synapomorphy of the Ostariophysi as a whole.

From the configuration found in first ostariophysans, two different evolutionary pathways seem to have been taken in the two major ostariophysan groups, that is, in gonorynchiforms and in otophysans. In gonorynchiforms, there is a somewhat mobile type of connection between the swimbladder and the back of the skull. The details of this connection were described in detail by Rosen and Greenwood (1970) and will thus be only briefly summarized here. As in otophysans, in gonorynchiforms there is usually an attachment between the swimbladder and the ribs of the third and fourth free vertebrae (see Fig. 73). However, although in gonorynchiforms the rib of the third free vertebra is usually enlarged relative to the ribs of the following vertebrae (see Fig. 73A), this rib is not as highly modified as in otophysans and is not integrated in the peculiar Weberian apparatus of these latter fishes (see Fig. 73B). Instead, in gonorynchiforms this rib is usually deeply associated, via ligaments, muscles, and/or connective tissue, with the pectoral girdle and with well-developed, rather mobile cephalic ribs anteriorly abutting bones of the posteroventral region of the skull, near the level of the inner ear, such as the basioccipital and/or exoccipital (e.g., Rosen and Greenwood, 1970: fig. 5). My dissections of the postcranial region of gonorynchiforms revealed that, when movements to the external layer of the swimbladder are provoked, the somewhat mobile rib of the third free vertebra does move, and this causes a corresponding movement of the dorsolateral surface of the pectoral girdle and of the also rather mobile cephalic ribs.

However, as pointed out by, for example, Chardon et al. (2003), because in gonorynchiforms the anterior portions of the cephalic ribs do not really penetrate any kind of opening/tunnel of the neurocranium allowing them to directly contact structures of the inner ear, it does not seem plausible that the movements of these cephalic ribs could be efficiently transmitted to the inner ear. That is why I agree that the indirect connection between the swimbladder and the bones of the posteroventral surface of the neurocranium in gonorynchiforms hardly constitutes a "true", functionally efficient otophysic connection between the swimbladder and the inner ear. In fact, such an indirect connection between the swimbladder and the bones of the posteroventral region of the neurocranium is also seen in otocephalan fishes other than gonorynchiforms, such as in derived clupeomorphs such as the pristigasteroid *Ilisha* and the engrauloid *Thryssa* (see Fig. 3). For instance, in specimens of the genus *Ilisha* the rib of the third free vertebra is modified into a rather mobile structure attached to the swimbladder, as in gonorynchiforms (see Fig. 73A). As in many gonorynchiforms, *Ilisha* specimens also have numerous intermuscular bones, muscles, and ligaments in the postcranial region, which may establish an indirect connection between this rib and the back of the skull. Moreover, the lateral surface of the rib of the third free vertebra of these specimens is deeply associated with the also rather mobile dorsolateral portion of the pectoral girdle. This latter is, in turn, connected to the back of the skull by two strong, thick ligaments: the Baudelot's ligament attaches anteriorly on the first free vertebra, which is in contact with the basioccipital; the "posttemporal-intercalar" ligament attaches anteriorly on the intercalar, which is a small bone situated on the back of the neurocranium.

As explained above, clupeiforms have a functional otophysic connection between the inner ear and the swimbladder, realized by the anterior extensions of this bladder (see Fig. 70A). Thus, it can be said that derived clupeiforms such as *Ilisha* and *Thryssa* have: (1) a direct connection between the swimbladder and the inner ear, which, according to, for example, Chardon et al. (2003), could have originated the Weberian connection between the swimbladder and the inner ear of otophysans; and (2) an indirect connection between the swimbladder and the back of the skull through connective tissue, muscle tendons, intermuscular bones, and/or the pectoral girdle, similar to that found in gonorynchiforms. As in gonorynchiforms, this latter indirect connection hardly seems to constitute a functional, efficient system allowing the inner ear to detect the changes of volume of the swimbladder. Such an indirect connection between the swimbladder and the back of the skull, found homoplasically in gonorynchiforms and in some derived clupeiforms, can thus seemingly be interpreted as an anatomical consequence eventually associated with the functioning of other systems (e.g., the movements of the pectoral girdle), or, less likely, as part of a system related to an unknown function.

Otophysan Clade

The evolutionary pathway followed by otophysans was quite different from that followed by gonorynchiforms. Unlike extant gonorynchiforms, extant Otophysi do have a direct, "functional" otophysic connection between the swimbladder and the inner ear, as described above (see Figs. 69, 70D, 71B, C, D, 73B, compare with Fig. 73A).

Some of the key morphological transformations that have lead to the characteristic Weberian apparatus of extant otophysans, and that differentiate the condition found in these fishes from that found in gonorynchiforms, were seemingly acquired in the node leading to the clade including cypriniforms, characiforms, gymnotiforms, and fossils such as †*Clupavus maroccanus*, †*Santanichthys diasii*, †*Chanoides macropoma*, and †*Lusitanichthys characiformis* (see Figs. 3, 4). As explained above, these fossils already exhibit a "rudimentary" tripus, a "rudimentary" os suspensorium, a "rudimentary" intercalarium and a "rudimentary" scaphium (e.g., Figs. 71A, 72). Some authors suggested that features such as the presence of the claustrum and the presence of a sinus impar are otophysan synapomorphies (e.g., Fink and Fink, 1981, 1996). However, because the data available does not allow us to discern whether or not these features were present in fossils such as †*Clupavus maroccanus*, †*Santanichthys diasii*, †*Chanoides macropoma*, and †*Lusitanichthys characiformis*, we cannot discern whether they are synapomorphies of the whole Otophysi or of the clade including only the four extant otophysan orders (see Figs. 3, 4).

Although fossils such as †*Clupavus maroccanus*, †*Santanichthys diasii*, †*Chanoides macropoma*, and †*Lusitanichthys characiformis* already had a "rudimentary" tripus, scaphium, intercalarium, and os suspensorium, there are major differences concerning the configuration of these ossicles in those fossils and in extant otophysans (see above). For instance, unlike in †*Santanichthys diasii* (see Fig. 72D), in extant otophysans there is a well-developed transformator tripodis (see Fig. 71B, C, D); unlike, for example, †*Clupavus maroccanus* (see Fig. 72C), in extant otophysans the tripus is a rather mobile element with a rather thin articulation with the centrum of the third free vertebra (e.g., Fig. 71B, C, D); in contrast to, for example, †*Santanichthys diasii* and †*Lusitanichthys characiformis* (see Fig. 72B, D), part of the tripus of extant otophysans is enclosed in a tunnel/bifurcation formed by the os suspensorium and the other ventral elements of the fourth free vertebra (see Fig. 71B, C, D); and, in contrast to, for example, †*Chanoides macropoma* (see Fig. 71A), in extant Otophysi the scaphium and intercalarium do not have foramina for dorsal and ventral nerve roots (e.g., Fig. 71A). A more detailed and complete list of the anatomical differences between the Weberian ossicles of these fossils and those of extant otophysans is given above. What is important to emphasize here is that the cladistic analysis of Chapter 3 supports the view that the "rudimentary" Weberian apparatus found in †*Clupavus maroccanus*, †*Santanichthys diasii*, †*Chanoides macropoma*, and †*Lusitanichthys characiformis* was seemingly acquired in the node leading to these fossils plus the extant otophysans, and that the characteristic Weberian apparatus of extant otophysans was acquired later in evolution, in the node leading to the four extant otophysan orders (see Figs. 3, 4).

General Conclusions

The remarkably complex Weberian apparatus of extant otophysans seems, in fact, to be the outcome of a functional integration of features acquired in basal otocephalans and in basal ostariophysans, which were very likely not directly related with the functioning of this apparatus, as well as of features acquired in the nodes leading to the Otophysi and to the clade including the four extant otophysan orders, which could well have been the result of a selection directly related to the functioning of this apparatus. This is, of course, just a scientific hypothesis, which should be, and hopefully will be, tested in future studies. There are interesting issues that should be addressed in future works in order to shed further light on the origin, evolution, and functioning of the Weberian apparatus. In my opinion, it would be, for instance, advisable to undertake works on the development and comparative anatomy of the various muscles and ligaments associated with the anterior vertebrae in otocephalans, the histological and mechanical properties of the tunica externa and of its fibers, the ontogeny of the ossicles, interossicular ligament, and surrounding structures of the Weberian apparatus of otophysans, the hydrostatic pressure sensitivity of the otophysan swimbladder and the possible transmission of that information through the Weberian ossicles to the labyrinths, and, if this is possible, the configuration of new fossil Weberian apparatuses. As explained above, of these and other possible lines of research for future studies, I consider that it is particularly important to investigate the development and comparative anatomy of the ligaments, muscles, and intermuscular bones of the postcranial region of otocephalans, with a particular focus on the ontogeny and homologies of the interossicular ligament of otophysans. It is hoped that the present work may stimulate, and somehow facilitate, such future studies.

4.6 MYOLOGICAL VERSUS OSTEOLOGICAL CHARACTERS IN PHYLOGENETIC RECONSTRUCTIONS: A NEW INSIGHT

The phylogenetic results of the present work provide background material for a discussion on an interesting, but unfortunately often neglected, issue concerning the use of different kinds of information in phylogenetic reconstructions: the use of *myological* versus *osteological* data. Regarding bony fishes, one of the few discussions on the utility of myological data for phylogenetic reconstructions was made by Borden (1999). In that paper, Borden described in detail the configuration and variation of 93 muscles in 15 species of the genus *Naso*, or unicornfishes, of the Acanthuridae (Teleostei: Percomorpha) and discussed the phylogenetic implication of the myological results obtained. As noted by Borden (1999: 191), very few studies in ichthyology focus on myology, for a variety of reasons: "investigators may be reluctant to use myology due, for example, to the plethora of names that have been used to describe the same muscles, to the realization that osteological proficiency is mandatory in order to identify muscle, leading them to concentrate only on osteology, or to the requirement of potentially finer dissection to preserve muscle bundles and nerves; furthermore, fossil fishes leave few if any myological clues, complicating hypotheses between extinct and extant fishes". In consequence, "of those studies using myology as a basis of information, most are functional works often analyzing the role of various muscles in feeding or locomotion or comparing a muscle or specific group across a number of taxa systematically and/or ecologically related" (Borden, 1999: 191). Explicit cladistic analyses based on myological data are, thus, rather rare. Within bony fishes, some of the most relevant examples of such analyses listed by Borden (1999) are those of Winterbottom (1974), using 74 myological characters to investigate the relationships of the tetraodontiforms, Winterbottom (1993), using 46 myological characters to reconstruct the relationships among the acanthurid genera, or Borden (1998), using about ten myological characters to investigate the phylogeny inside the acanthurid genus *Naso*. As these studies are essentially restricted to myology, they do not allow a direct comparison between the homoplasy and phylogenetic relevance of myological versus osteological data. Some works by other authors did include some myological characters together with some osteological ones, for example, Howes (1983a) and Schaefer (1990). However, the number of myological characters included in those works is rather small; moreover, those works do not provide a direct comparison between the homoplasy and retention levels found in the osteological and myological characters examined.

In order to provide such a comparison, I compared (Diogo, 2004b) the homoplasy levels and phylogenetic relevance of 91 myological and 303 osteological characters in the reconstruction of the higher-level phylogeny of a diverse and representative group of teleosts, the Siluriformes (see above). As explained in that paper, such a direct comparison of the relative contribution of these two types of data sets for phylogenetic reconstructions has the advantage that the homoplasy levels and the phylogenetic trees being compared refer to the same group and, more importantly, to the very same terminal taxa. In major lines, the overall analysis of the results presented in the paper seemingly indicated that: (1) osteological structures display a greater morphological variation than myological ones; (2) this difference (which is very likely over-enhanced by the fact that the phylogenetic variation of osteological structures has historically been the subject of more studies and descriptions than that concerning the myological ones) is particularly notable regarding lower taxa, such as genera or species; (3) myological characters provide, however, a high proportion of informative characters for disclosing the relationships between higher taxa, and, thus, for disclosing the phylogeny of the clades in which these taxa are included.

It is thus interesting to investigate whether or not the conclusions of that paper, based on a phylogenetic analysis of the Siluriformes, are supported by a comparison of the myological and osteological characters included in the cladistic analysis of Chapter 3 of the present book, which mainly deals with the higher-level phylogeny of osteichthyans.

The 122 characters included in the cladistic analysis of Chapter 3 that refer *exclusively* to the configuration of muscles are listed in Table 10; the 198 characters that refer *exclusively* to the configuration of the bones and associated cartilage constituting a certain osteological component (e.g., main body of the palatine plus the palatine cartilages), but not to the ligaments connecting different osteological components, are given in Table 11. For each myological (Table 10) and osteological (Table 11) character listed, these tables show its respective CI (Consistency Index) and RI (Retention Index). Those characters that are *autapomorphic for a single terminal taxon*, that is, for a particular genus included in the cladistic analysis, and, thus, that are not informative of the phylogenetic relationships *between* the different terminal taxa, are respectively indicated as "AUT".

As in the case study provided by siluriforms (Diogo, 2004b), the osteological components included in the cladistic analysis of Chapter 3 exhibit a greater morphological variation than the myological ones. The 81 different osteological components examined for that cladistic analysis (see Chapter 3) provided, as mentioned above, 198 phylogenetic characters (in an arithmetical mean of about *2.4 phylogenetic characters for each osteological component* examined). The 63 muscles examined for the analysis (see Chapter 3) provided 122 phylogenetic characters (in an arithmetical mean of about *1.9 phylogenetic characters for each muscle* examined). This seems to corroborate the opinion of Borden (1999), who pointed out that muscular characters are somewhat more "conservative" than osteological ones, these latter thus demonstrating a higher variability. However, contrary to what was suggested by Borden (1999) and supported by the data of Diogo (2004b), the proportion of autapomorphic characters within the osteological characters listed in Table 11 (49 out of a total of 198, i.e., about 25%) is not higher than that within the myological characters listed in Table 10 (33 out of a total of 122, i.e., about 27%).

All things being counted, it can thus be said that each osteological component examined for the cladistic analysis of Chapter 3 provided, on average, more informative characters: the 81 osteological components examined for that analysis provided 198 phylogenetic characters, of which 149 are informative (i.e., *1.8* informative characters per osteological component examined); the 63 muscles examined for the analysis provided 122 phylogenetic characters, of which 89 are informative (i.e., *1.4* informative characters per muscle examined). However, it is important to stress that although very likely indicating a true higher osteological variation, as suggested by Borden (1999), these numbers are probably also related to the fact that in the literature there are much fewer phylogenetic works based on myology than based on osteology (see above). In fact, a significant number of the osteological characters included in the cladistic analysis of Chapter 3 were inspired, in some way, by characters previously pointed out in phylogenetic studies by other authors. These studies thus called my attention to the variability exhibited by certain osteological

Table 10 List of 122 myological characters (Char.) included in the cladistic analysis of Chapter 3 and their respective CI and RI. "AUT" means that the character is autapomorphic for a single genus (thus, it is not informative for the inference of the phylogenetic relationships between the groups studied); the last cells, indicated with TN, MC and MR and marked with thick lines, represent, respectively, the sum total of the number of myological characters (TN), the arithmetical mean of the consistency indexes of these characters (MC), and the arithmetical mean of the retention index of these characters (MR) (for more details, see text).

Char.	*CI*	*RI*	*Char.*	*CI*	*RI*
1	14	40	**35**	AUT	AUT
2	AUT	AUT	**36**	100	100
3	33	85	**37**	100	100
4	AUT	AUT	**38**	100	100
5	100	100	**39**	100	100
6	AUT	AUT	**40**	100	100
7	100	100	**41**	100	100
8	AUT	AUT	**42**	100	100
9	AUT	AUT	**43**	100	100
10	100	100	**44**	100	100
11	100	100	**45**	100	100
12	20	77	**46**	100	100
13	20	69	**47**	100	100
14	AUT	AUT	**48**	100	100
15	50	50	**49**	100	100
16	100	100	**50**	AUT	AUT
17	50	88	**51**	AUT	AUT
18	100	100	**52**	100	100
19	AUT	AUT	**53**	100	100
20	8	60	**54**	100	100
21	100	100	**55**	100	100
22	100	100	**56**	100	100
23	100	100	**57**	100	100
24	100	100	**58**	100	100
25	100	100	**59**	100	100
26	50	0	**60**	100	100
27	40	85	**61**	AUT	AUT
28	25	86	**62**	AUT	AUT
29	14	81	**63**	AUT	AUT
30	50	80	**157**	8	64
31	50	66	**158**	16	0
32	AUT	AUT	**159**	100	100
33	100	100	**160**	100	100
34	100	100	**161**	25	25

Contd.

Table 10 Contd.

Char.	*CI*	*RI*	*Char.*	*CI*	*RI*
162	AUT	AUT	**190**	100	100
163	AUT	AUT	**191**	100	100
164	50	0	**192**	100	100
165	100	100	**193**	33	80
166	50	50	**194**	66	83
167	100	100	**195**	AUT	AUT
168	AUT	AUT	**196**	AUT	AUT
169	AUT	AUT	**197**	7	25
170	AUT	AUT	**198**	AUT	AUT
171	AUT	AUT	**199**	AUT	AUT
172	AUT	AUT	**200**	100	100
173	AUT	AUT	**201**	AUT	AUT
174	AUT	AUT	**202**	50	50
175	100	100	**203**	16	44
176	20	76	**204**	12	50
177	20	42	**205**	25	0
178	33	75	**206**	20	42
179	50	75	**207**	AUT	AUT
180	100	100	**208**	AUT	AUT
181	100	100	**209**	33	77
182	25	40	**210**	50	50
183	AUT	AUT	**211**	100	100
184	100	100	**212**	100	100
185	16	83	**213**	AUT	AUT
186	25	78	**214**	50	87
187	7	45	**347**	100	100
188	100	100	--------	--------	--------
189	AUT	AUT	**TN=122**	**MC=0.71**	**MR=0.82**

features within the groups included in the cladistic analysis of the present work. This historical bias should, therefore, be kept in mind when one discusses the "variability" of, and the consequent number of informative phylogenetic characters provided by, osteological versus myological structures.

The arithmetical mean of the RI of the informative muscular characters listed in Table 10 of the present work (= 0.82) is considerably higher than that of the informative osteological ones listed in Table 11 (= 0.71). It should be explained here that the CI of a character is simply related with the level of homoplasy exhibited by that character, while the RI also takes into account whether the character helps to retain a certain clade in the obtained cladogram of the group being studied. So, to give a simple example, if a

Table 11 List of 198 osteological characters (Char.) included in the cladistic analysis of Chapter 3 and their respective CI and RI. "AUT" means that the character is autapomorphic for a single genus (thus, it is not informative for the inference of the phylogenetic relationships between the groups studied); the last cells, indicated with TN, MC and MR and marked with thick lines, represent, respectively, the sum total of the number of osteological characters (TN), the arithmetical mean of the consistency indexes of these characters (MC), and the arithmetical mean of the retention index of these characters (MR) (for more details, see text).

Char.	*CI*	*RI*	*Char.*	*CI*	*RI*
64	36	66	**99**	25	72
65	50	50	**100**	25	25
66	16	79	**101**	50	92
67	16	79	**102**	AUT	AUT
68	11	72	**103**	AUT	AUT
69	AUT	AUT	**104**	33	89
70	100	100	**105**	100	100
71	AUT	AUT	**106**	25	76
72	100	100	**107**	100	100
73	33	0	**108**	100	100
74	100	100	**109**	14	73
75	25	25	**110**	11	33
76	50	88	**111**	100	100
78	6	61	**112**	33	93
79	AUT	AUT	**113**	100	100
80	50	0	**114**	50	92
81	11	68	**115**	100	100
82	11	69	**120**	33	75
83	100	100	**121**	20	75
84	100	100	**122**	AUT	AUT
85	12	63	**123**	50	94
86	33	80	**124**	14	75
87	AUT	AUT	**125**	50	50
88	AUT	AUT	**126**	33	33
89	100	100	**134**	50	66
90	100	100	**135**	AUT	AUT
91	100	100	**136**	25	70
92	50	75	**137**	33	80
93	100	100	**138**	13	62
94	AUT	AUT	**139**	12	0
95	100	100	**140**	50	85
96	16	73	**141**	14	64
97	16	84	**142**	25	40
98	100	100	**143**	33	60

Contd.

Table 11 Contd.

Char.	*CI*	*RI*	*Char.*	*CI*	*RI*
144	AUT	AUT	**251**	AUT	AUT
145	25	40	**252**	33	50
146	100	100	**253**	100	100
147	100	100	**254**	100	100
148	AUT	AUT	**255**	50	83
149	100	100	**259**	33	60
150	100	100	**260**	33	33
151	100	100	**261**	AUT	AUT
152	AUT	AUT	**262**	AUT	AUT
153	25	25	**263**	8	60
154	33	60	**264**	AUT	AUT
155	100	100	**265**	AUT	AUT
156	AUT	AUT	**266**	50	75
215	50	0	**267**	AUT	AUT
216	7	53	**268**	AUT	AUT
217	20	69	**269**	100	100
218	AUT	AUT	**270**	AUT	AUT
219	100	100	**271**	20	50
220	11	61	**272**	14	76
221	100	100	**273**	25	50
224	50	50	**274**	33	0
225	50	88	**275**	50	0
226	100	100	**276**	50	33
227	50	0	**277**	10	64
228	11	55	**278**	100	100
229	50	0	**279**	AUT	AUT
230	50	92	**280**	25	87
233	AUT	AUT	**281**	20	73
234	AUT	AUT	**282**	100	100
235	AUT	AUT	**283**	100	100
236	AUT	AUT	**284**	33	60
237	AUT	AUT	**285**	AUT	AUT
238	16	76	**286**	AUT	AUT
242	100	100	**287**	25	50
243	100	100	**288**	40	66
244	33	33	**289**	33	50
245	12	79	**290**	16	50
246	50	50	**291**	AUT	AUT
247	33	81	**292**	AUT	AUT
248	50	85	**294**	100	100
249	100	100	**296**	50	0
250	50	88	**297**	AUT	AUT

Contd.

Table 11 Contd.

Char.	*CI*	*RI*	*Char.*	*CI*	*RI*
298	AUT	AUT	**326**	50	88
299	100	100	**327**	AUT	AUT
300	AUT	AUT	**328**	33	50
301	100	100	**329**	33	60
302	12	58	**330**	16	50
303	AUT	AUT	**331**	20	82
305	100	100	**332**	33	60
308	AUT	AUT	**333**	100	100
309	AUT	AUT	**334**	50	0
310	AUT	AUT	**335**	50	50
311	50	50	**336**	AUT	AUT
312	AUT	AUT	**337**	AUT	AUT
313	11	11	**338**	AUT	AUT
314	100	100	**339**	AUT	AUT
315	33	75	**340**	AUT	AUT
316	100	100	**341**	33	66
317	20	50	**342**	50	91
318	AUT	AUT	**343**	100	100
320	33	66	**344**	AUT	AUT
321	100	100	**349**	100	100
322	20	33	**350**	100	100
323	16	73	**356**	100	100
324	25	25	--------	--------	--------
325	33	71	--------	--------	--------
			TN=198	**MC=0.52**	**MR=0.71**

certain feature X is found in the actinopterygian *Pimelodus* and in the distantly related sarcopterygian *Timon*, that character will have, according to the phylogenetic scenario shown in Fig. 3, a CI of 0.5 (it was independently acquired two times) and an RI of 0.0 (it does not help to retain any of the clades shown in that figure). However, if a certain feature Z is found, for instance, in *Pimelodus* and *Timon*, but also in all the elopomorph genera examined (i.e., clade 7 of Fig. 3), this feature Z will have, according to the scenario shown in Fig. 3, a CI of 0.33 and an RI of 0.75 (since it was independently acquired in three different, non-related groups, that is, *Pimelodus*, *Timon*, and the clade 7, but, unlike feature X, it helps to retain one of the clades shown in the cladogram of Fig. 3, the clade 7).

Taking this into account, it can thus be said that each of the 89 informative myological characters listed in Table 10 is on average more useful for the retention of the clades shown in Fig. 3 than each of the 149 osteological informative characters listed in Table 11. In the case study provided by catfishes (Diogo, 2004b), the arithmetical mean of the RI of the informative muscular characters was also higher than that of the informative osteological ones, but the difference was less pronounced (0.80 and 0.77, respectively). The arithmetical mean of the CI of the informative muscular characters listed in Table 10 is also considerably higher that that of the informative osteological characters listed in Table 11 (0.71 and 0.52, respectively). These results are different from those obtained in the case study of catfishes (Diogo, 2004b), in which the arithmetical mean of the CI of the informative osteological characters was a little higher than that of the informative myological characters (0.66 and 0.64, respectively).

As stressed by Diogo (2004b), another interesting way to analyze the phylogenetic contribution of myological versus osteological features is to simply compare the phylogenetic trees generated from these two different kinds of data. Figure 122 of the present work illustrates the "majority fools" cladogram obtained by the cladistic analysis of those 122 myological characters listed in Table 10; Figure 123 illustrates the "majority fools" cladogram resulting from the cladistic analysis of the 198 osteological characters listed in Table 11 (see above). The CI of the "myological" cladogram shown in Fig. 122 is higher than the RI of the "osteological" cladogram shown in Fig. 123 (0.48 and 0.37, respectively). The RI of the "myological" cladogram is also higher than the RI of the "osteological" cladogram, but the difference between these indexes is less pronounced (0.77 and 0.76, respectively). It is important to note that the "myological" cladogram does not include the seven fossil taxa included in the "osteological" cladogram, since the information available on the musculature of each of these taxa is very scarce, or even non-existent.

One way of comparing the results shown in the "myological" and "osteological" cladograms of Figs. 122 and 123 is to analyze which of the clades obtained in the cladistic analysis of Chapter 3 (Figs. 3, 4) are supported by these cladograms. Within the context of the cladistic analysis of Chapter 3, that is, of the higher-level phylogeny of the osteichthyan groups included in the present work, one can differentiate two main types of clades (Fig. 4). At a higher level, one has major clades including various orders, such as the Sarcopterygii, the non-actinistian sarcopterygians, the Tetrapoda, the Actinopterygii, the Actinopteri, the Neopterygii, the Halecostomi, the Teleostei, the Elopomorpha, the Osteoglossomorpha, the Clupeocephala, the Euteleostei, the Neoteleostei, the Otocephala, the Ostariophysi, and the Otophysi (Fig. 4). At a lower level on the cladogram obtained in the cladistic analysis of Chapter 3, one has for instance taxa such as orders (see below). It is thus interesting to examine how these two types of clades are supported in the "myological" and "osteological" cladograms of Figs. 122 and 123.

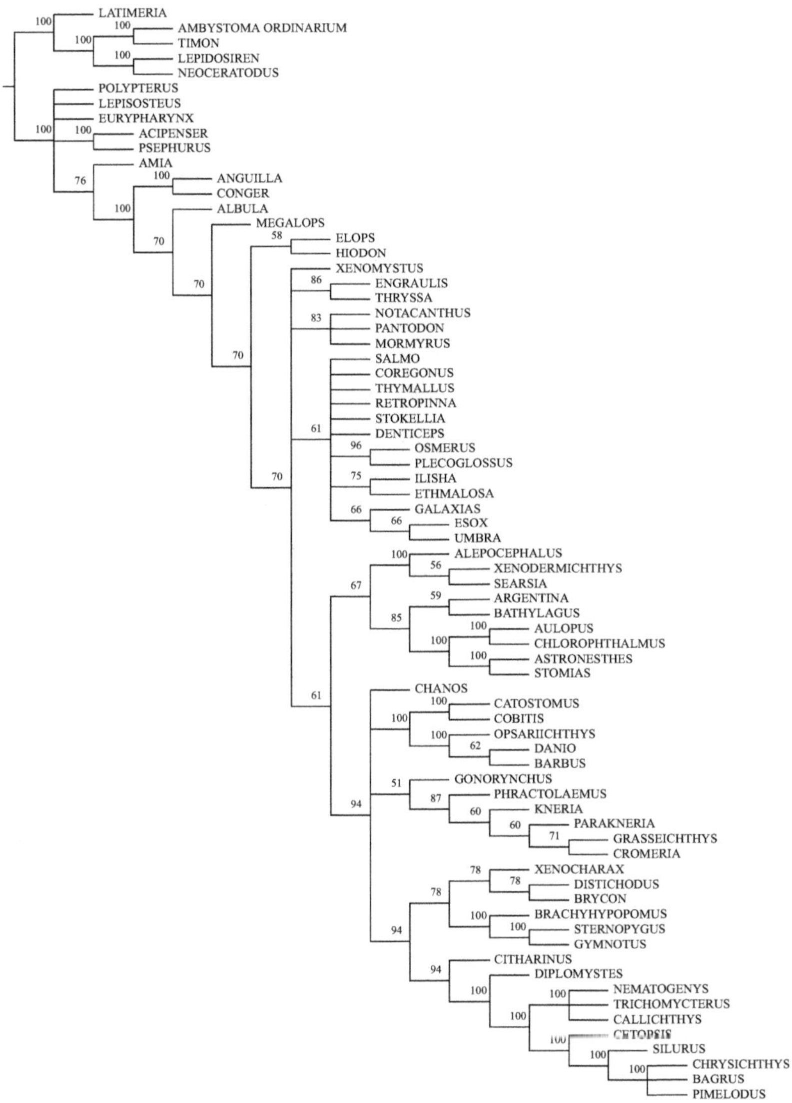

Figure 122 "Majority fools" tree (CI= 0.48; RI= 0.77) obtained by applying the "majority fools" option of Nona&Winclada to the 7,368 equally parsimonious trees (CI= 0.50; RI= 0.79) obtained when only those 122 myological characters listed in Table 10 were included in the cladistic analysis (note: fossil taxa were not included in the analysis). The numbers 51, 56, 58, 59, 60, 61, 62, 66, 67, 70, 71, 75, 76, 78, 83, 85, 86, 87, 94, 96, and 100 indicate that the respective clades were supported by 51%, 56%, 58%, 59%, 60%, 61%, 62%, 66%, 67%, 70%, 71%, 75%, 76%, 78%, 83%, 85%, 86%, 87%, 94%, 96%, and 100% of the equally parsimonious trees obtained, respectively (for more details, see text).

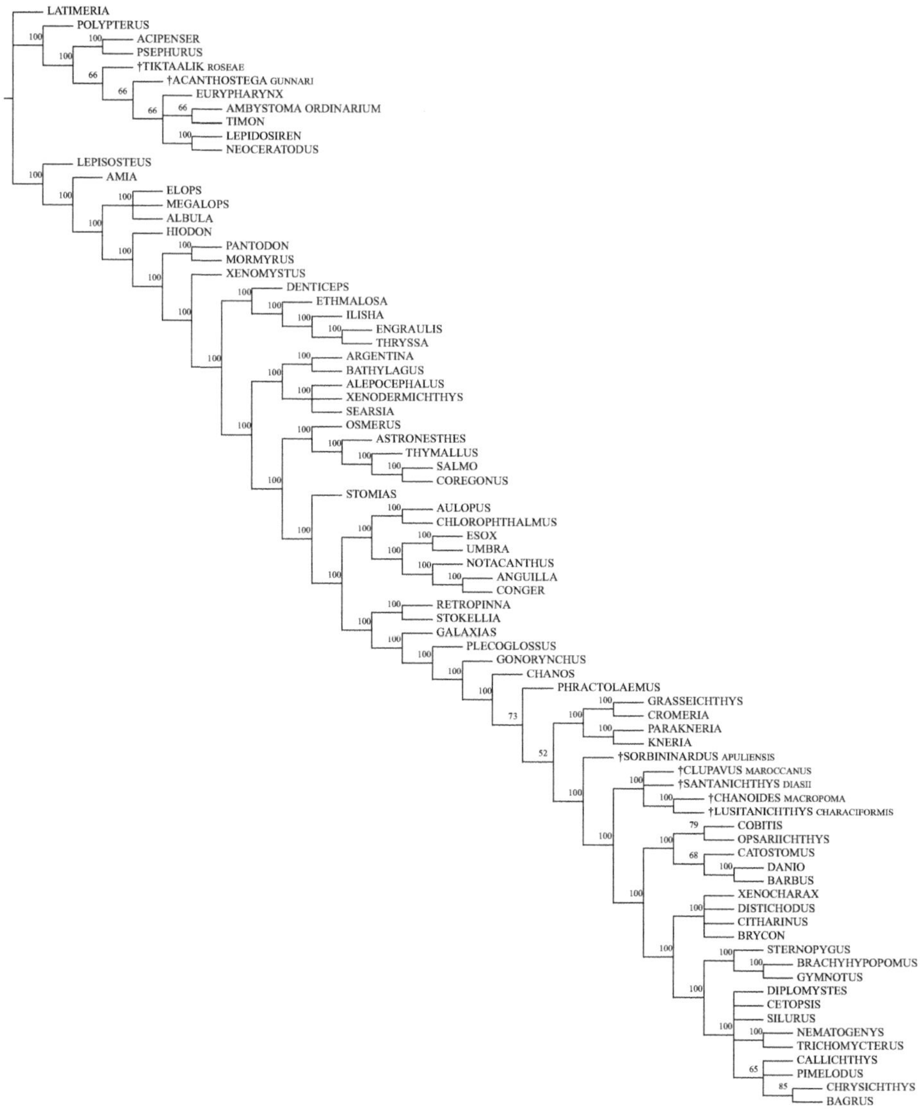

Figure 123 "Majority fools" tree (CI= 0.37; RI= 0.76) obtained by applying the "majority fools" option of Nona&Winclada to the 15,840 equally parsimonious trees (CI= 0.38; RI= 0.77) obtained when only those 198 osteological characters listed in Table 11 were included in the cladistic analysis (note: fossil taxa were included in the analysis). The numbers 52, 65, 66, 68, 73, 79, 85, and 100 indicate that the respective clades were supported by 52%, 65%, 66%, 68%, 73%, 79%, 85%, and 100% of the equally parsimonious trees obtained, respectively (for more details, see text).

Among the 16 major clades listed in the paragraph above, six appear as monophyletic in the "myological" cladogram obtained from the cladistic analysis including only those 122 osteological characters listed in Table 10: the Actinopterygii, the Sarcopterygii, the non-cladistian sarcopterygians, the Tetrapoda, the Neoteleostei, and the Ostariophysi (Fig. 122). It should be noted that some other clades of the "myological cladogram" strongly resemble other major clades of the cladogram shown in Fig. 4. For instance, except the highly peculiar *Eurypharynx*, all halecostomes are grouped into a monophyletic unit in the "myological" cladogram (Fig. 122). Likewise, excepting this peculiar genus, all teleosts are grouped together in that cladogram (Fig. 122).

Only two of the 16 major clades listed above appear as monophyletic in the "osteological" cladogram obtained from the cladistic analysis including only those 198 osteological characters listed in Table 11: the Ostariophysi and the Otophysi (Fig. 123). Again, one of the major problems in this "osteological" cladogram concerns the position of *Eurypharynx*, which appears in an unresolved trichotomy that also leads to the tetrapods and to the Dipnoi (Fig. 123). If one were to exclude this genus from that "osteological" cladogram, clades such as the non-cladistian sarcopterygians, the Neopterygii, the Halecostomi and the Teleostei would appear monophyletic (Fig. 123). However, even so, major clades such as the Actinopterygii, the Sarcopterygii, and the Neoteleostei, which appear as monophyletic in the "myological" cladogram (Fig. 122), would not constitute monophyletic units in the "osteological" cladogram (Fig. 123; note: the Tetrapoda would also not appear as a monophyletic unit in the "osteological" cladogram, but this would be due to the position of †*Acanthostega gunnari*, a fossil taxon that was not included in the "myological" cladogram: see above).

Some readers may consider it rather strange to see *Eurypharynx* appearing more closely related to certain sarcopterygians than to actinopterygians in the "osteological" cladogram of Fig. 122. However, as explained in the earlier chapters, the Saccopharyngiformes, which includes *Eurypharynx*, is a group of bizarre fishes that are anatomically very different from any other fishes, and in fact from any other vertebrates I have ever seen. One should remember that in some of the first descriptions of these fishes they were not even included in the Osteichthyes, and that their phylogenetic position has been, and continues to be, the subject of much controversy. As explained in Chapter 3, although the elopiforms, albuliforms, notacanthiforms, anguilliforms and saccopharyngiforms are considered to be elopomorphs in some general textbooks, no published morphological cladistic analysis has actually included representatives of all these groups and supported their grouping in a monophyletic unit. The only morphological cladistic analysis that has done so is thus the one provided in Chapter 3 of the present work, which includes a large number of both myological and osteological characters (Figs. 3 and 4). This, together with the fact that neither the "osteological" cladogram nor the "myological" cladogram of Figs. 122 and 123 support the monophyly of the Elopomorpha, stresses how important it is to use the complementary data provided by muscular and skeletal structures (as well by other type of structures, if possible) in morphological cladistic analyses.

As explained above, in opposition to the higher clades discussed in the previous paragraphs, at a lower level on the cladogram obtained in the cladistic analysis of Chapter 3 one has, for instance, clades such as orders (see Fig. 4). Among the 16 orders included in the cladistic analysis of Chapter 3 that were represented by more than one terminal taxon in that cladistic analysis, that is, the Acipenseriformes, Elopiformes, Anguilliformes, Osteoglossiformes, Argentiniformes, Esociformes, Osmeriformes, Salmoniformes, Aulopiformes, Stomiiformes, Clupeiformes, Gonorynchiformes, Cypriniformes, Characiformes, Gymnotiformes and Siluriformes, 8 appear as monophyletic units in the "myological" cladogram of Fig. 122. These are the Acipenseriformes, Anguilliformes, Esociformes, Aulopiformes, Stomiiformes, Cypriniformes, Gymnotiformes, and Siluriformes. With respect to the "osteological" cladogram of Fig. 123, 11 of these 16 orders appear as monophyletic units: the Acipenseriformes, Anguilliformes, Argentiniformes, Esociformes, Salmoniformes, Aulopiformes, Clupeiformes, Cypriniformes, Characiformes, Gymnotiformes, and Siluriformes.

The examination of the data discussed in the above paragraphs thus seems to support the hypothesis of Diogo (2004b). That is, in a higher level, within the 16 major clades obtained in the cladistic analysis of Chapter 3 (Figs. 3, 4), 6 appear as monophyletic in the "myological cladogram", while only 2 appear as monophyletic in the "osteological" cladogram. In contrast, in a lower level, within those 16 orders included in the cladistic analysis of Chapter 3 that were represented by more than one terminal taxon in that analysis, 11 appear as monophyletic in the "osteological" cladogram, while only 8 appear as monophyletic in the "myological" cladogram. An interesting case is that of the salmoniforms, which do not appear grouped in a monophyletic unit in the cladistic analysis of Chapter 3 including both osteological and myological characters (Figs. 3, 4) nor in the "myological" cladogram of Fig. 122, but appear grouped together in the "osteological" cladogram of Fig. 123. That is, in this specific case the myological characters included in the cladistic analysis of Chapter 3 have seemingly "corrupted" the phylogenetic signal provided by the osteological characters included in that analysis (if one considers that the salmoniforms are indeed monophyletic, as is usually accepted: see Chapter 3). However, one should also keep in mind that there is also a case in which one order, namely the Stomiiformes, appears as a monophyletic group in the "myological" cladogram of Fig. 122 but not in the "osteological" cladogram of Fig. 123. That is, in this particular case the support for a sister-group relationship between the stomiiform taxa *Astronesthes* and *Stomias* obtained in the cladistic analysis of Chapter 3 (see Figs. 3, 4) comes mainly from the myological characters included in that analysis. It should also be noted there are four orders that appear as monophyletic units in the cladogram obtained in the cladistic analysis of Chapter 3 (Figs. 3, 4) and that do not appear as monophyletic in the "osteological" cladogram of Fig. 123 or in the "myological" cladogram of Fig. 122: the Elopiformes, Osteoglossiformes, Osmeriformes, and Gonorynchiformes. Thus, all things being counted, it seems that the best option is to use the complementary data provided by both osteological and myological structures, as stressed above.

In summary, the overall examination of the data provided by Diogo (2004b) and by the present work indicates that osteological structures seemingly display a greater morphological variation than myological ones, providing more characters for phylogenetic analyses (this being very likely over-enhanced by a historical bias: see above). But this difference refers particularly to lower clades; myological characters are seemingly highly useful in providing informative characters to disclose the relationships between higher clades and, thus, to help in inferring the higher-level phylogeny of the groups being studied. As stressed by Diogo (2004b), other authors have defended the reliability of muscular characters in revealing the higher-level phylogeny of major groups such as birds (McKitrick, 1991), hadrosaurian dinosaurs (Dilkes, 2000), or primates (Shoshani et al., 1996; Collard and Wood, 2000; Gibbs et al., 2000, 2002). For instance, contrary to works mainly based on osteological characters, phylogenetic analyses of the primate superfamily Hominoidea mainly based on myological data support the now commonly accepted molecular cladograms indicating a sister-group relationship between the hominin tribes Panini and Hominini and between the clade formed by these two tribes and the subfamily Gorilliinae (see Gibbs et al., 2002).

The points discussed above thus raise interesting questions. For instance, what are the reasons for the seemingly greater morphological variation of osteological structures? Does natural selection eventually act, in certain cases, more on bones than on muscles? Is the development of myological structures more constrained than that of osteological features? Why do muscular structures seem to be particularly reliable in revealing the higher-level phylogeny of major clades? According to Gibbs et al. (2002), for example, one of the reasons for the apparently high reliability of muscular characters in revealing the phylogeny of higher taxa may be related to the results of the experiments that used rhombomeric quail-to-chick grafts to investigate the influence of hindbrain segmentation on craniofacial patterning (Köntges and Lumsden, 1996). Those experiments showed that each rhombomeric population remains coherent throughout ontogeny, with rhombomere-specific matching of muscle connective tissue and their attachment sites for all branchial and tongue muscles. As hypothesized by Gibbs et al. (2002: 13), if "a similar system operates elsewhere in the body, it would help explain how muscle gross morphology is conserved, whereas the shapes of the skeletal elements to which the muscles are attached are susceptible to changes that contrive to obscure phylogeny". In fact, at least with respect to teleost fishes, the principal points of muscular origin and insertion seem indeed to be rather stable, as pointed out by, for example, Takahasi (1925), Alexander (1965), and Diogo (2004a). Another contributory factor suggested by Gibbs et al. (2002: 13) to explain the seemingly high reliability of muscular structures in phylogenetic reconstructions of higher taxa is that these structures "are not as prone to homolology as skeletal characters". The term *homolology* refers to shared character states that are phylogenetically misleading and that result from similarities in the way that genotypes interact with the environment. Thus, because bone is a dynamic tissue, "many osseous morphologies would be homolologous", with "homolology playing a minor role in the generation of the phenotypes" referring to muscular and soft tissues, since "whereas the mass of a muscle may be affected by activity or inactivity, its attachments are unlikely to be" (Gibbs et al., 2002: 13). However, homolology cannot be the whole explanation for the difference in phylogenetic reliability between osteological and myological structures, since dental enamel, for example, does not remodel and, thus, is not prone to homolology (Gibbs et al., 2002). But some authors have suggested that function or developmental constraints may make tooth morphology particularly prone to homoplasy and, therefore, a poor guide to phylogenetic reconstructions (Hartman, 1988; Hunter and Jernvall, 1995; Jernvall, 2000).

More direct comparisons of the homoplasy levels and the reliability of muscular and osteological data for phylogenetic reconstructions, such as those provided by Diogo (2004b) and in the present work, are needed to infer whether the patterns found in these two studies are supported by other works and, thus, to help clarify some of the general questions formulated above. Anyway, these two studies stress that the inclusion of muscular characters, and particularly the conjugation of the complementary information provided by these and other types of morphological characters, allows more comprehensive and, very likely, more resolved phylogenetic reconstructions. Also, the inclusion of myological data on phylogenetic and evolutionary analyses allows a more comprehensive, integrative discussion on the functional aspects and the evolution of the structural complexes being examined and, thus, on the origin and macroevolution of the higher-clades being studied. As explained above, many of the synapomorphies characterizing higher clades such as the Actinopterygii and the Sarcopterygii, among others, refer to myological features. The analysis of these features may, thus, play a crucial role in the understanding of the origin and macroevolution of these and other higher clades. Therefore, despite the intrinsic difficulties involved, such as those related with museum availability or the arduousness of muscular dissections, it is hoped that the present volume will stimulate zoologists to take into account myological features in their phylogenetic reconstructions and in their evolutionary analyses.

References

Abdala V, Moro S. 2003. A cladistic analysis of ten lizard families (Reptilia: Squamata) based on cranial musculature. Russ J Herpetol 10:53-78.

Abdala V, Moro S. 2006. Comparative myology of the forelimb of *Liolaemus* sand lizards (Liolaemidae). Acta Zool (Stockh) 87:1-12.

Adamicka P, Ahnelt H. 1992. Two jaw articulations in *Latimeria chalumnae* (Actinistia, Coelacanthidae). Zool Jb Anat 122:107-112.

Adriaens D, Verraes W. 1996. Ontogeny of cranial musculature in *Clarias gariepinus* (Siluroidei: Clariidae): the adductor mandibulae complex. J Morphol 229:255-269.

Adriaens D, Decleyre D, Verraes W. 1993. Morphology of the pectoral girdle in *Pomatoschistus lozanoi* De Buen, 1923 (Gobiidae), in relation to pectoral fin adduction. Belg J Zool 123:135-157.

Aguilera O. 1988. La musculatura estriada en los peces Gymnotiformes (Teleostei-Ostariophysi): musculatura facial. Acta Biol Venez 12:13-23.

Ahlberg PE, Clack JA. 1998. Lower jaws, lower tetrapods—review based on the Devonian genus *Acanthostega*. Trans R Soc Edin (Earth Sci) 89:11-46.

Ahlberg PE, Clack JA. 2006. A firm step from water to land. Nature 440:747-749.

Albert JS. 2001. Species diversity and phylogenetic relationships of American knifefishes (Gymnotiformes, Teleostei). Misc Publ Mus Zool Univ Mich 190:1-127.

Albert JS, Campoz-da-Paz R. 1998. Phylogenetic systematics of Gymnotiformes with diagnoses of 58 clades: a review of available data. In: Malabarba LR, Reis RE, Vari RP, Lucena ZM, Lucena CAS (eds.). Phylogeny and Classification of Neotropical Fishes. Edipucrs, Porto Alegre, pp. 410-446.

Alexander R McN. 1961a. The physical properties of the swimbladders of some South American Cypriniformes. J Exp Biol 38:403-410.

Alexander R McN. 1961b. The structure of the Weberian apparatus in the Cyprini. Proc Zool Soc Lond 139:451-473.

Alexander R McN. 1964a. The evolution of the Weberian apparatus in the Cobitidae. Proc Zool Soc Lond 143:177-190.

Alexander R McN. 1964b. The structure of the Weberian apparatus in the Siluri. Proc Zool Soc Lond 142:419-440.

Alexander R McN. 1965. Structure and function in catfish. J Zool (Lond) 148:88-152.

Alexander R McN. 1973. Jaw mechanisms of the coelacanth *Latimeria*. Copeia 1973:156-158.

Allis EP. 1897. The cranial muscles and cranial nerves of *Amia calva*. J Morphol 12:487-737.

Allis EP. 1922. The cranial anatomy of *Polypterus*, with special reference to *Polypterus bichir*. J Anat 56:180-294.

Alves-Gomez JA, Orti G, Haygood M, Heiligenberg W, Meyer A. 1995. Phylogenetic analysis of the South American electric fishes and the evolution of their electrogenic system: a synthesis based on morphology, electrophysiology and mitochondrial data. Mol Biol Evol 12:298-318.

Andrews SM. 1977. The axial skeleton of the coelacanth, *Latimeria*. In: Andrews SM, Miles RS, Walker AD (eds.). Problems in Vertebrate Evolution. Academic Press, London, pp. 271-288.

Anthony J. 1980. Évocation des travaux français sur *Latimeria* notamment depuis 1972. Proc R Soc Lond B 208:349-367.

Arnason U, Gullberg A, Janke A. 2001. Molecular phylogenetics of gnathostome (jawed) fishes: old bones, new cartilage. Zool Scr 30:249-249.

Arnason U, Gullberg A, Janke A, Joss J, Elmerot C. 2004. Mitogenomic analyses of deep gnathostome divergences: a fish is a fish. Gene 333: 61-70.

Arratia G. 1987. Description of the primitive family Diplomystidae (Siluriformes, Teleostei, Pisces): morphology, taxonomy and phylogenetic implications. Bonn zool Monogr 24:1-120.

Arratia G. 1992. Development and variation of the suspensorium of primitive catfishes (Teleostei: Ostariophysi) and their phylogenetic relationships. Bonn zool Monogr 32:1-148.

Arratia G. 1997. Basal teleosts and teleostean phylogeny. Palaeo Ichthyologica 7:5-168.

Arratia G. 1999. The monophyly of Teleostei and stem-group teleosts: consensus and disagreements. In: Arratia G, Schultze H-P (eds.). Mesozoic Fishes 2: Systematic and Fossil Record. Verlag Pfeil, Munich, pp. 265-334.

Arratia G. 2000. Phylogenetic relationships of Teleostei—past and present. Estud Oceanol 19:19-51.

Arratia G, Schultze H-P. 1990. The urohyal: development and homology within osteichthyans. J Morphol 203:247-282.

Arratia G, Schultze H-P. 1991. Development and homology of the palatoquadrate within osteichthyans. J Morphol 208:1-8.

Baldwin CC, Johnson GD. 1996. Interrelationships of Aulopiformes. In: Stiassny MLJ, Parenti LR, Johnson GD (eds.). The Interrelationships of Fishes. Academic Press, San Diego, pp. 355-404.

Bamford TW. 1948. Cranial development of *Galeichthys felis*. Proc Zool Soc Lond 118:364-391.

Bartsch P. 1992. On the constructional anatomy of the jaw suspension and the cranial base in the larva of *Neoceratodus forsteri* (Krefft, 1870). Zool Jb Anat 122:113-127.

Bartsch P. 1993. Development of the snout in the Australian lungfish *Neoceratodus forsteri* (Krefft, 1870), with special reference to cranial nerves. Acta Zool (Stockh) 74:15-29.

Bartsch P. 1994. Development of the cranium of *Neoceratodus forsteri*, with a discussion of the suspensorium and the opercular apparatus in Dipnoi. Zoomorphology 114:1-31.

Bartsch P, Gemballa S. 1992. On the anatomy and development of the vertebral column and pterygiophores in *Polypterus senegalus* Cuvier, 1829 ("Pisces", Polypteriformes). Zool Jb Anat Ontog Tiere 122:497-529.

Bauchot M-L. 1987. Poissons osseux. In: Fischer W, Bauchot ML, Schneider M (eds.). Fiches FAO d'identification pour les besoins de la pêche. (rev. 1)—Méditerranée et mer Noire—Zone de pêche 37—Vol. II. Commission des Communautés Européennes and FAO, Rome, pp. 891-1421.

Bauer WJ. 1992. A contribution to the morphology of the M. interhyoideus posterior (VII) of Urodele Amphibia. Zool Jb Anat 122:129-139.

Bauer WJ. 1997. A contribution to the morphology of visceral jaw-opening muscles of urodeles (Amphibia: Caudata). J Morphol 233:77- 97.

Begle DP. 1991. Relationships of the osmeroid fishes and the use of reductive characters in phylogenetic analysis. Syst Zool 40:33-53.

Begle DP. 1992. Monophyly and relationships of the argentinoid fishes. Copeia 1992:350-366.

Belouze A. 2002. Compréhension morphologique et phylogénétique des taxon actuels et fossiles rapportés aux Anguilliformes ("poissons" téléostéens). Doc Lab Geo Lyon 158:1-401.

Bels VL, Goosse V, Kardong KV. 1993. Kinematic analysis of drinking by the lacertid lizard, *Lacerta viridis* (Squamates, Scleroglossa). J Zool Lond 229:659-682.

Bemis WE. 1986. Feeding mechanisms of living Dipnoi: anatomy and function. J Morphol, suppl 1:249-275.

Bemis WE, Lauder CV. 1986. Morphology and function of the feeding apparatus of the lungfish, *Lepidosiren paradoxa* (Dipnoi). J Morphol 187:81-108.

Bemis WE, Northcutt RG. 1991. Innervation of the basicranial muscle of *Latimeria chalumnae*. Environ Biol Fish 32:147-158.

Bemis WE, Schwenk K, Wake MH. 1983. Morphology and function of the feeding apparatus in *Dermophis mexicanus*. Zool J Linn Soc Lond 77:75-96.

Bemis WE, Findeis EK, Grande L. 1997. An overview of Acipenseriformes. Environ Biol Fish 48:25-71.

Benton MJ. 1985. Classification and phylogeny of the diapsid reptiles. Zool J Linn Soc Lond 84: 97-164.

Bernstein P. 2003. The ear region of *Latimeria chalumnae*: functional and evolutionary implications. Zoology 106:223-242.

Bischoff TLW. 1840 Description anatomique du *Lepidosiren paradoxa*. Ann Sci Nat 14, Ser 2:116-159.

Bishai RM. 1967. Cranial muscles of *Mormyrus caschive* (L.). Anat Anz Bd 121:12-25.

Bjerring HC. 1973. Relationships of coelacanthiforms. In: Greenwood PH, Miles RS, Patterson C (eds.). Interrelationships of Fishes. Zool J Linn Soc Lond 53, suppl 1, pp. 179-205.

Bjerring HC. 1993. Yet another interpretation of the coelacanthiform basicranial muscle and its innervation. Acta Zool (Stockh) 74:289-299.

Blagden CD, Currie PD, Infham PW, Hughes SM. 1997. Notochord induction of zebrafish slow muscle mediated by Sonic hedgehog. Genes & Dev 11:2163-2175.

Bogutskaia HG. 1991. Development of the Weberian apparatus during ontogeny of some species of Cyprinidae. J Ichthyol 31:363-372.

Bonebrake JE, Brandon RA. 1971. Ontogeny of cranial ossification in the small-mouthed salamander, *Ambystoma texanum*. J Morphol 133:189-204.

Borden WC. 1998. Phylogeny of the unicornfishes (*Naso*, Acanthuridae) based on soft anatomy. Copeia 1998:104-113.

Borden WC. 1999. Comparative myology of the unicornfishes, *Naso* (Acanthuridae, Percomorpha), with implications for phylogenetic analysis. J Morphol 239:191-224.

Bornbusch AH. 1991. Monophyly of the catfish family Siluridae (Teleostei: Siluriformes), with a critique of previous hypotheses of the family's relationships. Zool J Linn Soc Lond 101:105-120.

Bornbusch AH. 1995. Phylogenetic relationships within the Eurasian catfish family Siluridae (Pisces: Siluriformes), with comments on generic validities and biogeography. Zool J Linn Soc Lond 115:1-46.

Brazeau MD, Ahlberg PE. 2006. Tetrapod-like middle ear architecture in a Devonian fish. Nature 439:318-421.

Bridge TW. 1878. On the osteology of *Polyodon folium*. Philos Trans R Soc Lond 169:683-733.

Bridge TW, Haddon AC. 1893. Contributions to the anatomy of fishes, II—the air-bladder and Weberian ossicles in the siluroid fishes. Philos Trans R Soc Lond B 84:65-333.

Bridge TW, Haddon AC. 1894. Notes on the production of sounds by the air bladder of certain siluroid fishes. Proc R Soc Lond 55:439-441.

Brinkmann H, Venkatesh B, Brenner S, Meyer A. 2004. Nuclear protein-coding genes support lungfish and not the coelacanth as the closest living relatives of land vertebrates. Proc Natl Acad Sci US 101:4900-4905.

Brock GT. 1938. The cranial muscles of the *Gecko*—a general account with a comparison of muscles in other gnathostomes. Proc Zool Soc Lond B 108:735-761.

Brosseau AR. 1978a. The pectoral anatomy of selected Ostariophysi, II—the Characiformes. J Morphol 148:89-136.

Brosseau AR. 1978b. The pectoral anatomy of selected Ostariophysi, I—the Cypriniformes and Siluriformes. J Morphol 150:79-115.

Buckup PA. 1998. Relationships of the Characidiinae and phylogeny of characiform fishes (Teleostei: Ostariophysi). In: Malabarba LR, Reis RE, Vari RP, Lucena ZM, Lucena CAS (eds.). Phylogeny and Classification of Neotropical Fishes. Edipucrs, Porto Alegre, pp. 123-144.

Butler J. 1960. Development of the Weberian apparatus of catostomid fish. Proc Iowa Acad Sci 67:532-543.

Campbell KSW, Barwick RE. 1986. Paleozoic lungfishes—a review. J Morphol, suppl 1:93-131.

Carroll AM, Wainwright PC. 2003. Functional morphology of feeding in the sturgeon, *Scaphirhyncus albus*. J Morphol 256:270-284.

Carroll RL. 1977. The origin of lizards. In: Andrews SM, Miles RS, Walker AD (eds.). Problems in Vertebrate Evolution. Academic Press, London, pp. 359-396.

Carroll RL, Holmes R. 1980. The skull and jaw musculature as guides to the ancestry of salamanders. Zool J Linn Soc Lond 68:1-40.

Carroll SB. 2005. The New Science of Evo-Devo: Endless Forms Most Beautiful. WW Norton & Company, New York.

Carroll SB, Grenier JK, Weathrbee SD. 2005. From DNA to Diversity—Molecular Genetics and the Evolution of Animal Design (2nd ed.). Blackwell Science, Malden.

Cavin L. 1999. A new Clupavidae (Teleostei, Ostariophysi) from the Cenomanian of Daoura (Morocco). C R Acad Sci Paris 329:689-695.

Cavin L, Suteethorn V, Buffetaut E, Tong H. In press. A new Thai Mesozoic lungfish (Sarcopterygii, Dipnoi) with an insight into post-Palaeozoic dipnoan evolution. Zool J Linn Soc Lond.

Chang M-M, Maisey JG. 2003. Redescription of †*Ellimma branneri* and †*Diplomystus shengliensis*, and relationships of some basal clupeomorphs. Am Mus Novit 3404:1-35.

Chapman WM. 1934. The osteology of the haplimous fish *Novumbra hubbsi* Schultz with comparative notes on related species. J Morphol 56:371-405.

Chapman WM. 1941. The osteology and relationships of the isospondylous fish *Plecoglossus altivelis* Temminck and Schegel. J Morphol 68:425-455.

Chapman WM. 1942. The osteology and relationships of the Argentinidae, a family of oceanic fishes. J Wash Acad Sci 32:104-117.

Chardon M. 1968. Anatomie comparée de l'appareil de Weber et des structures connexes chez les Siluriformes. Ann Mus R Afr Centr 169:1-273.

Chardon M, De La Hoz E. 1973. Notes sur le squelette, les muscles, les tendons et le cerveau des Gymnotoidei. Ann Soc Nat Zool Paris 12, Ser 15:1-10.

Chardon M, Vandewalle P. 1997. Evolutionary trends and possible origin of the Weberian apparatus. Neth J Zool 47:383-403.

Chardon M, Parmentier E, Vandewalle P. 2003. Morphology, development and evolution of the Weberian apparatus in catfish. In: Arratia G, Kapoor BG, Chardon M, Diogo R (eds.). Catfishes. Science Publishers, Enfield, pp. 71-120.

Chranilov NS. 1927. Beiträge zur Kenntniss der Weber'schen Apparates bei Cypriniformes. Zool Jb Anat 49:501-597.

Chranilov NS. 1929. Beiträge zur kenntnis der Weber'schen apparates der Ostariophysi, 2—der Weber'schen apparat bei Siluroidea. Zool Jahrb Anat 51:323-462.

Clack JA. 1998. The neurocranium of *Acanthostega gunnari* Jarvik and the evolution of the otic region in tetrapods. Zool J Linn Soc Lond 122:61-97.

Clack JA. 2001. The occipital region: origin, ontogeny and the fish-tetrapod transition. In: Ahlberg PE (ed.). Major Events in Early Vertebrate Evolution, Paleontology, Phylogeny, Genetics and Development. Taylor and Francis, London, pp. 392-405.

Clack JA. 2002. Gaining Ground—the Origin and Evolution of Tetrapods. Indiana University Press, Bloomington.

Clack JA. 2006. The emergence of early tetrapods. Palaeogeogr Palaeoclimatol Palaeoecol 232 (2006) 167-189.

Clack JA, Coates M. 1995. *Acanthostega gunnari*, a primitive, aquatic tetrapod? Bull Mus Natl Hist Nat Paris 17:359-372.

Clement G. 2005. A new coelacanth (Actinistia, Sarcopterygii) from the Jurassic of France, and the question of the closest relative to *Latimeria*. J Vertebr Paleont 25:481-491.

Cloutier R. 1991. Patterns, trends, and rates of evolution of the Actinistia. In: Musick JA, Bruton MN, Balon EK (eds.). The Biology of *Latimeria chalumnae* and Evolution of Coelacanths. Environ Biol Fish 32, pp. 23-58

Cloutier R, Forey PL. 1991. Diversity of extinct and living actinistian fishes (Sarcopterygii). In: Musick JA, Bruton MN, Balon EK (eds.). The Biology of *Latimeria chalumnae* and Evolution of Coelacanths. Environ Biol Fish 32, pp. 59-74.

Cloutier R, Ahlberg PE. 1996. Morphology, characters, and the interrelationships of basal sarcopterygians. In: Stiassny MLJ, Parenti LR, Johnson GD (eds.). The Interrelationships of Fishes. Academic Press, San Diego, pp. 425-426.

Cloutier R, Arratia G. 2004. Early diversification of actinopterygians. In: Arratia G, Tintori A (eds.). Mesozoic Fishes 3—Systematics, Paleonvironments and Biodiversity. Verlag Pfeil, Munich, pp. 217-270.

Coates MI. 1994. The origin of vertebrate limbs. Development, suppl 1:169-180.

Coates MI. 1996. The Devonian tetrapod *Acanthostega gunnari* Jarvik: postcranial anatomy, basal tetrapod interrelationships and patterns of skeletal evolution. Trans R Soc Edin (Earth Sci) 87: 363-422.

Coates MI, Cohn MJ. 1999. Vertebrate axial and appendicular patterning: the early development of paired appendages. Am Zool 39:676-685.

Coates MI, Jeffery JE, Ruta M. 2002. Fins to limbs: what the fossils say. Evol & Dev 4:390-401.

Coburn MM, Futey LM. 1996. The ontogeny of supraneurals and neural arches in the cypriniform Weberian apparatus (Teleostei, Ostariophysi). Zool J Linn Soc Lond 116:333-346.

Coburn MM, Grubach PG. 1998. Ontogeny of the armored catfish *Corydoras paleatus* (Siluriformes: Callichthyidae). Copeia 1998:301-311.

Coburn MM, Chai P. 2003. The development of the anterior vertebrae in *Chanos chanos* (Ostariophysi: Gonorynchiformes). Copeia 2003:175-180.

Cohn MJ, Lovejoy CO, Wolpert L, Coates MI. 2002. Branching, segmentation and the metapterygial axis: pattern versus process in the vertebrate limb. Bioessays 24:460-465.

Collard M, Wood B. 2000. How reliable are human phylogenetic hypotheses? Proc Natl Acad Sci US 97:5003-5006.

Conant R. 1975. A Field Guide to Reptiles and Amphibians of Eastern and Central North America. Houghton Mifflin Company, Boston.

Couly GF, Coltey PM, Le Douarin NM. 1992. The developmental fate of the cephalic mesoderm in quail-chick chimeras. Development 114:1-15.

Cubbage CC, Mabee PM. 1996. Development of the cranium and paired fins in the zebrafish *Danio rerio* (Ostariophysi, Cyprinidae). J Morphol 229:121-160.

Daeschler EB, Shubin NH, Jenkins FA. 2006. A Devonian tetrapod-like fish and the evolution of the tetrapod body plan. Nature 440:757-763.

Daget J. 1950. Revision des afinités phylogénétiques des Polypterides. Mem Inst Fr Afr Noire 11:1-178.

Danforth CH. 1913. The myology of *Polyodon*. J Morphol 24:107-146.

D'Aubenton F. 1961. Morphologie du crane de *Cromeria nilotica occidentalis* Daget 1954. Bull Inst Fr Afr Noire A 23:187-249.

Davis MC, Shubin NH, Force A. 2004. Pectoral fin and girdle development in the basal actinopterygians *Polyodon spathula* and *Acipenser transmontanus*. J Morphol 262:606-628.

Dawkins R. 2004. The Ancestor's Tale—a Pilgrimage to the Dawn of Life. Weidenfeld & Nicolson, London.

De la Hoz E. 1974. Definition et classification des poissons Gymnotoidei sur la base de la morphologie comparée et fonctionnelle du squelette et des muscles. Unpublished PhD thesis. University of Liège.

De la Hoz E, Chardon M. 1975. Le squelette de la région ethmoïdienne chez les Gymnotoidei (Teleostei Ostariophysi). Ann Sci Nat Zool Biol Anim Paris 17, Ser 12:121-140.

De la Hoz E, Chardon M. 1984. Skeleton, muscles, ligaments and swim-bladder of a gymnotid fish, *Sternopygus macrurus* Bloch & Schneider (Ostariophysi: Gymnotoidei). Bull Soc R Sci Liège 53:9-53.

De Pinna MCC. 1991. Concepts and tests of homology in the cladistic paradigm. Cladistics 7:367-394.

De Pinna MCC. 1993. Higher-level phylogeny of Siluriformes, with a new classification of the order (Teleostei, Ostariophysi). Unpublished PhD thesis. City University of New York.

De Pinna MCC. 1996. Teleostean monophyly. In: Stiassny MLJ, Parenti LR, Johnson GD (eds.). The Interrelationships of Fishes. Academic Press, San Diego, pp. 147-162.

De Pinna MCC. 1998. Phylogenetic relationships of Neotropical Siluriformes (Teleostei: Ostariophysi): historical overview and synthesis of hypotheses. In: Malabarba LR, Reis RE, Vari RP, Lucena ZM, Lucena CAS (eds.). Phylogeny and Classification of Neotropical Fishes. Edipucrs, Porto Alegre, pp. 279-330.

De Pinna MCC, Vari RP. 1995. Monophyly and phylogenetic diagnosis of the family Cetopsidae, with synonymization of the Helogenidae (Teleostei: Siluriformes). Smiths Contrib Zool 571:1-26.

De Pinna MCC, Grande T. 2003. Ontogeny of the accessory neural arch in pristigasteroid clupeomorphs and its bearing on the homology of the otophysan claustrum (Teleostei). Copeia 2003:938-845.

Deufel A, Cundall D. 2003. Feeding in *Atractaspis* (Serpentes: Atractaspididae): a study in conflicting functional constraints. Zoology 106:43-61.

Di Dario F. 2002. Evidence supporting a sister-group relationship between Clupeoidea and Engrauloidea (Clupeomorpha). Copeia 2002:496-503.

Di Dario F. 2004. Homology between the recessus lateralis and cephalic sensory canals, with the proposition of additional synapomorphies for the Clupeiformes and the Clupeoidei. Zool J Linn Soc Lond 141:257-270.

Dilkes DW. 2000. Appendicular myology of the hadrosaurian dinosaur *Maiasaura peeblesorum* from the Late Cretaceous (Campanian) of Montana. Trans R Soc Edin (Earth Sci) 90:87-125.

Dimmick WW, Larson A. 1996. A molecular and morphological perspective on the phylogenetic relationships of the othophysan fishes. Mol Phylogenet Evol 6:120-133.

Diogo R. 2004a. Morphological Evolution, Aptations, Homoplasies, Constraints and Evolutionary Trends: Catfishes as a Case Study on General Phylogeny and Macroevolution. Science Publishers, Enfield.

Diogo R. 2004b. Muscles versus bones: catfishes as a case study for an analysis on the contribution of myological and osteological structures in phylogenetic reconstructions. Anim Biol 54:373-391

Diogo R. 2005. Evolutionary convergences and parallelisms: their theoretical differences and the difficulty of discriminating them in a practical phylogenetic context. Biol Philos 20:735-744.

Diogo R, Chardon M. 2000. Homologies between different adductor mandibulae sections of teleostean fishes, with a special regard to catfishes (Teleostei: Siluriformes). J Morphol 243:193-208.

Diogo R, Chardon M. 2001. Adaptive transformation of the palatine-maxillary system in catfish: increased mobility of the maxillary barbel. In: Kapoor BG, Hara TJ (eds.). Sensory Biology of Jawed Fishes—New Insights. Science Publishers, Enfield, pp. 367-383.

Diogo R, Chardon M. 2003. Homologies and evolutionary transformation of the skeletal elements of catfish (Teleostei: Siluriformes) suspensorium: a morphofunctional hypothesis. In: Val AL, Kapoor BG (eds.). Fish Adaptations. Science Publishers, Enfield, pp. 275-284.

Diogo R, Vandewalle P. 2003. Review of superficial cranial musculature of catfishes, with comments on plesiomorphic states. In: Arratia G, Kapoor BG, Chardon M, Diogo R (eds.). Catfishes. Science Publishers, Enfield, pp. 47-69.

Diogo R, Oliveira C, Chardon M. 2000. The origin and transformation of catfish palatine-maxillary system: an example of adaptive macroevolution. Neth J Zool 50:373-388.

Diogo R, Oliveira C, Chardon M. 2001a. On the osteology and myology of catfish pectoral girdle, with a reflection on catfish (Teleostei: Siluriformes) plesiomorphies. J Morphol 249:100-125.

Diogo R, Oliveira C, Chardon M. 2001b. On the homologies of the skeletal components of catfish (Teleostei: Siluriformes) suspensorium. Belg J Zool 131:93-109.

Diogo R, Chardon M, Vandewalle P. 2003. Functional morphology of catfishes: movements of barbels. In: Arratia G, Kapoor BG, Chardon M, Diogo R (eds.). Catfishes. Science Publishers, Enfield, pp. 203-220.

Easter SS, Nicola GN. 1996. The development of vision in the zebrafish (*Danio rerio*). Dev Biol 180:646-663.

Eaton TH. 1948. Form and function in the head of the channel catfish, *Ictalurus lacustris punctatus*. J Morphol 83:181-194.

Edgeworth FH. 1902. The development of the head muscles in *Scyllium canicula*. J Anat Physiol 37:73-88.

Edgeworth FH. 1911. On the morphology of the cranial muscles in some vertebrates. Q J Micr Sci N S 56:167-316.

Edgeworth FH. 1923. On the development of the hypobranchial, branchial and laryngeal muscles of *Ceratodus*, with a note on the development of the quadrate and epihyal. Q J Micr Sci N S 67:325-368.

Edgeworth FH. 1926a. On the hyomandibula of Selachii, Teleostomi and *Ceratodus*. J Anat Physiol 60:173-193.

Edgeworth FH. 1926b. On the development of the coraco-branchialis and cucullaris in *Scyllium canicula*. J Anat Physiol 60:298-308.

Edgeworth FH. 1926c. On the development of the cranial muscles in *Protopterus* and *Lepidosiren*. Trans R Soc Edinb 54:719-734.

Edgeworth FH. 1928. The development of some of the cranial muscles of ganoid fishes. Philos Trans R Soc Lond (Biol) 217:39-89.

Edgeworth FH. 1935. The Cranial Muscles of Vertebrates. Cambridge University Press, Cambridge.

Elmerot C, Arnason U, Gojobori T, Janke A. 2002. The mitochondrial genome of the pufferfish, *Fugu rubripes*, and ordinal teleostean relationships. Gene 295:163-172.

Ericsson R, Olsson L. 2004. Patterns of spatial and temporal visceral arch muscle development in the Mexican axolotl (*Ambystoma mexicanum*). J Morphol 261:131-140.

Ericsson R, Cerny R, Falck P, Olsson L. 2004. Role of cranial neural crest cells in visceral arch muscle positioning and morphogenesis in the Mexican axolotl (*Ambystoma mexicanum*). Dev Dynam 231:237-247.

Farris JS. 1988. Hennig 86, version 1.5. Distributed by the author. Port Jefferson Station, New York.

Filleul A. 2000. Analyse critique des synapomorphies des Élopomorphes et analyse phylogénétique du groupe. Cybium 24, suppl:75-83.

Filleul A, Lavoué S. 2001. Basal teleosts and the question of elopomorph monophyly: morphological and molecular approaches. C R Acad Sci Paris 324:393-399.

Filleul A, Maisey JG. 2004. Redescription of *Santanichthys diasii* (Otophysi, Characiformes) from the Albian of the Santana formation and comments on its implications for Otophysan relationships. Am Mus Novit 3455:1-21.

Findeis EK. 1997. Osteology and phylogenetic relationships of recent sturgeons. Environ Biol Fish 48:73-126.

Fink SV, Fink WL. 1981. Interrelationships of the ostariophysan fishes. Zool J Linn Soc Lond 72:297-353.

Fink SV, Fink WL. 1996. Interrelationships of ostariophysan fishes (Teleostei). In: Stiassny MLJ, Parenti LR, Johnson GD (eds.). The Interrelationships of Fishes. Academic Press, San Diego, pp. 209-249.

Fink SV, Greenwood PH, Fink W. 1984. A critique of recent works on fossil ostariophysan fishes. Copeia 4:1033-1041.

Fink WL. 1984. Basal euteleosts: relationships. In: Moser HG, Richards WJ, Cohen DM, Fahay MP, Kendall AW, Richardson SL (eds.). Ontogeny and Systematics of Fishes. American Society of Ichthyologists and Herpetologists, Lawrence, Special Publication number 1, pp. 202-206.

Fink WL. 1985. Phylogenetic interrelationships of the stomiid fishes (Teleostei: Stomiiformes). Misc Publ Mus Zool Univ Mich 171:1-127.

Fink WL, Weitzman SH. 1982. Relationships of the stomiiform fishes (Teleostei), with a description of *Diplophos*. Bull Mus Comp Zool 150:3l-93.

Fink WL, Fink SV. 1986. A phylogenetic analysis of the genus *Stomias*, including the synonymization of *Macrostomias*. Copeia 1986:494-503.

Fitzhugh K. 2006. The philosophical basis of character coding for the inference of phylogenetic hypotheses. Zool Scr 35:261-286.

Forey PL. 1973a. A revision of the elopiform fishes, fossil and recent. Bull Br Mus Nat Hist (Geol), suppl, 10:1-222.

Forey PL. 1973b. Relationships of Elopomorpha. In: Greenwood PH, Miles RS, Patterson C (eds.). Interrelationships of Fishes. Zool J Linn Soc Lond 53, suppl 1, pp. 351-368.

Forey PL. 1980. *Latimeria*, a paradoxical fish. Proc R Soc Lond B 208: 369-384.

Forey PL. 1986. Relationships of lungfishes. J Morphol, suppl 1:75-91.

Forey PL. 1991. *Latimeria chalumnae* and its pedigree. In: Musick JA, Bruton MN, Balon EK (eds.). The biology of *Latimeria chalumnae* and evolution of coelacanths. Environ Biol Fish 32, pp. 75-97.

Forey PL. 1998. History of the Coelacanth Fishes. Chapman and Hall, London.

Forey PL, Littlewood DTJ, Ritchie P, Meyer A. 1996. Interrelationships of elopomorph fishes (Teleostei). In: Stiassny MLJ, Parenti LR, Johnson GD (eds.). The Interrelationships of Fishes. Academic Press, San Diego, pp. 175-191.

Fox H. 1963. The hyoid of *Neoceratodus* and a consideration of its homology in urodele Amphibia. Proc Zool Soc Lond 141:803-810.

Fox H. 1965. Early development of the head and pharynx of *Neoceratodus* with a consideration of its phylogeny. J Zool (Lond) 146:470–554.

Franz G. 1937. Über den Reflex des Gaspuckens bei Fischen und die Funktion des Weberschen Apparates. Z Vergl Physiol 25:193-200.

Frazzetta TH. 1962. A functional consideration of cranial kinesis in lizards. J Morphol 111:287-320.

Fritzsch B. 1987. The inner ear of the coelacanth fish *Latimeria* has tetrapod affinities. Nature 327:153-154.

Fritzsch B. 2003. The ear of *Latimeria chalumnae* revisited. Zoology 106:243-248.

Fukushima M, Kohno H, Fujita K, Taki Y. 1992. Ontogenic development of the Weberian apparatus in the bitterling, *Rhodeus ocellatus ocellatus*. Tokyo Univ Fisheries 79:196-200.

Galis F. 1996. The application of functional morphology to evolutionary studies. Tree 11:124-129.

Gans C, Carrier D, De Vree F. 1985. Usage pattern of the complex masticatory muscles in the shingleback lizard, *Trachydosaurus rugosus*: a model for muscle placement. Am J Anat 173:219-240.

Gardiner BG. 1984. The relationships of the palaeoniscid fishes, a review based on new specimens of *Mimia* and *Moythomasia* from the Upper Devonian of Western Australia. Bull Br Mus Nat Hist (Geol) 37:173-428.

Gardiner BG, Schaeffer B. 1989. Interrelationships of lower actinopterygian fishes. Zool J Linn Soc Lond 97:135-187.

Gardiner BG, Maisey JG, Littlewood DTJ. 1996. Interrelationships of basal neopterygians. In: Stiassny MLJ, Parenti LR, Johnson GD (eds.). The Interrelationships of Fishes. Academic Press, San Diego, pp. 117-146.

Gardiner BG, Schaeffer B, Masserie JA. 2005. A review of the lower actinopterygian phylogeny. Zool J Linn Soc Lond 144:511-525.

Garvey JM, Johanson Z, Warren A. 2005. Redescription of the pectoral fin and vertebral column of the rhizodontid fish *Barameda decipiens* from the Lower Carboniferous of Australia. J Vertebr Paleontol 25:8-18.

Gasc J-P. 1968. Contribution à l'osteologie et à la myologie de Dibamus novaeguineae Gray (Sauria, Reptilia)—discussion systématique. Ann Sci Nat Zool 10:127-150.

Gayet M. 1981. Contribution à l'étude anatomique et systématique de l'ichthyofaune Cénomanienne du Portugal, deuxième partie: les ostariophysaires. Comun Serv Geol Portugal 67:173-190.

Gayet M. 1985. Contribution à l'étude anatomique et systématique de l'ichthyofaune Cénomanienne du Portugal, troisième partie: complèment à l'étude des ostariophysaires. Comun Serv Geol Portugal 71:91-118.

Gayet M. 1986a. About ostariophysan fishes: a reply to S.V. Fink, P.H. Greenwood and W.L. Fink's criticisms. Bull Mus Nat Hist, Sec C, 8:393-409.

Gayet M. 1986b. *Ramallichthys* Gayet du Cénomanien inférieur marin de Ramallah (Judée), une introduction aux relations phylogénétiques des Ostariophysi. Mém Mus Natl Hist Nat C 51:1-81.

Gayet M. 1993. Relations phylogénétiques des Gonorynchiformes (Ostariophysi). Belg J Zool 123:165-192.

Gayet M, Chardon M. 1987. Possible otophysic connections in some fossil and living ostariophysan fishes. Proc Vth Congr Eur Ichthyol Stockh:31-42.

Gayet M, Meunier FJ, Kirschbaum F. 1994. *Ellisella kirschbaumi* Gayet & Meunier, 1991, gymnotiforme fossile de Bolivie et ses relations phylogenetiques au sein des formes actuelles. Cybium 18:273-306.

Gegenbaur C. 1878. Ueber das Kopfskelet von *Alepocephalus rostratus* Risso. Morphol Jahrb 4:1-42.

George JC, Berger AJ. 1966. Avian Myology. New York: Academic Press.

Gibbs S, Collard M, Wood BA. 2000. Soft-tissue characters in higher primate phylogenetics. Proc Natl Acad Sci US 97:11130-11132.

Gibbs S, Collard M, Wood BA. 2002. Soft-tissue anatomy of the extant hominoids: a review and phylogenetic analysis. J Anat 200:3-49.

Goode GB (ed.). 1884-1887. The Fisheries and Fishery Industries of the United States, 7 volumes. Govt Print Off, Washington.

Gorniak GC. 1985. Trends in the action of mammalian masticatory muscles. Am Zool 25:331-337.

Gosline WA. 1969. The morphology and systematic position of the alepocephaloid fishes. Bull Br Mus Nat Hist (Zool) 18:183-218.

Gosline WA. 1973a. Functional Morphology and Classification of Teleostean Fishes. University Press of Hawaii, Honolulu.

Gosline WA. 1973b. Considerations regarding the phylogeny of cypriniform fishes, with special reference to structures associated with feeding. Copeia 1973:761-776.

Gosline WA. 1975. The palatine-maxillary mechanism in catfishes with comments on the evolution and zoogeography of modern siluroids. Occas Pap Calif Acad Sci 120:1-31.

Gosline WA. 1977. The structure and function of the dermal pectoral girdle in bony fishes with particular reference to ostariophysines. J Zool (Lond) 183:329-338.

Gosline WA. 1980. The evolution of some structural systems with reference to the interrelationships of modern lower teleostean fish groups. Jap J Ichthyol 27:1-27.

Gosline WA. 1986. Jaw musculature configuration in some higher teleostean fishes. Copeia 1986:705-713.

Gosline WA. 1989. Two patterns of differentiation in the jaw musculature of teleostean fishes. J Zool (Lond) 218:649-661.

Gould SJ. 1977. Ontogeny and Phylogeny. Harvard University Press, Cambridge.

Gould SJ. 2002. The Structure of Evolutionary Theory. Belknap, Harvard.

Gouréne G, Teugels GG. 1994. Synopsis de la classification et phylogénie des Pellonulinae de l'Afrique Occidentale et Centrale (Teleostei: Clupeidae). J Afr Zool 108:77-91.

Graham A. 2005. Vertebrate evolution: turning heads. Curr Biol 15:764-766.

Grande L. 1985a. Recent and fossil clupeomorph fishes with materials for revision of the subgroups of clupeoids. Bull Amer Mus Nat Hist 181:231-372.

Grande L. 1985b. A revision of the fossil genus †*Diplomystus*, with comments on the interrelationships of clupeomorph fishes. Am Mus Novit 2728:1-34.

Grande L. 2005. Phylogenetic study of gars and closely related species, based mostly on skeletal morphology—the resurrection of Holostei. In: Poyato-Ariza FJ (ed.). Abstracts of the Fourth International Meeting on

Mesozoic Fishes—Systematics, Homology, and Nomenclature. Ediciones Universidad Autónoma de Madrid, Madrid, pp. 119-121.

Grande L, Bemis WE. 1991. Osteology and phylogenetic relationships of fossil and recent paddlefishes (Polyodontidae) with comment on the interrelationships of Acipenseriformes. J Vertebr Paleontol, Special Memoir Number 1 (suppl to Vol 11):1-121.

Grande L, Bemis WE. 1996. Interrelationships of Acipenseriformes, with comments on "Chondrostei". In: Stiassny MLJ, Parenti LR, Johnson GD (eds.). The Interrelationships of Fishes. Academic Press, San Diego, pp. 85-115.

Grande L, Bemis WE. 1998. A comprehensive phylogenetic study of amiid fishes (Amiidae) based on comparative skeletal anatomy—an empirical search for interconnected patterns of natural history. J Vertebr Paleontol 18, suppl:1-681.

Grande T. 1994. Phylogeny and paedomorphosis in an African family of freshwater fishes (Gonorynchiformes: Kneriidae). Fieldiana (Zool) 78:1-20.

Grande T, Poyato-Ariza FJ. 1999. Phylogenetic relationships of fossil and recent gonorynchiform fishes (Teleostei: Ostariophysi). Zool J Linn Soc Lond 125:197-238.

Grande T, Braun C. 2002. Evolution of the Weberian apparatus. Bioacoustics 12:120-122.

Grande T, Shardo JD. 2002. Morphology and development of the postcranial skeleton in the channel catfish *Ictalurus punctatus* (Ostariophysi: Siluriformes). Fieldiana (Zool) 1518:1-30.

Grande T, De Pinna MCC. 2004. The evolution of the Weberian apparatus: a phylogenetic perspective. In: Arratia G, Tintori A (eds.). Mesozoic Fishes 3—Systematics, Paleonvironments and Biodiversity. Verlag Pfeil, Munich, pp. 429-448.

Grande T, Young B. 2004. The ontogeny and homology of the Weberian apparatus in the zebrafish *Danio rerio* (Ostariophysi: Cypriniformes). Zool J Linn Soc Lond 140:241-254.

Grande T, Laten H, Lopez JA. 2004. Phylogenetic relationships of extant esocid species (Teleostei: Salmoniformes) based on morphological and molecular characters. Copeia 2004:743-757.

Greenwood PH. 1968. The osteology and relationships of the Denticipitidae, a family of clupeomorph fishes. Bull Br Mus Nat Hist (Zool) 16:215-273.

Greenwood PH. 1970. Skull and swimbladder connections in fishes of the family Megalopidae. Bull Br Mus Nat Hist (Zool) 19:119-135.

Greenwood PH. 1971. Hyoid and ventral gill arch musculature in osteoglossomorph fishes. Bull Br Mus Nat Hist (Zool) 22: 1-55.

Greenwood PH. 1973. Interrelationships of osteoglossomorphs. In: Greenwood PH, Miles RS, Patterson C (eds.). Interrelationships of Fishes. Zool J Linn Soc Lond 53, suppl 1, pp. 307-332.

Greenwood PH. 1977. Notes on the anatomy and classification of elopomorph fishes. Bull Br Mus Nat Hist (Zool) 32:65-103.

Greenwood PH. 1986. The natural history of African lungfishes. J Morphol, suppl 1:163-179.

Greenwood PH, Thomson KS. 1960. The pectoral anatomy of *Pantodon buchholzi* Peters (a freshwater flying fish) and the related Osteoglossidae. Proc Zool Soc Lond 135:283-301.

Greenwood PH, Rosen DE. 1971. Notes on the structure and relationships of the alepocephaloid fishes. Am Mus Novit 2473: 1-41.

Greenwood PH, Lauder GV. 1981. The protractor pectoralis muscle and the classification of teleost fishes. Bull Br Mus Nat Hist (Zool) 41:213-234.

Greenwood PH, Rosen DE, Weitzman SH, Meyers GS. 1966. Phyletic studies of teleostean fishes, with a provisional classification of living forms. Bull Am Mus Nat Hist 131:339-456.

Günther K, Deckert K. 1959. Morphologie und Funktion des Kiefer- und Kiemenapparates von Tiefseefischen der Gattungen *Malacosteus* und *Photostomias* (Teleostei, Isospondyli, Stomiatoidea, Malacosteidae). Dana-Rep (Copenhagen) 49:1-54.

Haas A. 2001. The mandibular arch musculature of anuran tadpoles with comments on the homologies of amphibian jaw muscles. J Morphol 247: 1-33.

Harold AS, Weitzman SH. 1996. Interrelationships of the Stomiiformes. In: Stiassny MLJ, Parenti LR, Johnson GD (eds.). The Interrelationships of Fishes. Academic Press, San Diego, pp. 333-353.

Hartel KE, Stiassny MLJ. 1986. The identification of larval *Parasudis* (Teleostei, Chlorophthalmidae): with notes on the relationships of aulopiform fishes. Breviora 487:1-23.

Hartman SE. 1988. A cladistic analysis of hominoid molars. J Human Evol 17: 489-502.

Hatta K, Schilling TF, Bremiller R, Kimmel CB. 1990. Specification of jaw muscle identity in zebrafish: correlation with engrailed-lhomeoprotein expression. Science 250:802-805.

Hatta K, Bremiller R, Westerfield M, Kimmel CB. 1991. Diversity of expression of engrailed-like antigens in zebrafish. Development 112:821-832.

Hernandez LP, Barresi MJF, Devoto SH. 2002. Functional morphology and developmental biology of zebrafish: reciprocal illumination from an unlikely couple. Integr Comp Biol 42:222-231.

Hernandez LP, Patterson SE, Devoto SH. 2005. The development of muscle fiber type identity in zebrafish cranial muscles. Anat Embryol 209:323-334.

Herrel A, Meyers JJ, Nishikawa KC, De Vree F. 2001. The evolution of feeding motor patterns in lizards: modulatory complexity and possible constraints. Amer Zool 41:1311-1320.

Herrel A, Canbek M, Özelmas Ü, Uyanoglu M, Karakaya M. 2005. Comparative functional analysis of the hyolingual anatomy in lacertid lizards. Anat Rec A 284:561-573.

Heyd A, Pfeiffer W. 2000. Über die Lauterzeitung der Welse (Siluroidei, Ostariophysi, Teleostei) und ihren Zusammenhang mit der Phylogenie und der Schreckreaktion. Rev Suisse Zool 107:165-211.

Hilton EJ. 2003. Comparative osteology and phylogenetic systematics of fossil and living bony-tongue fishes (Actinopterygii, Teleostei, Osteoglossomorpha). Zool J Linn Soc Lond 137:1-100.

Hilton EJ, Bemis WE. 1999. Skeletal variation in shortnose sturgeon (*Acipenser brevirostrum*) from the Connecticut River: implications for the comparative osteological studies of fossil and living fishes. In: Arratia G, Schultze H-P (eds.). Mesozoic Fishes 2: Systematic and Fossil Record. Verlag Pfeil, Munich, pp. 69-94.

Hinchliffe JR, Vorobyeva EI, Géraudie J. 2001. Is there a tetrapod developmental bauplan underlying limb evolution? Evidence from a teleost fish and from urodele and anuran amphibians. In: Ahlberg PE (ed.). Major Events in Early Vertebrate Evolution, Paleontology, Phylogeny, Genetics and Development. Taylor and Francis, London, pp.377-391.

Hoedemann J. 1960a. Studies on callichthyid fishes—3, notes on the development of *Callichthys*—1 (Pisces Siluriformes). Bull Aquat Biol 19:53-72.

Hoedemann J. 1960b. Studies on callichthyid fishes—5, Development of the skull of *Callichthys* and *Hoplosternum*—2 (Pisces Siluriformes). Bull Aquat Biol 2:21-36.

Howell AB. 1933a. Morphogenesis of the shoulder architecture, Part I—general considerations. Quart Rev Biol 8:247-259.

Howell AB. 1933b. Morphogenesis of the shoulder architecture, Part II—Pisces. Quart Rev Biol 8:434-456.

Howell AB. 1935. Morphogenesis of the shoulder architecture, Part III—Amphibia. Quart Rev Biol 10:397-431.

Howell AB. 1936a. Morphogenesis of the shoulder architecture, Part IV—Reptilia. Quart Rev Biol 11:183-208.

Howell AB. 1936b. The phylogenetic arrangement of the muscular system. Anat Rec 66:295-316.

Howes GJ. 1978. The anatomy and relationships of the cyprinid fish *Luciobrama macrocephalus* (Lacepède). Bull Br Mus Nat Hist (Zool) 34:1-64.

Howes GJ. 1979. Notes on the anatomy of *Macrochirichthys macrochirus* (Valenciennes), 1844, with comments on the Cultrinae (Pisces, Cyprinidae). Bull Br Mus Nat Hist (Zool) 36:147-200.

Howes GJ. 1983a. The cranial muscles of the loricarioid catfishes, their homologies and value as taxonomic characters. Bull Br Mus Nat Hist (Zool) 45:309-345.

Howes GJ. 1983b. Problems in catfish anatomy and phylogeny exemplified by the Neotropical Hypophthalmidae (Teleostei: Siluroidei). Bull Br Mus Nat Hist (Zool) 45:1-39.

Howes GJ. 1985a. Cranial muscles of gonorynchiform fishes, with comments on generic relationships. Bull Br Mus Nat Hist (Zool) 49: 273-303.

Howes GJ. 1985b. The phylogenetic relationships of the electric family Malapteruridae (Teleostei: Siluroidei). J Nat Hist 19:37-67.

Howes GJ, Sanford CPJ. 1987a. The phylogenetic position of the Plecoglossidae (Teleostei, Salmoniformes), with comments on the Osmeridae and Osmeroidei. Proc 5th Congr Eur Ichthyol (Stockholm): 17-30.

Howes GJ, Sanford CPJ. 1987b. Oral ontogeny of the ayu, *Plecoglossus altivelis* and comparison with the jaws of other salmoniform fishes. Zool J Linn Soc Lond 89:133-169.

Hunter JP, Jernvall J. 1995. The hypocone as a key innovation in mammalian evolution. Proc Natl Acad Sci US 92:10718-10722.

Hunter MP, Prince VE. 2002. Zebrafish Hox Paralogue group 2 genes function reduntantly as selector genes to pattern the second pharyngeal arch. Dev Biol 247: 367-389.

Huxley TH. 1876. Contributions to morphology: Ichthyopsida No. 1—on *Ceratodus forsteri*, with observations on the classification of fishes. PZS of London 1876:24-59.

Ichiyanagi T, Kohno H, Fujita K. 1993. Ontogenetic development of the Weberian apparatus in the silurid catfish, *Silurus asotus.* J Tokyo Univ Fisheries 80:205-211.

Ichiyanagi T, Kohno H, Fujita K, Taki Y. 1996. Ontogenetic development of the Weberian ossicles in two cyprinids, *Triblodon hakonensis* and *Zacco platypus.* J Tokyo Univ Fisheries 82:119-124.

Ichiyanagi T, Kohno H, Fujita K. 1997. Ontogenetic development of the Weberian ossicles in the bagrid catfish, *Pseudobagrus ichikawai.* J Tokyo Univ Fisheries 84:93-97.

Inoue JG, Miya M, Tsukamoto K, Nishida M. 2001. A mitogenomic perspective on the basal teleostean phylogeny: resolving higher-level relationships with longer DNA sequences. Mol Phylogenet Evol 20:275-285.

Inoue JG, Miya M, Tsukamoto K, Nishida M. 2003. Basal actinopterygian relationships: a mitogenomic perspective on the phylogeny of the "ancient fish". Mol Phylogenet Evol 26:110-120.

Inoue JG, Miya M, Tsukamoto K, Nishida M. 2004. Mitogenomic evidence for the monophyly of elopomorph fishes (Teleostei) and the evolutionary origin of the leptocephalus larva. Mol Phylogenet Evol 32:274-286.

Iordansky NN. 1992. Jaw muscles of the Urodela and Anura: some features of development, functions, and homology. Zool Jb Anat 122:225-232.

Ishiguro NB, Miya M, Nishida M. 2003. Basal euteleostean relationships: a mitogenomic perspective on the phylogenetic reality of the "Protacanthopterygii". Mol Phylogenet Evol 27:476-488.

Janvier P. 1996. Early Vertebrates. Clarendon Press, Oxford.

Jarvik E. 1944. On the exoskeletal shoulder-girdle on teleostomian fishes, with special reference to *Eusthenopteron foordi* Whiteaves. Kungl Sven Veten Handl 21:1-32.

Jarvik E. 1963. The composition of the intermandibular division of the head in fishes and tetrapods and the diphyletic origin of the tetrapod tongue. Kungl Sven Veten Handl 9:1-74.

Jarvik E. 1965. On the origin of girdles and paired fins. Israel J Zool 14:141-172.

Jarvik E. 1980. Basic Structure and Evolution of Vertebrates. Academic Press, London.

Jeffery JE. 2003. Mandibles of rhizodontids: anatomy, function and evolution within the tetrapod stem-group. Trans R Soc Edin (Earth Sci) 93:255-276.

Jernvall J. 2000. Linking development with generation of novelty in mammalian teeth. Proc Natl Acad Sci US 97:2641-2645.

Jessen H. 1968. The gular plates and branchiostegal rays in *Amia, Elops* and *Polypterus*. In: Ørvig T (ed.). Current Problems of Lower Vertebrate Phylogeny. Nobel Symposium 4, pp. 429-438.

Jessen H. 1972. Schultergürtel und Pectoralflosse bei Actinopterygiern. Fossils Strata 1:1-101.

Johanson Z. 2003. Placoderm branchial and hypobranchial muscles and origins in jawed vertebrates. J Vertebr Paleontol 23:735-749.

Johanson Z, Joss J, Wood D. 2004. The scapulocoracoid of the Queensland lungfish *Neoceratodus forsteri* (Dipnoi: Sarcopterygii): morphology, development and evolutionary implications for bony fishes (Osteichthyes). Zoology 107:93-109.

Johanson Z, Sutija M, Joss J. 2005. Regionalization of axial skeleton in the lungfish *Neoceratodus forsteri* (Dipnoi). J Exp Zool 304B:1-9.

Johnson GD. 1992. Monophyly of the euteleostean clades—Neoteleostei, Eurypterygii, and Ctenosquamata. Copeia 1992:8-25.

Johnson GD, Patterson C. 1996. Relationships of lower euteleostean fishes. In: Stiassny MLJ, Parenti LR, Johnson GD (eds.). The Interrelationships of Fishes. Academic Press, San Diego, pp. 251-332.

Jollie M. 1975. Development of the head skeleton and pectoral girdle in *Esox*. J Morphol 147:61-88.

Jollie M. 1980. Development of head and pectoral girdle skeleton and scales in *Acipenser*. Copeia 1980:226-249.

Jollie M. 1982. Ventral branchial musculature and synapomorphies questioned. Zool J Linn Soc Lond 75:35-47.

Jollie M. 1984a. Development of the head and pectoral skeleton of *Amia* with a note on the scales. Gegenbaurs Morphol Jahrb 130:315-51.

Jollie M. 1984b. Development of head and pectoral girdle bones of *Lepisosteus* with a note on scales. Copeia 476-502.

Jollie M. 1984c. Development of head and pectoral skeleton of *Polypterus* with notes on scales (Pisces, Actinopterygii). J Zool (Lond) 204:469-507.

Jollie M. 1986. A primer of bone names for the understanding of the actinopterygian head and pectoral girdle skeleton. Can J Zool 64:365-379.

Joss J, Longhurst T. 2001. Lungfish paired fins. In: Ahlberg PE (ed.). Major Events in Early Vertebrate Evolution, Paleontology, Phylogeny, Genetics and Development. Taylor and Francis, London, pp. 370-376.

Kardong KV. 2002. Vertebrates: Comparative Anatomy, Function, Evolution (3rd ed.). McGraw-Hill, New York.

Kardong KV, Zalisko EJ. 1998. Comparative Vertebrate Anatomy—a Laboratory Dissection Guide. McGraw-Hill, New York.

Kaseda Y, Nomura S-I. 1975. Electromyographical studies on the swimming movement of carp—II, pectoral fin movement. Jap J Vet Sci 37:75-81.

Kemp A. 1999. Ontogeny of the skull of the Australian lungfish *Neoceratodus forsteri* (Osteichthyes: Dipnoi). J Zool (Lond) 248:97-137.

Kershaw DR. 1970. The cranial osteology of the 'Butterfly Fish', *Pantodon buchholzi* Peters. Zool J Linn Soc Lond 49:5-19

Kershaw DR. 1976. A structural and functional interpretation of the cranial anatomy in relation to the feeding of osteoglossoid fishes and a consideration of their phylogeny. Trans Zool Soc Lond 33:179-252.

Kesteven HL. 1942-1945. The evolution of the skull and the cephalic muscles. Mem Aust Mus 8:1-361.

Kikugawa K, Katoh K, Kuraku S, Sakurai H, Ishida O, Iwabe N, Miyata T. 2004. Basal jawed vertebrate phylogeny inferred from multiple nuclear DNA-coded genes. BMC Biol 2:1-11.

Kirschner MW, Gerhart JC. 2005. The Plausibility of Life—Resolving Darwin's Dilemma. Yale University Press, London.

Kisia SM, Onyango DW. 2005. Muscular System of Vertebrates. Science Publishers, Enfield.

Kleerekoper H, Roggenkamp P. 1959. An experimental study of the effect of the swimbladder on hearing sensitivity in *Ameiurus nebulosus nebulosus* (Lesieur). Can J Zool 37:1-8.

Köntges G, Lumsden A. 1996. Rhombencephalic neural crest segmentation is preserved throughout craniofacial ontogeny. Development 122:3229-3242.

Kulshrestha SK. 1977. Development of the Weberian apparatus in the major carp *Labeo rohita* (Ham.). Anat Anz 141:433-444.

Ladich F. 2001. Sound-generating and -detecting motor system in catfish: design of swimbladder muscles in doradids and pimelodids. Anat Rec 263:297-306.

Larsen JH, Guthrie DJ. 1975. The feeding system of terrestrial tiger salamanders (*Ambystoma tigrinum melanostictum* Baird). J Morphol 147:137-154.

Lauder GV. 1980a. Evolution of the feeding mechanism in primitive actinopterygian fishes: a functional anatomical analysis of *Polypterus*, *Lepisosteus*, and *Amia*. J Morphol 163:283-317.

Lauder GV. 1980b. On the evolution of the jaw adductor musculature in primitive gnathostome fishes. Breviora 460:1-10.

Lauder GV. 1980c. The role of the hyoid apparatus in the feeding mechanism of the coelacanth *Latimeria chalumnae*. Copeia 1980:1-9.

Lauder GV, Liem KF. 1983. The evolution and interrelationships of the actinopterygian fishes. Bull Mus Comp Zool 150:95-197.

Lauder GV, Shaffer HB. 1985. Functional morphology of the feeding mechanism in aquatic ambystomatid salamanders. J Morphol 185:297-326.

Lauder GV, Shaffer HB. 1988. Ontogeny of functional design in tiger salamanders (*Ambystoma tigrinum*): are motor patterns conserved during major morphological transformations? J Morphol 197:249-268.

Lauder GV, Reilly SM. 1990. Metamorphosis of the feeding mechanism in tiger salamanders (*Ambystoma tigrinum*): the ontogeny of cranial muscle mass. J Zool (Lond) 222:59-74.

Lavoué S, Sullivan JP. 2004. Simultaneous analysis of five molecular markers provides a well-supported phylogenetic hypothesis for the living bony-tongue fishes (Osteoglossomorpha: Teleostei). Mol Phylogenet Evol 33:171-185.

Lavoué S, Miya M, Inoue JG, Saitoh K, Ishiguro NB, Nishida M. 2005. Molecular systematics of the gonorynchiform fishes (Teleostei) based on whole mitogenome sequences: implications for higher-level relationships within the Otocephala. Mol Phylogenet Evol 37:165-177.

Lecointre G. 1995. Molecular and morphological evidence for a Clupeomorpha-Ostariophysi sister-group relationship (Teleostei). Geobios 19:205-210.

Lecointre G, Le Guyader H. 2001. Classification Phylogénétique du Vivant (2nd ed.). Éditions Belin, Paris.

Lecuru S. 1968a. Étude des variations morphologiques du sternum, des clavicules et de l'interclavicule des lacertiliens. Ann Sci Nat Zool 10:511-544.

Lecuru S. 1968b. Remarques sur le scapulo-coracoide des lacertiliens. Ann Sci Nat Zool 10:475-510.

Le HL, Lecointre G, Perasso R. 1993. A 28S rRNA-based phylogeny of the Gnathostomes: first step in the analysis of conflict and congruence with morphologically based cladograms. Mol Phylogenet Evol 2:31-51.

Le Danois Y. 1967. Quelques figures descriptives de l'anatomie de *Pantodon buchholzi* Peters. Bull Inst Fr Afr Noire A 29:1051-1096.

Lee MSY. 1998. Convergent evolution and character correlation in burrowing reptiles—towards a resolution of squamate relationships. Biol J Linn Soc Lond 65:369-453.

Lekander B. 1949. The sensory line system and the canal bones in the head of some Ostariophysi. Acta Zool (Stockh) 30:1-131.

Le Lièvre C, Le Douarin NM. 1975. Mesenchymal derivatives of the neural crest: analysis of chimaeric quail and chick embryos. J Embryol Exp Morphol 34:125-154.

Lenglet G. 1974. Contribution a l'étude ostéologique des Kneriidae. Ann Soc R Zool Belg 103:239-270.

Levet Y. 1987. Comparative anatomy of cutaneous muscles of the face. Aesth Plast Surg 11:177-179.

Li G-Q. 1996. A new species of Late Cretaceous osteoglossid (Teleostei) from the Oldman Formation of Alberta, Canada, and its phylogenetic relationships. In: Arratia G, Viohl G (eds.). Mesozoic Fishes—Systematic and Paleoecology. Verlag Pfeil, Munich, pp. 285-298.

Li G-Q, Wilson MVH. 1996. Phylogeny of Osteoglossomorpha. In: Stiassny MLJ, Parenti LR, Johnson GD (eds.). The Interrelationships of Fishes. Academic Press, San Diego, pp. 163-174.

Liem KF, Woods LP. 1973. A probable homologue of the clavicle in the holostean fish *Amia calva*. J Zool Lond 170:521-531.

Lightoller GHS. 1939. Probable homologues: a study of the comparative anatomy of the mandibular and hyoid arches and their musculature. Trans R Soc London 24:349-445.

Liu H. 2004. Phylogenetic relationships of the Cypriniformes tested by mtDNA 12S rRNA sequence variations. Acta Gen Sinica 31:137-142.

Liu H, Tzeng C-S, Teng H-Y. 2002. Sequence variations in the mitochondrial DNA control region and their implications for the phylogeny of the Cypriniformes. Can J Zool 80:569-581.

López JA, Chen W-J, Orti G. 2004. Esociform phylogeny. Copeia 2004:449-464.

Lubosch W. 1914. Vergleischende Anatomie der Kaumusculatur der Wirbeltiere, in fünf Teilen: 1. Die Kausmukulatur der Amphibien. Jen Z Naturwiss 53:51-188.

Lund R. 2000. The new actinopterygian order Guildayichthyiformes from the Lower Carboniferous of Montana (USA). Geodiversitas 22:171-206.

Lundberg JG. 1975. Homologies of the upper shoulder girdle and temporal region bones in catfishes (Order Siluriformes), with comments on the skull of the Helogeneidae. Copeia 1975:66-74.

Luther A. 1913. Über die vom N trigeminus versorgte Muskulatur des Ganoiden and Dipneusten. Acta Soc Scient Fenn 41:1-72.

Luther A. 1914. Über die vom N trigeminus versorgte Muskulatur der Amphibien, mit einem vergleichenden Aublick über deu Adductor mandibulae der Gnathostomen, und cinem Beitrag zum Verständnis der Organisation der Anurenlarven. Acta Soc Scient Fenn 44:1-151.

Mabee PM. 1989a. An empirical rejection of the ontogenetic polarity criterion. Cladistics 5:409-416.

Mabee PM. 1989b. Assumptions underlying the use of ontogenetic sequences for determining character state order. Trans Amer Fish Soc 118:151-158.

Mabee PM. 1993. Phylogenetic interpretation of ontogenic change: sorting out the actual and artefactual in an empirical case study of centrarchid fishes. Zool J Linn Soc Lond 107:175-291.

Mabee PM. 2000. Developmental data and phylogenetic systematics: evolution of the vertebrate Limb. Amer Zool 40:789-800.

Mabee PM, Noordsy. 2004. Development of the paired fins in the Paddlefish, *Polyodon spathula*. J Morphol 261:334-344.

Mallat J. 1997. Shark pharyngeal muscles and early vertebrate evolution. Acta Zool (Stockh) 78:279-294.

Markle NR, Merrett R. 1980. The abyssal alepocephalid, *Rinoctes nasutus* (Pisces, Salmoniformes), a redescription and an evaluation of its systematic position. J Zool (Lond) 190:225-239.

Marshall CR. 1986. Lungfish: phylogeny and parsimony. J Morphol, suppl 1:151-162.

Marshall NB. 1962. Observations on the Heteromi, an order of teleost fishes. Bull Br Mus Nat Hist (Zool) 9:249-270.

Matsui T, Rosenblatt RH. 1987. Review of the deep-sea fish family Platytroctidae (Pisces: Salmoniformes). Bull Scripps Inst Oceanogr Univ Calif 26: 1-159.

Matthes H. 1963. A comparative study of the feeding mechanisms of some African Cyprinidae (Pisces, Cypriniformes). Bijdr Dierk 33:3-35.

Mayhew RL. 1924. The skull of *Lepisosteus ptatostomus*. J Morphol 38:315-346.

McAllister DE. 1968. The evolution of branchiostegals and associated opercular, gular, and hyoid bones and the classification of teleostome fishes, living and fossil. Bull Nat Mus Can 221:1-239.

McDowall RM. 1969. Relationships of galaxioid fishes with a further discussion of salmoniform classification. Copeia 1969:796-824.

McGonnell IM. 2001. The evolution of the pectoral girdle. J Anat 199:189-194.

McKitrick MC. 1991. Phylogenetic analysis of avian hindlimb musculature. Misc Publ Mus Zool Univ Mich 179:1-85.

McMahon BR. 1969. A functional analysis of the aquatic and aerial respiratory movements of an African lungfish, *Protopterus aethiopicus*, with reference to the evolution of lung ventilation movements in vertebrates. J Exp Biol 51:407-430.

McMurrich JP. 1884. The myology of *Amiurus catus* (L.) Gill. Proc Can Inst Toronto 2, N Ser:311-351.

Meunier FJ, Geistdoerfer P. 1991. Anatomie de la ceinture pectorale de *Thermobiotes mytilogeiton* (Synaphobranchidae), anguilliforme des sources hydrothermales du Pacifique Occidental. Cybium 15:83-87.

Meyer A, Zardoya R. 2003. Recent advances in the (molecular) phylogeny of Vertebrates. Annu Rev Ecol Evol Syst 34:311-338.

Miles R. 1977. Dipnoan (lungfish) skulls and the relationships of the group: a study based on new species from the Devonian of Australia. Zool J Linn Soc Lond 61:1-328.

Millot J, Anthony J. 1958. Anatomie de *Latimeria chalumnae*—I, squelette, muscles, et formation de soutiens. CNRS, Paris.

Miyake T, McEachran JD, Hall BK. 1992. Edgeworth's legacy of cranial muscle development with an analysis of muscles in the ventral gill arch region of batoid fishes (Chondrichthyes: Batoidea). J Morphol 212:213-256.

Mo T. 1991. Anatomy, relationships and systematics of the Bagridae (Teleostei: Siluroidei) with a hypothesis of siluroid phylogeny. Theses Zoologicae 17:1-216.

Monath T. 1965. The opercular apparatus of salamanders. J Morphol 116:149-170.

Monod T. 1963. Sur quelques points de l'anatomie de *Gonorhynchus gonorhynchus* (Linné 1766). Melanges Ichthyol 68:255-313.

Montero R, Moro SA, Abdala V. 2002. Cranial anatomy of *Euspondylus acutirostris* (Squamata: Gymnophthalmidae) and its placement in a modern phylogenetic hypothesis. Russ J Herpetol 9:215-228.

Moritz T, Britz R. 2005. Ontogeny and homology of the basipterygoid articulation in *Pantodon buchholzi* (Teleostei: Osteoglossomorpha). Zool J Linn Soc Lond 144:1-13.

Moro S, Abdala V. 2004. Análisis descriptivo de la miología flexora y extensora del miembro anterior de *Polychrus acutirostris* (Squamata, Polychrotidae). Pap Avulsos Zool (São Paulo) 44:81-89.

Moro SA, Abdala V. 2000. Cladistic analysis of Teiidae (Squamata) based on myological characters. Russ J Herpetol 7:87-102.

Musick JA, Bruton MN, Balon EK (eds.) (1991). The biology of *Latimeria chalumnae* and evolution of coelacanths. Environ Biol Fish 32:1-446.

Nelson CE, Tabin C. 1995. Footnote on limb evolution. Nature 375:630-631.

Nelson GJ. 1967. Epibranchial organs in lower teleostean fishes. J Zool (Lond) 153:71-89.

Nelson GJ. 1969. Gill arches and the phylogeny of fishes, with notes on the classification of vertebrates. Bull Amer Mus Nat Hist 141:475-552.

Nelson GJ. 1970. Gill arches of some teleostean fishes of the families Salangidae and Argentinidae. Jap J Ichthyol 17:61-66.

Nelson GJ. 1973. Relationships of clupeomorphs, with remarks on the structure of the lower jaw in fishes. In: Greenwood PH, Miles RS, Patterson C (eds.). Interrelationships of fishes. Zool J Linn Soc Lond 53, suppl 1, pp. 333-349.

Nelson JS. 1994. Fishes of the World (3rd ed.). John Wiley & Sons, New York.

Nelson JS. 2006. Fishes of the World (4th ed.). John Wiley & Sons, New York.

Nixon KC. 2002. Nona&Winclada, version 1.00.08. Published by the author, Ithaca, New York.

Noden DM. 1983. The embryonic origins of avian cephalic and cervical muscles and associated connective tissues. Am J Anat 168:257-276.

Noden DM. 1984. Craniofacial development: new views on old problems. Anat Rec 208:1-13.

Noden DM. 1986. Patterning of avian craniofacial muscles. Dev Biol 116:347-356.

Norden CR. 1961. Comparative osteology of representative salmonid fishes, with particular reference to the grayling (*Thymallus articus*) and its phylogeny. J Fish Res Bd Can 18:679-791.

Northcutt RG, Bemis WE. 1993. Cranial nerves of the coelacanth, *Latimeria chalumnae* (Osteichthyes, Sarcopterygii, Actinistia), and comparison with other Craniata. Brain Behav Evol 42, suppl 1:1-76.

Nüsslein-Volhard C, Gilmour DT, Dahm R. 2002. Introduction: zebrafish as a system to study development and organogenesis. In: Nüsslein-Volhard C, Dahm R (eds.). Zebrafish: A Practical Approach. Oxford University Press, New York, pp. 1-5.

Obermiller LE, Pfeiler E. 2003. Phylogenetic relationships of elopomorph fishes inferred from mitochondrial ribosomal DNA sequences. Mol Phylogenet Evol 26:202-214.

O'Connell CP. 1955. The gasbladder and its relation to the inner ear in *Sardinops caerula* and *Engraulis mordax*. Fish Bull US 56:503-533.

Oliveira C, Diogo R, Vandewalle P, Chardon M. 2002. On the myology of the cephalic region and pectoral girdle of three ariid species, *Arius heudeloti*, *Genidens genidens* and *Bagre marinus*, with a comparison with other catfishes (Teleostei: Siluriformes). Belg J Zool 59:243-266.

Olsson L, Falck P, Lopez K, Cobb J, Hanken J. 2001. Cranial neural crest cells contribute to connective tissue in cranial muscles in the anuran amphibian, *Bombina orientalis*. Dev Biol 237:354-367.

Orti G, Meyer A. 1997. The radiation of characiform fishes and the limits of resolution of mitochondrial ribosomal DNA sequences. Syst Biol 46:75-100.

Orton GL. 1963. Notes on larval anatomy of fishes of order Lyomeri. Copeia 1963:6-15.

Osse JWM. 1969. Functional morphology of the head of the perch (*Perca fluviatis* L.): an electromyographic study. Neth J Zool 19:289-392.

Owen R. 1841. Description of the *Lepidosiren annectens*. Trans Linn Soc Lond 18: 327-361.

Parker WK. 1882. On the structure and development of the skull in sturgeons (*Acipenser ruthenus* and *A. sturio*). Philos Trans R Soc Lond 173:139-185.

Parr AE. 1951. Preliminary revision of Alepocephalidae, with the introduction of a new family, Searsidae. Am Mus Novit 1531:1-21.

Parr AE. 1960. The fishes of the family Searsidae. Dana-Rep (Copenhagen) 51:1-109.

Pasleau F. 1974. Recherches sur la position phylétique des téleostéens Gonorynchiformes, basées sur l'étude de l'osteologie et de myologie cephalique. Unpublished Bachelor's thesis. University of Liège.

Patterson C. 1973. Interrelationships of holosteans. In: Greenwood PH, Miles RS, Patterson C (eds.). Interrelationships of fishes. Zool J Linn Soc Lond 53, suppl 1, pp. 233-305.

Patterson C. 1975. The braincase of pholidophorid and leptolepid fishes, with a review of the actinopterygian braincase. Philos Trans R Soc Lond B 269:275-579.

Patterson C. 1977a. The contribution of paleontology to teleostean phylogeny. In: Hecht PC, Goody PC, Hecht BM (eds). Major Patterns in Vertebrate Evolution. Plenum Press, New York, pp. 579-643.

Patterson C. 1977b. Cartilage bones, dermal bones and membrane bones, or the exoskeleton versus the endoskeleton. In: Andrews SM, Miles RS, Walker AD (eds.). Problems in Vertebrate Evolution. Academic Press, London, pp. 77-121.

Patterson C. 1981. Significance of fossils in determining evolutionary relationships. Annu Rev Ecol Evol Syst 12:195-223.

Patterson C. 1982. Morphology and interrelationships of primitive actinopterygian fishes. Amer Zool 22:241-259.

Patterson C. 1984. *Chanoides*, a marine Eocene otophysan fish (Teleostei: Ostariophysi). J Vertebr Paleontol 4:430-456.

Patterson C, Rosen DE. 1977. Review of ichthyodectiform and other Mesozoic fishes and the theory and practice of classifying fossils. Bull Amer Mus Nat Hist 158:85-172.

Patterson C, Johnson GD. 1995. The intermuscular bones and ligaments of teleostean fishes. Smiths Contrib Zool 599:1-83.

Peng Z, Shunping H, Wang J, Wang W, Diogo R. 2006. Mitochondrial molecular clocks and the origin of the major otocephalan clades (Pisces, Teleostei): a new insight. Gene 370:113-124.

Poggendorf D. 1952. Die absoluten Hörschwellen des Zwergwelses (*Amiurus nebulosus*) und Beiträge zur Physik des Weberschen Apparates der Ostariophysen. Z Vergl Physiol 34:222-257.

Poll M. 1967. Contribution à la faune icthyologique de l'Angola. Diamang Publ Cult 75:1-391.

Pollard HB. 1892. On the anatomy and phylogenetic position of *Polypterus*. Zool Jb 5:387-428.

Popper AN. 1971. The morphology of the Weberian ossicles of two species of the genus *Astyanax* (Ostariophysi : Characidae). J Morphol 133:179-188.

Pough FH, Heiser JB, McFarland WN. 1996. Vertebrate Life (4th ed.). Prentice-Hall, New Jersey.

Poyato-Ariza FJ. 1996. A revision of the ostariophysan fish family Chanidae, with special reference to the Mesozoic forms. Palaeo Ichthyologica 6:1-52.

Radermaker F, Surlemont C, Sanna P, Chardon M, Vandewalle P. 1989. Ontogeny of the Weberian apparatus of *Clarias gariepinus* (Pisces Siluriformes). Can J Zool 67:2090-2097.

Ramaswami LS. 1952a. Skeleton of cyprinoid fishes in relation to phylogenetic studies—I, the systematic position of the genus *Gyrinocheilus* Vaillant. Proc Nat Inst Sci India B (Biol Sci) 18:125-140.

Ramaswami LS. 1952b. Skeleton of cyprinoid fishes in relation to phylogenetic studies—II, the systematic position of *Psilorhynchus* Mclelland. Proc Nat Inst Sci India B (Biol Sci) 18:141-150.

Ramaswami LS. 1952c. Skeleton of cyprinoid fishes in relation to phylogenetic studies—III, the skull and other skeletal structures of homalopterid fishes. Proc Nat Inst Sci India B (Biol Sci) 18:495-517.

Ramaswami LS. 1952d. Skeleton of cyprinoid fishes in relation to phylogenetic studies—IV, the skull and other skeletal structures of gastromyzonid fishes. Proc Nat Inst Sci India B (Biol Sci) 18:519-538.

Ramaswami LS. 1953. Skeleton of cyprinoid fishes in relation to phylogenetic studies—V, the skull and gasbladder capsule of the Cobitidae. Proc Nat Inst Sci India B (Biol Sci) 19:323-347.

Ramaswami LS. 1955a. Skeleton of cyprinoid fishes in relation to phylogenetic studies—VI, the skull and Weberian apparatus in the subfamily Gobioninae (Cyprinidae). Acta Zool (Stockh) 36: 127-158.

Ramaswami LS. 1955b. Skeleton of cyprinoid fishes in relation to phylogenetic studies—VII, the skull and Weberian apparatus of Cyprininae (Cyprinidae). Acta Zool (Stockh) 36:199-242.

Ramaswami LS. 1957. Skeleton of cyprinoid fishes in relation to phylogenetic studies—VIII, the skull and Weberian ossicles of Catostomidae. Proc Zool Soc Calcutta, Mookerjee Mem Vol 1957:293-303.

Rasmussen A-S, Arnason U. 1999a. Molecular studies suggest that cartilaginous fishes have a terminal position in the piscine tree. Proc Natl Acad Sci US 96:2177-2182.

Rasmussen A-S, Arnason U. 1999b. Phylogenetic studies of complete mitochondrial DNA molecules place cartilaginous fishes within the tree of bony fishes. J Mol Evol 48:118-123.

Regan CT. 1911a. The classification of the teleostean fishes of the order Ostariophysi: 1, Cyprinoidea. Ann Mag Nat Hist 8:13-32.

Regan CT. 1911b. The classification of the teleostean fishes of the order Ostariophysi: 2, Siluroidea. Ann Mag Nat Hist 8:35-57.

Regan CT. 1923. The skeleton of *Lepisosteus*, with remarks on the origin and evolution of the lower neopterygian fishes. Proc Zool Soc Lond 1923:445-461.

Reilly SM, Lauder GV. 1989. Kinetics of tongue projection in *Ambystoma tigrinum*: quantitative kinematics, muscle function and evolutionary hypotheses. J Morphol 199:223-243.

Reilly SM, Lauder GV. 1990. The evolution of tetrapod feeding behavior: kinematic homologies in prey transport. Evolution 44:1542-1557.

Reilly SM, Lauder GV. 1991. Prey transport in the tiger salamander: quantitative electromyography and muscle function in tetrapods. J Exp Zool 260:1-17.

Reis RE. 1998. Anatomy and phylogenetic analysis of the Neotropical callichthyid catfishes (Ostariophysi, Siluriformes). Zool J Linn Soc Lond 124:105-168.

Ridewood WG. 1904a. On the cranial osteology of the clupeoid fishes. Proc Zool Soc Lond 29:448-493.

Ridewood WG. 1904b. On the cranial osteology of the fishes of the families Elopidae and Albulidae, with remarks on the morphology of the skull in the lower teleostean fishes generally. Proc Zool Soc Lond 2:35-81.

Ridewood WG. 1904c. On the cranial osteology of the fishes of the families Mormyridae, Notopteridae and Hiodontidae. J Linn Soc Lond (Zool) 29:188-217.

Ridewood WG. 1905a. On the cranial osteology of the fishes of the families Osteoglossidae, Pantodontidae, Phractolaemidae. J Linn Soc Lond (Zool) 29:252-282.

Ridewood WG. 1905b. On the skull of *Gonorynchus greyi*. Ann Mag Nat Hist 7:361-372.

Rieppel O. 1992. Studies on skeleton formation in reptiles—III, patterns of ossification in the skeleton of *Lacerta vivipara* Jacquin (Reptilia, Squamata). Fieldiana (Zool) 68:1-25.

Roberts TR. 1973. Interrelationships of ostariophysans. In: Greenwood PH, Miles RS, Patterson C (eds.). Interrelationships of fishes. Zool J Linn Soc Lond 53, suppl 1, pp. 373-395.

Robineau D. 1987. Sur la signification phylogénétique de quelques caractères anatomiques remarquables du coelacanthe *Latimeria chalumnae* Smith, 1939. Ann Sci Nat Zool Paris 8:43-60.

Romer AS. 1922. The locomotor apparatus of certain primitive and mammal-like reptiles. Bull Am Mus Nat Hist 46:517-606.

Romer AS. 1924. Pectoral limb musculature and shoulder-girdle structure in fish and tetrapods. Anat Rec 27:119-143.

Romer AS. 1944. The development of tetrapod limb musculature—the shoulder region of *Lacerta*. J Morphol 74:1-41.

Rosen DE. 1973. Interrelationships of higher euteleostean fishes. In: Greenwood PH, Miles RS, Patterson C (eds.). Interrelationships of fishes. Zool J Linn Soc Lond 53, suppl 1, pp. 397-513.

Rosen DE. 1974. Phylogeny and zoogeography of salmoniform fishes and relationships of *Lepidogalaxias salamandroides*. Bull Am Mus Nat Hist 153:265-326.

Rosen DE. 1985. An essay on euteleostean classification. Am Mus Novit 2827:1-45.

Rosen DE, Greenwood PH. 1970. Origin of the Weberian apparatus and the relationships of the ostariophysan and gonorynchiform fishes. Am Mus Novit 2428:1-25.

Rosen DE, Forey PL, Gardiner BG, Patterson C. 1981. Lungfishes, tetrapods, paleontology and plesiomorphy. Bull Am Mus Nat Hist 167:159-276.

Ruta M, Coates MI, Quicke DLJ. 2003. Early tetrapod relationships revisited. Biol Rev 78:251-345.

Sagemehl M. 1885. Beiträge zur vergleichenden Anatomie der Fische, III—das cranium der characiniden nebst allgemeinen über die mit einem Weber'schen apparat versehenen Physostomenfamilien. Morphol Jahr 10:1-119.

Saitoh K, Miya M, Inoue JG, Ishiguro NB, Nishida M. 2003. Mitochondrial genomics of ostariophysan fishes: perspectives on phylogeny and biogeography. J Mol Evol 56:464-472.

Saitoh K, Sado T, Mayden RL, Hanzawa N, Nakamura K, Nishida M, Miya M. 2006. Mitogenomic evolution and interrelationships of the Cypriniformes (Actinopterygii: Ostariophysi): the first evidence toward resolution of higher-level relationships of the world's largest freshwater fish clade based on 59 whole mitogenome sequences. J Mol Evol 63:826-841.

Sanford CPJ. 2000. Salmonoid fish osteology and phylogeny (Teleostei: Salmonoidei). Theses Zoologicae 33:1-264.

Sanford CPJ. 2001a. Kinematic analysis of a novel feeding mechanism in the brook trout *Salvelinus fontinalis* (Teleostei: Salmonidae): behavior modulation of a functional novelty. J Exp Biol 204:3905-3916.

Sanford CPJ. 2001b. The novel tongue-bite apparatus in the Notopteridae (Teleostei: Osteoglossomorpha): do kinematic patterns vary within a clade? Zool J Linn Soc Lond 132:259-275.

Sanford CPJ, Lauder GV. 1989. The functional morphology of the tongue-bite in the osteoglossomorph fish *Notopterus*. J Morphol 202:379-408.

Sanford CPJ, Lauder GV. 1990. Kinematics of the tongue bite apparatus in osteoglossomorph fishes. J Exp Biol 154:137-162.

Sato T, Nakabo T. 2002. Paraulopidae and *Paraulopus*, a new family and genus of aulopiform fishes with revised relationship within the order. Ichthyol Res 49:25-46.

Schaefer SA. 1990. Anatomy and relationships of the scoloplacid catfishes. Proc Acad Nat Sci (Phil) 142:167-210.

Schaeffer B. 1973. Interrelationships of chondrosteans. In: Greenwood PH, Miles RS, Patterson C (eds.). Interrelationships of fishes. Zool J Linn Soc Lond 53, suppl 1, pp. 207-226.

Schaeffer B, Rosen DE. 1961. Major adaptive levels in the evolution of the actinopterygian feeding mechanisms. Amer Zool 1:187-204.

Schilling TF. 2002. The morphology of larval and adult zebrafish. In: Nüsslein-Volhard C, Dahm R (eds.). Zebrafish: A Practical Approach. Oxford University Press, New York, pp. 59-94.

Schilling TF, Kimmel CB. 1997. Musculoskeletal patterning in the pharyngeal segments of the zebrafish. Development 124:2945-2960.

Schreiber B. 1935. La funzione dell'apparato de Weber—ricerche sperimentali sui Cyprinidi. Arch Zool Torino 21:1-10.

Schultze H-P. 1986. Dipnoans as sarcopterygians. J Morphol, suppl 1:39-74.

Schultze H-P. 2004. Mesozoic sarcopterygians. In: Arratia G, Tintori A (eds.). Mesozoic Fishes 3—Systematics, Paleonvironments and Biodiversity. Verlag Pfeil, Munich, pp. 463-492.

Schultze H-P, Campbell KSW. 1986. Characterization of the Dipnoi, a monophyletic group. J Morphol, suppl 1:25-37.

Schultze H-P, Cloutier R. 1991. Computed tomography and magnetic resonance imaging studies of *Latimeria chalumnae*. In: Musick JA, Bruton MN, Balon EK (eds.). The biology of *Latimeria chalumnae* and evolution of coelacanths. Environ Biol Fish 32, pp. 159-182.

Schultze H-P, Cumbaa SL. 2001. *Dialipina* and the characters of basal actinopterygians. In: Ahlberg PE (ed.). Major Events in Early Vertebrate Evolution, Paleontology, Phylogeny, Genetics and Development. Taylor and Francis, London, pp. 315-332.

Sewertzoff AN. 1928. The head skeleton and muscles of *Acipenser ruthenus*. Acta Zool (Stockh) 9:193-319.

Shaffer HB, Lauder GV. 1985a. Aquatic prey capture in ambystomatid salamanders: patterns of variation in muscle activity. J Morphol 183:273-284.

Shaffer HB, Lauder GV. 1985b. Patterns of variation in aquatic ambystomatid salamanders: kinematics of the feeding mechanism. Evolution 39, 83-92.

Shaposhnikova G. 1967. Comparative morphology of the whitefishes (Coregoninae) from the USSR. Trudy Zool Inst Akad 46:207-256.

Shellswell GB, Wolpert L. 1977. The pattern of muscle and tendon development in the chick wing. In: Ede D, Hinchcliffe R, Balls M (eds.). Vertebrate Limb and Somite Morphogenesis. Cambridge University Press, Cambridge, pp. 71-86.

Shen M. 1996. Fossil "osteoglossomorphs" in East Asia and their implications in teleostean phylogeny. In: Arratia G, Viohl G (eds.). Mesozoic Fishes—Systematic and Paleoecology. Verlag Pfeil, Munich, pp. 261-272.

Shoshani J, Groves CP, Simons EL, Gunnell GF. 1996. Primate phylogeny: morphological vs. molecular results. Mol Phylogenet Evol 5:101-153.

Shubin NH, Alberch P. 1986. A morphogenetic approach to the origin and basic organization of the tetrapod limb. Evol Biol 20:319-387.

Shubin NH, Tabin CT, Carroll SB. 1997. Fossils, genes and the evolution of animal limbs. Nature 388:638-648.

Shubin NH, Daeschler EB, Coates MI. 2004. The early evolution of the tetrapod humerus. Science 304:90-93.

Shubin NH, Daeschler EB, Jenkins FA. 2006. The pectoral fin of *Tiktaalik roseae* and the origin of the tetrapod limb. Nature 440:764-771.

Siebert DJ. 1987. Interrelationships among families of the order Cypriniformes (Teleostei). Unpublished PhD thesis. City University of New York.

Slevin JR. 1928. The amphibians of western North America. Occas Pap Calif Acad Sci 16:1-152.

Smith DG . 1989a. Family Anguillidae. In: Fishes of the Western North Atlantic. Mem Sears Found Mar Res 1:25-47.

Smith DG . 1989b. Family Congridae. In: Fishes of the Western North Atlantic. Mem Sears Found Mar Res 1:460-567.

Smith KK. 1988. Form and function of the tongue in agamid lizards with comments on its phylogenetic significance. J Morphol 196:157-71.

Smith MM. 1986. Latimeriidae. In: Smith MM, Heemstra PC (eds.). Smith's Sea Fishes. Springer-Verlag, Berlin, pp. 152-153.

Springer VG, Johnson GD. 2004. Study of the dorsal gill-arch musculature of teleosteome fishes, with special reference to the Actinopterygii. Bull Biol Soc Wash 11:1-236.

Stiassny MLJ. 1986. The limits and relationships of the acanthomorph fishes. J Zool (Lond.) B 1:411-460.

Stiassny MLJ. 1996. Basal ctenosquamate relationships and the interrelationships of the myctophiform (Scopelomorph) fishes. In:

Stiassny MLJ, Parenti LR, Johnson GD (eds.). The Interrelationships of Fishes. Academic Press, San Diego, pp. 405-436.

Stiassny MLJ. 2000. Gross functional anatomy: muscular system. In: Bullock G, Bunton TE (eds.). The Handbook of Experimental Animals. Academic Press, London, pp. 119-128.

Stiassny MLJ, Wiley EO, Johnson GD, De Carvalho MR. 2004. Gnathostome fishes. In: Donaghue MJ, Cracraft J (eds.). Assembling the Tree of Life. Oxford University Press, New York, pp. 410-429.

Sulak KJ. 1977. The systematics and biology of *Bathypterois* (Pisces, Chlorophthalmidae) with a revised classification of benthic myctophiform fishes. Galathea Rep 14:49-108.

Sullivan GE. 1962. Anatomy and embryology of the wing musculature of the domestic fowl (*Gallus*). Austr J Zool 10:458-518.

Swinnerton HH. 1903. The osteology of *Cromeria nilotica* and *Galaxias attenuatus*. Zool Jahr 18:58-70.

Takahasi N. 1925. On the homology of the cranial muscles of the cypriniform fishes. J Morphol 40:1-109.

Taverne L. 1972. Ostéologie des genres *Mormyrus* Linné, *Mormyrops* Muller, *Hyperopisus* Gill, *Myomyrus* Boulenger, *Stomatorhinus* Boulenger et *Gymnarchus* Cuvier. Ann Mus R Afr Centr 200:1-194.

Taverne L. 1974. Ostéologie d'*Elops* Linné, C., 1766 (Pisces, Elopiformes) et son intérêt phylogénétique. Acad R Belg, Mem Classe Sci 41:1-96.

Taverne L. 1977a. Ostéologie de *Clupavus maroccanus* (Crétacé Supérieur du Maroc) et considérations sur la position systématique et des relations des Clupavidae au sein de l'ordre des Clupeiformes sensu stricto (Pisces, Teleostei). Geobios 10:697-722.

Taverne L. 1977b. Ostéologie, phylogénèse, et systématique des téléostéens fossiles et actuels du super-ordre des ostéoglossomorphes, première partie. Ostéologie des genres *Hiodon*, *Eohiodon*, *Lycoptera*, *Osteoglossum*, *Scleropages*, *Heterotis* et *Arapaima*. Acad R Belg, Mem Classe Sci 41:1-235.

Taverne L. 1978. Ostéologie, phylogénèse, et systématique des téléostéens fossiles et actuels du super-ordre des ostéoglossomorphes, deuxième partie. Ostéologie des genres *Phareodus*, *Phareoides*, *Brychaetus*, *Musperia*, *Pantodon*, *Singida*, *Notopterus*, *Xenomystus* et *Papyrocranus*. Acad R Belg, Mem Classe Sci 42:1-212.

Taverne L. 1979. Ostéologie, phylogénèse, et systématique des téléostéens fossiles et actuels du super-ordre des ostéoglossomorphes, troisième partie. Évolution des structures ostéologiques et conclusions génerales relatives à la phylogénèse et à la systèmatique du super-ordre. Addendum. Acad R Belg, Mem Classe Sci 43:1-168.

Taverne L. 1995. Description de l'appareil de Weber du téléosteen crétacé marin *Clupavus maroccanus* et ses implications phylogénétiques. Belg J Zool 125:267-282.

Taverne L. 1999. Les poissons cretaces de Nardò, 8—*Sorbininardus apuliensis*, gen. nov., sp. nov. (Teleostei, Ostariophysi, Anatophysi, Sorbininardiformes, nov. ord.). Studi e Richerche sui giacimenti terziari di Bolca VIII, Spec Vol L Sorbiri, Mus Civ Storia Nat Verona 23:77-103.

Taverne L. 2004. *Libanechelys bultyncki* gen. et sp. nov., une nouvelle anguille primitive (Teleostei, Anguilliformes) du Cenomanien marin du Liban. Bull Inst R Sci Nat Belg, Sci Terre 74:73-87.

Taverne L. 2005. Les poissons cretaces de Nardo, 20, *Chanoides chardoni* sp. nov. Teleostei, Ostariophysi, Otophysi). Boll Mus Civ Storia Nat Verona 29:39-54.

Taverne L, De Vos L. 1997. Ostéologie et morphologie d'un bariliné nouveau du bassin de la Malagarasi (système du Lac Tanganyika): *Opsaridium splendens* sp. n. (Teleostei, Cyprinidae). J Afr Zool 111:281-300.

Tchernavin VV. 1947a. Six specimens of Lyomeri in the British Museum (with notes on the skeleton of the Lyomeri). J Linn Soc Lond (Zool) 41:287-350.

Tchernavin VV. 1947b. Further notes on the structure of the bony fishes of the order Lyomeri (Eurypharynx). J Linn Soc Lond (Zool) 41:387-393.

Tchernavin VV. 1953. The feeding mechanisms of a deep-sea fish, *Chauliodus sloani* Schneider. Br Mus Nat Hist (Zool) 1953:1-101.

Thomson KS. 1967. Mechanisms of intracranial kinetics in fossils rhipidistian fishes (Crossopterygii) and their relatives. J Linn Soc Lond (Zool) 46:223-253.

Thorsen DH, Hale ME. 2005. Development of zebrafish (*Danio rerio*) pectoral fin musculature. J Morphol 266:241-55.

Thorsen DH, Westneat MW. 2005. Diversity of pectoral fin structure and function in fishes with labriform propulsion. J Morphol 263:133-150.

Thys van den Audenaerde DFE. 1961. L'anatomie de *Phractolaemus ansorgei* Blgr. et la position systématique des Phractolaemidae. Ann Mus R Afr Centr 103:101-167.

Trewavas E. 1932. A contribution to the classification of the fishes of the order Apodes, based on the osteology of some rare eels. Proc Zool Soc Lond 1932:639-659.

Triques ML. 1993. Filogenia dos géneros de Gymnotiformes (Actinopterygii, Ostariophysi), com base em caracteres esqueléticos. Comun Mus Ciênc PUCRS (Zool) 6:85-130.

Vandewalle P. 1975. Des formes aux fonctions: une étude de morphologie fonctionnelle et comparée chez trois poissons cyprinidés. Unpublished PhD thesis. University of Liège.

Vandewalle P. 1977. Particularités anatomiques de la tête de deux Poissons Cyprinidés *Barbus barbus* (L.) et *Leuciscus leuciscus* (L). Bull Acad R Belg 5:469-479.

Vandewalle P. 1978. Analyse des mouvements potentiels de la region céphalique du Goujon, *Gobio gobio* (L.) (Poisson, Cyprinidae). Cybium 3:15-33.

Vandewalle P, Victor D, Sanna P, Surlemont C. 1989. The Weberian apparatus of a 18.5 mm fry of *Barbus barbus*. In: Splechtna H, Hilgers G (eds.). Trends in Vertebrate Morphology (35th vol). Fischer Verlag, Stuttgard, pp. 363-366.

Vandewalle P, Radermaker F, Surlemont C, Chardon M. 1990. Apparition of the Weberian characters in *Barbus barbus* (Pisces Cyprinidae). Z Anz 225:262-376.

Vari RP. 1979. Anatomy, relationships and classification of the families Citharinidae and Distichodontidae (Pisces, Characoidea). Bull Br Mus Nat Hist (Zool) 36:261-344.

Venkatesch B, Erdmann MV, Brenner S. 2001. Molecular synapomorphies resolve evolutionary relationships of extant jawed vertebrates. Proc Natl Acad Sci US 98:11382-11387.

Vetter B. 1878. Untersuchungen zur vergleichenden anatomie der kiemen- und kiefer-musculatur der Fische: II—Thiel. Jena Z Naturw 12:431-550.

Vickaryous MK, Hall BK. 2006. Homology of the reptilian coracoid and a reappraisal of the evolution and development of the amniote pectoral apparatus. J Anat 208:263-85.

Vrba ES. 1968. Contributions to the functional morphology of fishes—part V, the feeding mechanism of *Elops saurus* Linnaeus. Zool Afr 3:211-236.

Walthall JC, Ashley-Ross MA. 2006. Postcranial myology of the California newt, *Taricha torosa*. Anat Rec A 288:46-57.

Wang C-H, Kuo C-H, Mok H-H, Lee S-C. 2003. Molecular phylogeny of elopomorph fishes inferred from mitochondrial 12S ribosomal RNA sequences. Zool Scr 32:231-241.

Waters JM, Lopez JA, Wallis GP. 2000. Molecular phylogenetics and biogeography of galaxiid fishes (Osteichthyes: Galaxiidae): dispersal, vicariance, and the position of *Lepidogalaxias salamandroides*. Syst Biol 49:777-795.

Watson JM. 1939. The development of the Weberian ossicles and anterior vertebrae in the goldfish. Proc R Soc Lond B 127:452-472.

Weber EH. 1820. De Aure et Auditu Hominis et Animalium: Pars I—De Aure Animalium aquatilium. Gehard Fleischer, Leipzig.

Weisel GF. 1960. The osteocranium of the catostomid fish, *Catostomus macrocheilus*—a study in adaptation and natural relationship. J Morphol 106:109-129.

Weiss B, Strother W, Hartig G. 1969. Auditory sensitivity in the Bullhead Catfish (*Ictalurus nebulosus*). Proc Nat Acad Sci US 64:552-556.

Weitzman SH. 1962. The osteology of *Brycon meeki*, a generalized characid fish, with an osteological definition of the family. Stanford Ichthyol Bull 8:1-77.

Weitzman SH. 1964. Osteology of and relationships of South American characid fishes of subfamilies Lebiasininae and Erythrinae with special reference to subtribe Nannostomina. Proc US Natl Mus 116:127-170.

Weitzman SH. 1967a. The origin of stomiatoid fishes with comment on the classification of salmoniform fishes. Copeia 1967:507-540.

Weitzman SH. 1967b. The osteology and relationships of the Astronesthidae, a family of oceanic fishes. Dana-Rep (Copenhagen) 71:1-54.

West-Eberhard MJ. 2003. Developmental Plasticity and Evolution. Oxford University Press, New York.

Westneat MW, Thorsen DH, Walker JA, Hale ME. 2004. Structure, function, and neural control of pectoral fins in fishes. IEEE J Oceanic Engen 29:674-683.

Wiley EO. 1976. The phylogeny and biogeography of fossil and recent gars (Actinopterygii: Lepisosteidae). Univ Kansas Misc Publ (Nat Hist) 64:1-111.

Wiley EO. 1979a. Ventral gill arch muscles and the interrelationships of gnathostomes, with a new classification of the Vertebrata. J Linn Soc Lond (Zool) 67:149-179.

Wiley EO. 1979b. Ventral gill arch muscles and the phylogenetic interrelationships of *Latimeria*. Occ Pap Calif Acad Sci 134:56-67.

Wilga CD, Wainwright PC, Motta PJ. 2000. Evolution of jaw depression mechanisms in aquatic vertebrates: insights from Chondrichthyes. Biol J Linn Soc Lond 71:165-185.

Wilson MVH, Veilleux P. 1982. Comparative osteology and relationships of the Umbridae (Pisces: Salmoniformes). Zool J Linn Soc Lond 76:321-325.

Winterbottom R. 1974. A descriptive synonymy of the striated muscles of the Teleostei. Proc Acad Nat Sci (Phil) 125:225-317.

Winterbottom R. 1993. Myological evidence for the phylogeny of recent genera of surgeonfishes (Percomorpha, Acanthuridae), with comments on Acanthuroidei. Copeia 1993:21-39.

Wu K-Y, Shen S-C. 2004. Review of the teleostean adductor mandibulae and its significance to the systematic position of the Polymixiiformes, Lampridiformes, and Triacanthoidei. Zool Stud 43:712-736.

Zaragüeta-Bagils R, Lavoué S, Tillier A, Bonillo C, Lecointre G. 2002. Assessment of otocephalan and protacanthopterygian concepts in the light of multiple molecular phylogenies. C R Biologies 325:1191-1207.

Zardoya R, Meyer A. 1996. Evolutionary relationships of the coelacanth, lungfishes, and tetrapods based on the 28S ribosomal RNA gene. Proc Natl Acad Sci US 93:5449-5454.

Zardoya R, Cao Y, Hasegawa M, Meyer A. 1998. Searching for the closest living relative(s) of tetrapods through evolutionary analysis of mitochondrial and nuclear data. Mol Biol Evol 15:506-517.

Zhu M, Schultze H-P. 1997. The oldest sarcopterygian fish. Lethaia 30:293-304.

Zhu M, Schultze H-P. 2001. Interrelationships of basal osteichthyans. In: Ahlberg PE (ed.). Major Events in Early Vertebrate Evolution, Paleontology, Phylogeny, Genetics and Development. Taylor and Francis, London, pp. 289-314.

Index

For Product Safety Concerns and Information please contact our EU representative GPSR@taylorandfrancis.com Taylor & Francis Verlag GmbH, Kaufingerstraße 24, 80331 München, Germany

Batch number: 10726388

Printed by Printforce, the Netherlands